THE ELEMENTS OF CONTINUUM BIOMECHANICS

THE ELEMENTS OF CONTINUUM BIOMECHANICS

Marcelo Epstein

University Of Calgary, Canada

A John Wiley & Sons, Ltd., Publication

Library of Congress Cataloging-in-Publication Data

Epstein, M. (Marcelo)
   The elements of continuum biomechanics / Marcelo Epstein.
      p. cm.
   Includes bibliographical references and index.
   ISBN 978-1-119-99923-2 (cloth)
   1. Biomechanics. 2. Biomechanics–Case studies. I. Title.
   QH513.E77 2012
   573.7'9343–dc23

                                        2012007486

A catalogue record for this book is available from the British Library.

Print ISBN: 9781119999232

Set in 10/12pt Times by Laserwords Private Limited, Chennai, India
Printed and bound in Malaysia by Vivar Printing Sdn Bhd

1   2012

To my mother
Catalina Ana Blejer de Epstein
with love and admiration

Contents

Preface

> *Euclides a Ptolemaeo interrogatus an non esset methodus discendae Geometriae methodo sua facilior: Non est regia, inquit Euclides, ad Geometriam via.*
>
> (PROCLUS, *Commentary on Euclid's Book 1*, Latin translation of the Greek original.)

There is no *via regia*, no royal road, to continuum mechanics. All one can hope for is a more comfortable ride. Many of us, trained in the good old days of pure thought, driven by intellectual curiosity, did actually enjoy the hardships of the journey and our lives were ennobled by it, as a pilgrimage to a sacred Source. This highly personal transformative phenomenon continues to take place, to be sure, as I have had the chance to observe in the lives of more than a few of my former and present students. Today, on the other hand, we face a more formidable and exciting challenge, since, as it turns out, some of the most interesting applications of continuum mechanics have arisen from the field of biology. Materials of interest are no longer inert or passive systems upon which one can apply the classical paradigms of thought: mass is conserved, the history of the deformation and the temperature suffice to determine the material response, and so on. Instead, biological systems are alive, open to mass exchanges, controlled by external agents, subject to chemical reactions with outside sources, highly structured at various phenomenological levels. Moreover, those most interested in using the rigorous and elegant cutting tools of continuum mechanics in biological applications are not only the applied mechanicians and mathematicians of yore. They are, in rapidly increasing numbers, members of the younger generation working in human performance laboratories, kinesiology departments, faculties of medicine and engineering schools the world over.

One of the objectives of this book is to allow the newcomer to the field to get involved in important issues from the very beginning. To achieve this aim, the policy in the first part of the book is to present everything that can be presented in a one-dimensional spatial context. The polar decomposition of the deformation gradient and the divergence theorem become trivial in one dimension. Nevertheless, important topics such as fading memory and mixture theory retain most of the features of the general formalism. It is thus possible to deal with these important topics rather early in the book. To emphasize the potential of these material models, the landscape is dotted with case studies motivated by biomedical applications. Some examples of these case studies are: vibration of air in the ear canal, hyperthermia treatment of tumours, striated-muscle memory and cartilage mechanics.

The Elements of Continuum Biomechanics, First Edition. Marcelo Epstein.
© 2012 John Wiley & Sons, Ltd. Published 2012 by John Wiley & Sons, Ltd.

Upon completion of the first four chapters, the reader should have acquired a fairly good idea of the scope and method of continuum mechanics. In the second part of the book, the three-dimensional context is introduced and developed over two chapters. A further chapter is devoted to the modern theories of growth and remodelling. Finally, a chapter on the finite-element method, although not strictly part of continuum mechanics, is also included, and the opportunity is not wasted to talk about the calculus of variations and the notion of weak formulation of balance equations.

As in so many other realms, when considering the material to be included in a technical book, choices have to be made. Every educator has a hidden agenda. Mine has been to open the eyes of the student to the enchanted forest of truly original ideas which, long after having been learned and seemingly forgotten, continue to colour our view of the world.

Part I

A One-dimensional Context

1

Material Bodies and Kinematics

1.1 Introduction

Many important biological structures can be considered as continuous, and many of these can be regarded as one-dimensional and straight. Moreover, it is not uncommon to observe that, whenever these structures deform, grow, sustain heat and undergo chemical reactions, they remain straight. Let us look at some examples.

Tendons. One of the main functions of tendons is to provide a connection between muscles (made of relatively soft tissue) and bone (hard tissue). Moreover, the deformability of tendinous tissue and its ability to store and release elastic energy are important for the healthy performance of human and animal activities, such as walking, running, chewing and eye movement. Tendons are generally slender and straight. Figure 1.1 shows a human foot densely populated by a network of tendons and ligaments. The Achilles tendon connects the calcaneus bone with the gastrocnemius and soleus muscles located in the lower leg.

Muscle components. Most muscles are structurally too complex to be considered as one-dimensional entities. On the other hand, at some level of analysis, muscle fibres and their components down to the myofibril and sarcomere level can be considered as straight one-dimensional structural elements, as illustrated schematically in Figure 1.2.

Hair. Figure 1.3 shows a skin block with follicles and hair. When subjected to tensile loads, hair can be analysed as a one-dimensional straight structure.

One of the questions that continuum mechanics addresses for these and more complex structures is the following: what is the mechanism of transmission of load? The general answer to this question is: *deformation*. It took millennia of empirical familiarity with natural and human-made structures before this simple answer could be arrived at. Indeed, the majestic Egyptian pyramids, the beautiful Greek temples, the imposing Roman arches, the overwhelming Gothic cathedrals and many other such structures were conceived, built and utilized without any awareness of the fact that their deformation, small as it might be, plays a crucial role in the process of transmission of load from one part of the structure to another. In an intuitive picture, one may say that the deformation of a continuous structure is the reflection of the change in atomic distances at a deeper level, a change that results in the development of internal forces in response to the applied external loads. Although this

The Elements of Continuum Biomechanics, First Edition. Marcelo Epstein.
© 2012 John Wiley & Sons, Ltd. Published 2012 by John Wiley & Sons, Ltd.

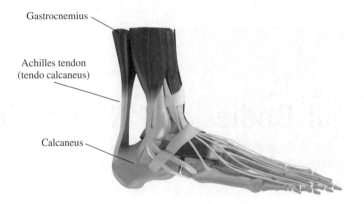

Figure 1.1 Tendons and ligaments in the human foot

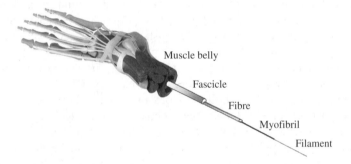

Figure 1.2 A skeletal muscle and its components

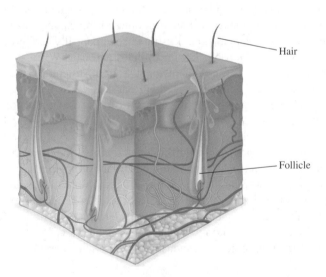

Figure 1.3 Skin cube with follicles and hair

naïve model should not be pushed too far, it certainly contains enough physical motivation to elicit the general picture and to be useful in many applications.

Once the role of the deformation has been recognized, continuum mechanics tends to organize itself in a tripartite fashion around the following questions:

1. How is the deformation of a continuous medium described mathematically?
2. What are the physical laws applicable to all continuous media?
3. How do different materials respond to various external agents?

This subdivision of the discipline is not only paedagogically useful, but also epistemologically meaningful. The answers to the three questions just formulated are encompassed, respectively, under the following three headings:

1. continuum kinematics;
2. physical balance laws;
3. constitutive theory.

From the mathematical standpoint, *continuum kinematics* is a direct application of the branch of mathematics known as differential geometry. In the one-dimensional context implied by our examples so far, all that needs to be said about differential geometry can be summarily absorbed within the realm of elementary calculus and algebra. For this to be the case, it is important to bear in mind not only that the structures considered are essentially one-dimensional, but also that they remain straight throughout the process of deformation.

The *physical balance laws* that apply to all continuous media, regardless of their material constitution, are mechanical (balance of mass, linear momentum, angular momentum) and thermodynamical (balance of energy, entropy production). In some applications, electromagnetical, chemical and other laws may be required. The fact that all these laws are formulated over a continuous entity, rather than over a discrete collection of particles, is an essential feature of continuum mechanics.

Finally, by not directly incorporating the more fundamental levels of physical discourse (cellular, molecular, atomic, subatomic), continuum mechanics must introduce phenomenological descriptors of material behaviour. Thus, tendon responds to the application of forces differently from muscle or skin. In other words, geometrically identical pieces of different materials will undergo vastly different deformations under the application of the same loads. One may think that the only considerations to be borne in mind in this respect are purely experimental. Nevertheless, there are some principles that can be established *a priori* on theoretical grounds, thus justifying the name of *constitutive theory* for this fundamental third pillar of the discipline. In particular, the introduction of *ideal material models*, such as elasticity, viscoelasticity and plasticity, has proven historically useful in terms of proposing material responses that can be characterized by means of a relatively small number of parameters to be determined experimentally.

1.2 Continuous versus Discrete

A pendulum, an elastic spring, a shock absorber: are they to be considered as continuous entities? The answer to this question depends ultimately on the level of description adopted. Leaving aside the deeper fact that bodies are made of a very large, though finite, number of particles, it is clear that, at a more mundane phenomenological level, the three entities just mentioned could be considered as fully-fledged three-dimensional continuous bodies. Moreover, if a spring is slender and straight, it might be appropriate to analyse it as a one-dimensional structure of the kind discussed in the previous section. A tendon, in fact, can be regarded as an example of a spring-like biological structure.

On the other hand, as we know from many encounters with an elastic spring attached to a mass in physics textbooks, there is a different sense in which words such as 'spring' or 'damper' can be used. To highlight the main difference between a 'real' spring and the physics textbook spring (which may be called a 'spring element'), we observe that in the former the mass and the elastic quality are smoothly distributed over the length of the spring. In other words, these properties are specified as some *functions of position* along the length of the spring. Concomitantly, the deformation is expressed in terms of some *displacement field* over the same domain. Physically, this implies that phenomena such as the propagation of sound waves become describable in this context. In contrast, the physics textbook spring element is characterized by just three numbers: its total mass (concentrated at one end), its relaxed length and its stiffness. The deformation is given by a single number, namely, the total instantaneous length. In a model of this kind, therefore, the description of any phenomenon distributed throughout the structure has been irrevocably sacrificed.

Example 1.2.1   Sonomicrometry. Introduced in the 1950s in biomechanical applications, *sonomicrometry* is a length-measuring technique based on the implantation of piezoelectric crystals at two points of an organ and the measurement of the time elapsed between the emission and the reception of an ultrasound signal from one crystal to the other. In particular, sonomicrometry has been used to measure the real-time elongation of muscle fibres during locomotion. Clearly, although for many purposes the muscle fibre can be considered as a simple spring element, the technique of sonomicrometry relies on the fact that sound waves propagate along the fibre and, therefore, on the fact that its properties are smoothly distributed.

From the foregoing, it appears that a fundamental distinction should be drawn between discrete and continuous mechanical systems. Although the intuitive understanding of the meaning of these terms can easily be illustrated by means of examples, as shown in Figure 1.4, the details are surprisingly technical. A cursory reading of the following formal, yet not completely rigorous, characterization should be sufficient for our purposes.

Definition   1.2.2 A *mechanical system* consists of an *underlying set M* and a finite number n of (real-valued) *state variables*:

$$\psi_i = \psi_i(p,t), \quad i = 1,2,\ldots,n, \tag{1.1}$$

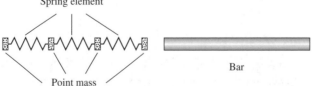

Figure 1.4   Discrete and continuous mechanical systems

where $p \in M$ and t denotes time. The underlying set is also called the *material set* of the system. Its elements are called *material points* or *particles*. (Some basic terminology on sets is reviewed in Window 1.1.)

Window 1.1   Some terminology and notation

Regarding a *set* informally as a collection of objects, called *elements* or *members* of the set, the symbol '$\in$' stands for the expression 'belongs to' or 'is an element of'. Thus, '$p \in A$' means 'p is an element of the set A', or 'p belongs to A'. A set C is a *subset* of the set A if every element of C is also an element of A. This relation is indicated as $C \subset A$. In particular, every set is a subset of itself.

The *Cartesian product* $A \times B$ of two sets, A and B, is the set whose elements are all the ordered pairs (a, b), where $a \in A$ and $b \in B$. The symbol $\mathbb{R}$ represents the *real line*, which is the collection of all real numbers. The set $\mathbb{R}^2 = \mathbb{R} \times \mathbb{R}$ is the *real plane*, consisting of all ordered pairs of real numbers. The *n-dimensional real space* $\mathbb{R}^n = \underbrace{\mathbb{R} \times \ldots \times \mathbb{R}}_{n}$ is constructed similarly by induction.

The physical meaning of the state variables ψ_i depends on the system at hand. They may represent displacements, velocities, temperature, stress, enthalpy, and so on. The time interval of interest, $t_0 \leq t \leq t_1$, is a subset of the real line $\mathbb{R}$. We assume that, within this interval, each state variable is a continuous (and usually also differentiable) function of the time variable t for each fixed point p of the underlying set M. It is only the nature of this set that differentiates between discrete and continuous systems.

Definition 1.2.3   A mechanical system is said to be *discrete* if its underlying set is finite or countable.

Recall that a collection is finite (countable) if it can be put in a one-to-one correspondence with a finite (infinite) subset of the natural numbers. The evolution in time of a discrete mechanical system is usually described by a system of ordinary differential equations (ODEs).

If a system is not discrete, its underlying set M is infinite and non-countable. This result, in and of itself, is not enough for a non-discrete system to be continuous. A classical counterexample is described in Window 1.2. To make matters more precise, let us declare that all the mechanical systems of interest in our present context have a material set M that is a subset of a one-dimensional *material universe* consisting of a copy $\mathcal{M}$ of the real line $\mathbb{R}$ (Figure 1.5). In this primeval material universe there are no pre-assigned length scales. The only properties that matter are the topological ones, such as the notion of open interval.

Definition 1.2.4   A *continuous one-dimensional mechanical system* is a system whose underlying material subset is an open interval $\mathcal{B}$ of the material universe $\mathcal{M}$. Such a material set $\mathcal{B}$ is called a *one-dimensional body* or a *one-dimensional material continuum*.

Figure 1.5   The material universe and a material body

Window 1.2   A non-discrete set that is not a continuum

The *Cantor set*, named after Georg Cantor (1845–1916), one of the founders of modern set theory, is obtained by starting from a closed finite interval of the line and deleting its open middle third, thus obtaining two separate closed intervals, each of length equal to one-third of the original segment. The process is repeated on each of these intervals, and so on *ad infinitum*. The result of the first few steps is shown in Figure 1.6.

It can be shown that what is left is a collection of points as numerous as the original segment. In other words, there exists a one-to-one correspondence between the Cantor set and the set of real numbers (which is not countable).

Figure 1.6   Generation of the Cantor set

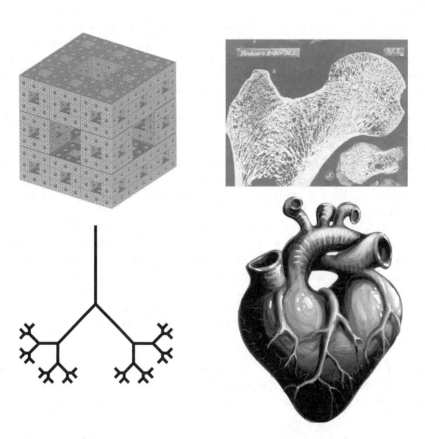

Figure 1.7   Fractals and biological applications

The Cantor set is an example of a self-similar *fractal*. Although originally introduced as curious mathematical entities, self-similar fractals (such as the Koch snowflake, the Sierpiński carpet, the Menger sponge and the fractal tree) can be treated as viable mechanical systems, called *fractal bodies*, to model biological structures, as suggested in Figure 1.7. Fractal bodies are neither discrete systems nor material continua according to our definitions.

A similar approach will be used in Section 5.5 to define a three-dimensional material body $\mathcal{B}$ as a connected open subset of $\mathbb{R}^3$, namely, of the three-dimensional material universe. The material universe does not presuppose any particular values for the mass or for any material properties whatsoever. Thus, a material body $\mathcal{B}$ can be regarded as just a continuous collection of particles. In contrast with the discrete counterpart, we observe that the state variables ψ_i now become *fields* defined over $\mathcal{B}$. For this reason it may be said that continuum mechanics belongs to the realm of physical *field theories*. The governing equations are, in general, partial differential equations (PDEs) rather than ODEs, as was the case in discrete systems.

Remark 1.2.5   We have followed the traditional definition of a continuous material body as an *open* subset of the material universe, thus automatically excluding its boundary (which, in the one-dimensional case, consists of at most two points). In other words, we are just concerned with the interior points of a physical material body. The inclusion of the boundary plays an important role when effects such as surface tension are of interest.

1.3   Configurations and Deformations

The material body itself, defined as a subset of the abstract material universe, is inaccessible to measurement. It acts as some kind of Platonic idea which manifests itself imperfectly in the world of experience. These manifestations, in the case of continuum mechanics, are called *configurations*. Thus, a configuration is a map κ of the material body into the physical space (which, in our one-dimensional context, is the real line):[1]

$$\kappa : \mathcal{B} \to \mathbb{R}. \tag{1.2}$$

To qualify as a configuration, however, this map cannot be arbitrary. The first restriction we will impose is *continuity*. Physically, continuity means that we do not want the body to be broken into separate pieces, but to remain as a connected unit. The second condition that we are going to impose corresponds to the physical notion of *impenetrability of matter*. In other words, we want to avoid a situation whereby two different material points occupy simultaneously the same position in space. From the mathematical point of view, this condition is equivalent to the requirement that the map be *injective*, or *one-to-one*. We summarize these conditions in the following definition.

Definition 1.3.1   A *configuration* of a (one-dimensional) body $\mathcal{B}$ is a continuous one-to-one map of $\mathcal{B}$ into $\mathbb{R}$.

---

[1] Some basic terminology on maps is reviewed in Window 1.3.

Window 1.3   Functions and maps

Given two sets, A and B, a *function* f from A (the *domain*) to B (the *codomain*),

$$f : A \to B,$$

assigns to *every* element of A some element of B. If $p \in A$, we say that the function f *acts* on p to produce the element $f(p) \in B$. A common notation for the action of a function is

$$p \mapsto f(p).$$

We also say that p is *mapped* by f to $f(p)$, the *image* of p. Depending on the context, functions are also called *maps* or *mappings*.

The *range* of the function $f : A \to B$ is the subset of the codomain B consisting of all elements $q \in B$ such that, for at least one $p \in A$, $f(p) = q$. For example, the range of the function $f : \mathbb{R}^2 \to \mathbb{R}$ defined by $(x, y) \mapsto x^2 + y^2$ is the closed positive semi-axis of the real line. If $C \subset A$, we denote by $f(C)$ the subset of B consisting of all the images of elements of C. Thus, the range of a function is precisely $f(A)$.

A function is *onto* or *surjective* if its range equals its codomain. In other words, every element of the codomain is the image of at least one element of the domain. A function $f : A \to B$ is *one-to-one* or *injective* if, for every pair $p_1, p_2 \in A$, $p_1 \neq p_2$ implies $f(p_1) \neq f(p_2)$. Thus, no two different elements of the domain are assigned the same element of the range. A function is *bijective* if it is both one-to-one (injective) and onto (surjective). If $f : A \to B$ is bijective, and only in this case, we can define the *inverse function* $f^{-1} : B \to A$ by assigning to each element $q \in B$ the unique element $p \in A$ such that $f(p) = q$. Noting that every function is clearly surjective over its range $f(A) \subset B$, we can always define the restricted inverse $f^{-1} : f(A) \to A$ of any injective function f.

If $f : A \to B$ and $g : B \to C$ are functions, the *composition* of g with f is the function $g \circ f : A \to C$ defined by

$$p \mapsto (g \circ f)(p) = g(f(p)),$$

for each $p \in A$. Notice that the codomain of f must be a subset of the domain of g for the composition to be defined. The composition of functions enjoys the associative property: $h \circ (g \circ f) = (h \circ g) \circ f$, which justifies the less precise notation $h \circ g \circ f$.

We note that a configuration can also be understood as the introduction of a coordinate system in the body. Indeed, a formula such as equation (1.2), under the conditions of continuity and injectivity, assigns to each point of the body a unique real number within an open interval of the real line. Based on this observation, it turns out to be convenient to arbitrarily choose a configuration as a reference, so that quantities defined over the material body can be explicitly expressed in terms of coordinate formulas. We will denote this *reference configuration* by some special symbol such as κ_0.

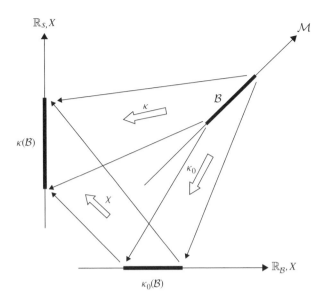

Figure 1.8   Configurations and deformation

Definition 1.3.2   Given two configurations, κ_0 and κ, the *deformation* of the second relative to the first is defined as the composition:

$$\chi = \kappa \circ \kappa_0^{-1}. \tag{1.3}$$

It is important to notice that the inverse function κ_0^{-1} is well defined over the image $\kappa_0(\mathcal{B})$, since configurations are by definition one-to-one. Figure 1.8 depicts the notion of deformation. It is customary to refer to a configuration other than the reference as the *current* or *spatial* configuration. The deformation is thus the map that assigns to each point of the reference configuration a point in the current configuration.

Conceptually, it is convenient to distinguish between the copy of $\mathbb{R}$ where the reference configuration lives and the copy of $\mathbb{R}$ where the current configurations take place. In Figure 1.8 we denote these two copies by $\mathbb{R}_B$ and $\mathbb{R}_S$, respectively.

A nice way to visualize the difference between the reference space $\mathbb{R}_B$ and the physical space $\mathbb{R}_S$ in a biomechanical context is to imagine that the former represents the space where samples are prepared for experiment. For, example, a tendon or a muscle fibre may initially lie on a table where markers and electrodes are placed. This is the world of $\mathbb{R}_B$. Once the preparation has been accomplished, the specimen is placed between, say, the jaws of a tensioning device. This is the world of $\mathbb{R}_S$, where deformations take place.

The usefulness of a fixed (albeit arbitrarily chosen) reference configuration resides in the fact that all configurations can be referred to it so that the notion of configuration can be advantageously replaced by the notion of deformation, thereby making coordinate expressions explicitly available to describe the physical phenomena at hand. Denoting the running coordinate in the reference configuration κ_0 by X and in the current configuration κ by x, the deformation χ is expressed in coordinates by some function

$$x = x(X). \tag{1.4}$$

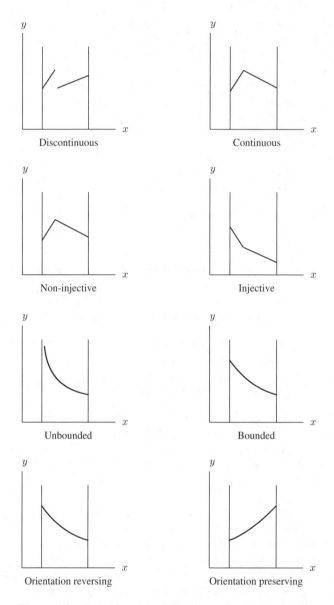

Figure 1.9   Some features of functions of a single variable

This function is continuous and one-to one. For physical reasons, we will impose two further restrictions. The first restriction stems from the fact that we would like deformations to be *orientation preserving* (so that what we call left and right in the reference configuration is consistent with what we call left and right in all other configurations). The second restriction is slightly more technical. We will assume that all deformations (and their inverses) are differentiable functions and that the derivatives are bounded.[2] Figure 1.9 illustrates various properties of functions that have been mentioned above.

---

[2] This condition can be further relaxed by not requiring differentiability and replacing the boundedness by a Lipschitz condition.

Although in principle we could adopt different measuring units in $\mathbb{R}_B$ and $\mathbb{R}_S$, we will assume that the same units (inches, millimetres, etc.) are used in both. As far as the material universe $\mathcal{M}$ is concerned, it is devoid of units since, as may be recalled, it just represents all possible collections of particles and their topological properties (such as the notions of open and closed intervals) without any metric connotation. For practical purposes, one may identify the body with any one of its reference configurations, thus avoiding the need for the material universe, with the proviso that the reference configuration is an arbitrary choice.

1.4 The Deformation Gradient

Having assumed the deformation to be differentiable, it is natural to consider the meaning of its derivative.

Definition 1.4.1 Given a (differentiable) deformation $x = x(X)$, the *deformation gradient* at a body point X_0 is defined as the derivative

$$F = \left.\frac{\partial x}{\partial X}\right|_{X=X_0}. \tag{1.5}$$

The reason for using partial derivatives is that, as we will soon establish, the deformation is in general considered also a function of time. Since the derivative can be evaluated at each and every body point, the deformation gradient can be considered as a function of X. We note that by the assumption of preservation of orientation the deformation is always a monotonically increasing function, whence we obtain

$$F > 0. \tag{1.6}$$

Notice that, by virtue of its definition, the deformation gradient is non-dimensional.

We now explore the physical meaning of the deformation gradient. In this regard, it is useful to adopt Leibniz's view of the notion of derivative (at a point X_0) as a linear function that maps increments dX of the independent variable into increments dx of the dependent variable according to the formula

$$dx = \left.\frac{\partial x}{\partial X}\right|_{X=X_0} dX. \tag{1.7}$$

The derivative is thus a local linear approximation to the function, as shown in Figure 1.10. In our context, we write

$$dx = F\ dX. \tag{1.8}$$

If we regard the differential dX as a small piece of the body around a point X_0 in the reference configuration, the deformation gradient tells us the (approximate) size $dx = F\ dX$ of the piece made up of the same material points in the current configuration, as shown in Figure 1.10. The two pieces will be of the same size if, and only if, it so happens that, at the point in question, $F = 1$. We may say, therefore, that a unit value of the deformation gradient corresponds to a rigid deformation of a small neighbourhood of the point. If the deformation gradient is greater than 1, we have an increase in length (or *extension*). A deformation gradient less than 1 corresponds to a decrease in length (or *contraction*).

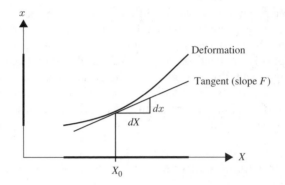

Figure 1.10   The deformation gradient

1.5   Change of Reference Configuration

It is important to bear in mind that the deformation gradient depends on *both* the current and the reference configuration. We want to investigate what happens to the deformation gradient when a different choice of reference configuration is made. The conceptual scheme is illustrated in Figure 1.11. Let κ_0 and κ_1 denote two different reference configurations and let X and Y denote the coordinates in the respective referential spaces $\mathbb{R}_{\mathcal{B}_0}$ and $\mathbb{R}_{\mathcal{B}_1}$. Then, the *change of reference configuration* κ_{10} from κ_0 to κ_1 is defined by the composition

$$\kappa_{10} = \kappa_1 \circ \kappa_0^{-1}, \tag{1.9}$$

or, in coordinates, by some function

$$Y = Y(X). \tag{1.10}$$

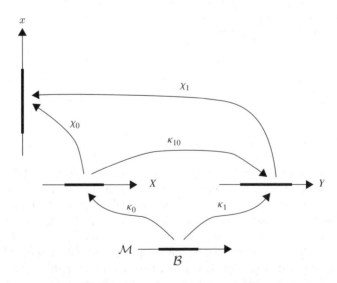

Figure 1.11   Change of reference configuration

This function is smooth, orientation preserving and smoothly invertible. As such, we may calculate the *gradient of the change of reference* at X as the (positive) scalar

$$H = \frac{dY}{dX}. \tag{1.11}$$

Let us denote the deformations relative to κ_0 and κ_1 by χ_0 and χ_1, respectively. They are related by the composition

$$\chi_0 = \chi_1 \circ \kappa_{10}, \tag{1.12}$$

as can be inferred from Figure 1.11. In terms of coordinate representations, these deformations are given by some functions

$$x = x(X) \tag{1.13}$$

and

$$x = \hat{x}(Y), \tag{1.14}$$

respectively, where we have placed a hat over the second function to indicate a different functional relationship. Using equation (1.10), we may write

$$x = \hat{x}(Y) = \hat{x}(Y(X)) = x(X), \tag{1.15}$$

as the relation between the two deformations of the same current configuration. We are interested in the relation between the deformation gradients F and $\hat{F}$ of the deformations $x(X)$ and $\hat{x}(Y)$, respectively. Taking the derivative of equation (1.15) with respect to X, we obtain

$$F = \frac{\partial x}{\partial X} = \frac{\partial \hat{x}}{\partial Y} \frac{dY}{dX} = \hat{F}H. \tag{1.16}$$

In short, the chain rule of differentiation provides us with the result that the deformation gradients with respect to different reference configurations are related by multiplication with the gradient of the change of reference configuration. We notice that, since H cannot vanish (it is, in fact, always positive according to our assumption of preservation of orientation), it makes sense to write equation (1.16) alternatively as

$$\hat{F} = FH^{-1}. \tag{1.17}$$

We may thus establish that the arbitrariness in the choice of reference configuration is governed by equation (1.15) in the sense that all kinematical measurements (such as the deformation gradient) will be necessarily related by a strict transformation rule (such as equation (1.16)).

1.6 Strain

A notion closely related to the deformation gradient is that of *strain*. One of the reasons for introducing this notion is that we would like to have a quantity that vanishes whenever the neighbourhood of a point undergoes a rigid deformation (rather than attaining a unit value, as the deformation gradient does). An easy way to tackle this issue is to define the strain as

$$e = F - 1. \tag{1.18}$$

An alternative and illuminating way to write this definition is obtained by using Leibniz's notation for the derivative. We have

$$e = F - 1 = \frac{dx - dX}{dX}. \tag{1.19}$$

In other words, the strain is a measure of the relative elongation, namely, the change in length divided by the original length (in the reference configuration) of a small piece of material. Notice that, since F is positive, the range of variation of e is the interval $(-1, \infty)$, where the lower limit corresponds to the collapse of a segment to a point.

The measure of strain e just defined, also known as the *engineering strain*, is not the only possibility. In three-dimensional problems (when the deformation gradient is represented not by just a number but by a matrix) it turns out that a more convenient and legitimate measure of strain is obtained by considering first a squared measure of the deformation gradient. In our one-dimensional context this measure, called the (right) *Cauchy–Green* scalar, is defined as

$$C = F^2. \tag{1.20}$$

Note that this number preserves the above mentioned property of the deformation gradient in the sense that it attains a unit value if and only if the neighbourhood preserves its length (i.e., undergoes a rigid deformation). Thus, we can define a new measure of strain by means of the formula:

$$E = \frac{1}{2}(C - 1). \tag{1.21}$$

This strain measure is called the *Lagrangian strain*. As desired, it vanishes for rigid deformations. The presence of the factor $\frac{1}{2}$ requires some further justification. It stems from the following condition to be imposed on all strain measures.

Condition 1.6.1 Small strains. For *small strains*, all strain measures must reduce to the engineering strain e.

Recalling that all strain measures are non-dimensional (or imposing this as a restriction) and that for rigid deformations all strain measures must vanish, a *small strain* is, by definition, a number with an absolute value that is very small compared to the number 1 or, more precisely, a number whose square is very small when compared with its absolute value. Typically, a small strain is of the order of 10^{-2} or smaller. Small strains in biomechanical applications are typical of hard tissue, such as bone, tooth and hair under tension. Soft tissues, such as skin, muscle and connective tissue, can undergo much larger strains. Let us check whether the condition just proposed is verified by the Lagrangian strain given by equation (1.21). Using this equation in conjunction with equations (1.18) and (1.20), we can write

$$E = \frac{1}{2}(C - 1) = \frac{1}{2}(F^2 - 1) = \frac{1}{2}(F + 1)(F - 1) = \frac{1}{2}(e + 2)e = e + \frac{1}{2}e^2. \tag{1.22}$$

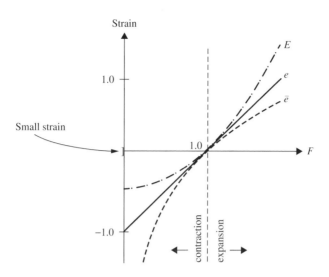

Figure 1.12   Various strain measures

For small strains the last (quadratic) term is negligible compared to e, so that we have that $E \approx e$. In this derivation, the factor $\frac{1}{2}$ plays a decisive role.

Another useful measure of strain is the *logarithmic strain*, given by

$$\bar{e} = \ln F, \tag{1.23}$$

where 'ln' denotes the natural logarithm. Notice that this definition makes sense, since F is strictly positive. Moreover, the logarithmic strain vanishes for rigid deformations, as expected for a measure of strain. The range of variation of the logarithmic strain is the whole line $(-\infty, \infty)$. To check the fulfilment of the small-strain condition, we recall that the natural logarithm function has a convergent Taylor expansion around the unit argument given by

$$\ln F = F - \frac{1}{2}F^2 + O(F^3), \tag{1.24}$$

which proves that the small-strain condition is indeed satisfied by the logarithmic strain. Clearly, an equivalent way to formulate the small-strain condition is to require that all measures of strain (defined analytically in terms of the deformation gradient) must have a unit derivative at $F = 1$. There are many other (in fact, infinite) possible measures of strain satisfying this condition, the choice being a matter of convenience. Figure 1.12 illustrates this fact for the three measures of strain discussed in this section.

1.7 Displacement

It may seem surprising that the concept of displacement, sometimes thought of as being the fundamental kinematic field, has not arisen yet in our treatment and, more particularly, that it was not necessary to invoke it in order to define the important notion of strain. What we have called a configuration may also be called a *placement*, namely a positioning of each point of a body in space. A deformation is thus the passage from one placement

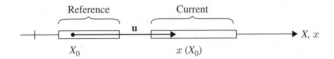

Figure 1.13 Displacement vector

(the reference configuration) to another (the current configuration) and one might be tempted, if only for linguistic reasons, to call it a *dis-placement*. But the usual concept of displacement requires the introduction of additional geometrical structure into the picture, a structure that, as already remarked, does not need to be assumed at all, except as a matter of convenience.

More specifically, to introduce the notion of displacement, we need to immerse both the reference space $\mathbb{R}_\mathcal{B}$ and the physical space $\mathbb{R}_\mathcal{S}$ in a common all-encompassing space. Alternatively, we may choose to identify $\mathbb{R}_\mathcal{B}$ with $\mathbb{R}_\mathcal{S}$ and adopt a common coordinate origin and the same orientation. When so doing, the referential and spatial coordinate axes, X and x, become identical, while keeping the distinction of their roles as labels, respectively, of points in the reference and in the current configurations. Adopting this policy, the displacement of a point X_0 is, by definition, the vector $\mathbf{u}$ issuing from X_0 and ending at $x(X_0)$, as shown in Figure 1.13.

Since in a one-dimensional context (recalling that $\mathbb{R}_\mathcal{B}$ has been identified with $\mathbb{R}_\mathcal{S}$) a vector is represented by a single scalar, positive if pointing forward and negative if backward, we conclude that the displacement associated with a given deformation is the scalar field $u(X)$ defined by the formula

$$u(X) = x(X) - X. \tag{1.25}$$

Denoting the *displacement gradient* by ∇u, we obtain the following equations:

$$\nabla u = \frac{\partial u}{\partial X} = F - 1, \quad e = \nabla u, \quad E = \nabla u + \frac{1}{2}(\nabla u)^2. \tag{1.26}$$

In particular, we observe that the engineering strain is identical to the displacement gradient.[3]

1.8 Motion

A *motion* is a collection of configurations smoothly parametrized by time. In terms of coordinates, a motion is represented by a function of two variables,

$$x = x(X, t), \tag{1.27}$$

where the first argument extends over a fixed reference configuration and the second over some closed time interval $[t_0, t_1]$. The assumption of smoothness can be relaxed, but we will always assume that the motion is at least twice differentiable with respect to the time variable.

---

[3] For those even superficially acquainted with the more general setting, it hardly needs to be remarked that, in a three-dimensional context, the situation is more involved and that the very definition of large engineering strains is cast into doubt (see Chapter 5).

Recalling that X plays the role of labelling body points, we define the *trajectory of a body point* X_0 as the curve

$$x_0(t) = x(X_0, t). \tag{1.28}$$

We are using the term 'curve' in the sense of a *parametrized* curve. In our spatially one-dimensional setting the image of the curve is clearly contained within the straight line $\mathbb{R}_S$.

Motions can be regarded both as collections of configurations and as collections of trajectories. Graphically speaking, if we represent the motion in equation (1.27) as a surface, this surface can be combed in two ways, namely, by means of trajectories or by means of configurations. Notice that, while the configurations are monotonically increasing curves, this is not necessarily the case for the trajectories (a particle may reverse the sense of its motion, for example). Figure 1.14 is a Mathematica®-generated plot of the motion

$$x = 0.4X(X+1)(1 + \sin t) + X, \quad 0 < X < 1, \quad 0 < t < 2\pi. \tag{1.29}$$

This might be the motion of a piece of tendon of unit reference length, one of whose ends is fixed while the other is attached to a reciprocating motor. The figure clearly shows how the surface representing the motion may be spanned by either one of two families of curves, representing trajectories ($X = $ constant) or configurations ($t = $ constant).

We define the *velocity* and the *acceleration* at a point X and at time t as the derivatives

$$v(X,t) = \frac{\partial x(X,t)}{\partial t} \tag{1.30}$$

and

$$a(X,t) = \frac{\partial v(X,t)}{\partial t} = \frac{\partial^2 x(X,t)}{\partial t^2}. \tag{1.31}$$

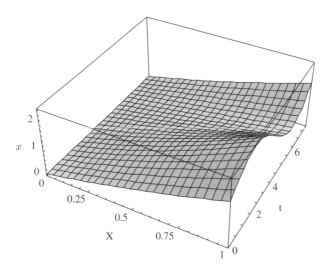

Figure 1.14   A motion $x = x(X,t)$

As before, since the variable X plays the role of labelling body points, we may also say that these expressions represent the velocity and acceleration of the material particle X. Letting X and t vary over their domains of definition, we obtain the *velocity* and *acceleration fields*. These fields are defined so far over the reference configuration. Notice that in these definitions the motion can be replaced by the displacement as a function of X and t.

Exercise 1.8.1   Even when we identify the referential and spatial axes, the reference configuration need not be a configuration actually occupied by the body at any point of time. (a) Show that, for the motion defined by equation (1.29), the spatial configuration does indeed coincide with the reference configuration at specific times. What are these times? (b) Propose a small modification of equation (1.29) to produce a different motion for which the spatial configuration never coincides with the reference configuration.

Exercise 1.8.2   For the motion defined by equation (1.29): (a) Obtain the engineering strain and the Lagrangian strain. (b) Find the time ranges for which these two measures of strain agree over the whole range of the body within a tolerance of 1% . (c) What is the maximum absolute value of the strain in those time ranges?

Exercise 1.8.3   For the motion (1.29), it is proposed to change the reference configuration to that occupied by the body at time $t = 0$. (a) Denoting by Y the corresponding reference coordinate, rewrite the equation of motion with respect to the new reference configuration. What is the range of Y occupied by the body? (b) Repeat Exercise 1.8.2 for the new description of the motion. Are the time ranges the same? Why, or why not?

Exercise 1.8.4   Yet another change of reference configuration is suggested. It consists of stretching the original reference configuration by means of a constant engineering strain of 0.2. Denoting the new reference coordinate by Z, rewrite the equation of motion with respect to the new reference configuration. Comment on the possible agreement of the engineering and the Lagrangian strains in this case. Does the body ever occupy the suggested reference configuration?

1.9   The Lagrangian and Eulerian Representations of Fields

We have already introduced several kinematic fields: the motion itself, the deformation gradient, the various strain fields, the displacement, the velocity and the acceleration. At any fixed instant of time, these fields are defined over the body in the reference configuration, namely, over the open interval $\kappa_0(\mathcal{B})$ of $\mathbb{R}_{\mathcal{B}}$. Letting time vary over its domain of interest $[t_0, t_1]$, we obtain what may be called a body–time field, that is, a function of X and t defined over the rectangle with sides $\kappa_0(\mathcal{B})$ and $[t_0, t_1]$ or, more precisely, over the Cartesian product $\kappa_0(\mathcal{B}) \times [t_0, t_1]$. Any such function $\psi(X, t)$ will be called a *Lagrangian* (or *referential*) *field*. In addition to the fields already introduced, we will encounter other fields of interest, such as temperature fields, stress fields, and so on.

A Lagrangian field assigns a value of some physical quantity to each material particle at each instant of time. But suppose now that, instead of adopting this point of view, we place measurement devices (such as thermometers) at many fixed locations spread over the physical space $\mathbb{R}_S$ and record their readings as time passes. Collecting the results of these measurements at each spatial location x and at each time t, we obtain a function $\psi(x, t)$. A field defined in this manner is called an *Eulerian* (or *spatial*) *field*.

Remark 1.9.1   The space-time domain of an Eulerian field. There are two possibilities that may arise as the body moves. At any given location x of the measuring device and at any given instant of time t, there either happens to be a material particle or not. If the former is the case, we obtain a reading $\psi(x,t)$. Otherwise, we either do not obtain any reading or the reading is physically meaningless, since there is no particle associated with that place and time. Except for the case in which the body is the whole real line, this situation is likely to arise at some points and at some instants. Collecting the results of the measuring devices, at each instant of time t we certainly obtain a field, namely a function of x defined over the current configuration, that is, the open interval $\chi(\kappa_0(\mathcal{B}),t)$ of $\mathbb{R}_S$. This open interval, however, is a function of time. In other words, the field $\psi(x,t)$ is no longer defined over a rectangle, but over a more sinuous two-dimensional figure, as illustrated in Figure 1.15 for the motion described in equation (1.29).

Our immediate interest is in establishing a precise connection between the Lagrangian and Eulerian versions of the field associated with the same physical variable. In the case of temperature, for instance, the Lagrangian version would correspond to having thermocouples glued to various body particles which, thus, accompany the body in its motion. The Eulerian version corresponds to placing the thermocouples at fixed spatial positions. Leaving aside the practical details of such experimental arrangements, it is clear that both thought experiments make perfect physical sense. Given a Lagrangian field $\psi = \psi(X,t)$, we denote the corresponding Eulerian field by $\hat{\psi}(x,t)$. This notational distinction is not always necessary, since it is usually clear in each case which the independent variables are. Nevertheless, we want to emphasize the different functional dependence for the sake of clarity at this junction. The functions $\psi = \psi(X,t)$ and $\hat{\psi}(x,t)$ are related via the motion by the following equation:

$$\hat{\psi}(x,t) = \hat{\psi}(x(X,t),t) = \psi(X,t). \qquad (1.32)$$

A field is called *steady* or *stationary* if the function $\hat{\psi}(x,t)$ happens to be independent of time.

It is often necessary to evaluate the gradient of a given field, namely, the derivative of the field with respect to the space-like variable. It follows from equation (1.32) that the

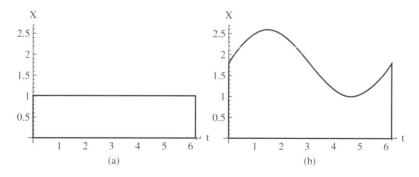

Figure 1.15   Domains of definition of a (a) Lagrangian field (b) and corresponding Eulerian field for the motion of Figure 1.14

referential gradient and the *spatial gradient* are related by

$$\frac{\partial \psi(X,t)}{\partial X} = \frac{\partial \hat{\psi}(x,t)}{\partial x}\frac{\partial x(X,t)}{\partial X} = \frac{\partial \hat{\psi}(x,t)}{\partial x}F(X,t). \tag{1.33}$$

In terms of the gradient notation already used in equation (1.26), we can rewrite equation (1.33) as follows:

$$\nabla \psi = \nabla \hat{\psi}\, F. \tag{1.34}$$

Alternatively, the designations Grad and grad are often used to denote, respectively, the referential and spatial gradients of fields. In terms of this notation, equation (1.34) can be written as

$$\text{Grad }\psi = (\text{grad }\psi)F, \tag{1.35}$$

where the 'hat' has been dropped, since there should not be any room for confusion.

Exercise 1.9.2   An unsteady temperature field has been measured by means of thermo-couples placed at fixed spatial positions. Over the domain represented in Figure 1.15(b), the experimental results have been approximated by the formula

$$\theta = (300 + 20x)(1 + 0.1\cos 0.2t). \tag{1.36}$$

Knowing the motion (1.29), express the temperature as a Lagrangian field over the reference configuration.

Exercise 1.9.3   Calculate the Eulerian velocity field of the motion (1.29) by first obtaining the Lagrangian field and then explicitly inverting the equation of motion as $X = X(x,t)$.

1.10   The Material Derivative

A similar treatment of derivatives with respect to the time variable gives rise to the important notion of the *material derivative* or, more precisely, *material time derivative*. Using, as before, the chain rule of differentiation on equation (1.33), we obtain

$$\frac{\partial \psi(X,t)}{\partial t} = \frac{\partial \hat{\psi}(x,t)}{\partial x}\frac{\partial x(X,t)}{\partial t} + \frac{\partial \hat{\psi}(x,t)}{\partial t} = \frac{\partial \hat{\psi}(x,t)}{\partial t} + \frac{\partial \hat{\psi}(x,t)}{\partial x}v(X,t), \tag{1.37}$$

where $v(X,t)$ is the referential version of the velocity field. Equation (1.37) can also be written as

$$\frac{\partial \psi(X,t)}{\partial t} = \frac{\partial \hat{\psi}(x,t)}{\partial t} + (\text{grad }\psi)\, v(X,t). \tag{1.38}$$

Notice that on the left-hand side of this equation, it is the particle that is held fixed while taking the limit, while on the right-hand side it is the spatial position that is held fixed. To avoid the need to explicitly write the independent variables and/or indicate by a hat the different functions involved, the following convention is used for time derivatives: if the particle is held fixed, then the time derivative is denoted by $\frac{D}{Dt}$, while the notation $\frac{\partial}{\partial t}$ is retained whenever the spatial position is kept fixed. In terms of this convenient notation, equation (1.38) can be written as

$$\frac{D\psi}{Dt} = \frac{\partial\psi}{\partial t} + (\text{grad } \psi)\, v. \tag{1.39}$$

The derivative $\frac{D\psi}{Dt}$ of a field ψ with respect to time while keeping the material particle fixed (whether calculated directly from the Lagrangian representation or indirectly by the right-hand side of equation (1.39)) is known as the material derivative of the field. For a stationary field, equation (1.39) yields the following interesting result:

$$\left.\frac{D\psi}{Dt}\right|_{\text{stationary}} = (\text{grad } \psi)\, v. \tag{1.40}$$

This means that the 'correction' term between the spatial and referential partial derivatives has to do with the spatial gradient and with the velocity. This corrective term is also known as *advection*.

It may prove instructive to provide an intuitive picture of the meaning of the advection term appearing in the calculation of the material derivative out of purely spatial data. Imagine that two observers are placed at two nearby fixed positions in the laboratory separated by a small space interval Δx, as shown in Figure 1.16. They are charged with observing some phenomenon, such as the flow of blood through an artery narrowed by atherosclerosis, and are supposed to record the readings (perhaps temperatures or velocities) of instruments fixed at their respective stations as time goes on. If the field happens to be stationary, the readings will be constant in time. Let these constant readings be ψ and $\psi + d\psi$. A particle (a blood cell, say) that passes in front of the first observer at time t with velocity v will pass in front of the second observer at time $t + \frac{\Delta x}{v}$. This means that, in going from the first observer to the second in an interval of time $\Delta t = \frac{\Delta x}{v}$, this particle undergoes an increase in the quantity ψ precisely equal to the value $\Delta\psi$. The rate of change of ψ at the particle in question is therefore given by

$$\left.\frac{\Delta\psi}{\Delta t}\right|_{\text{particle}} = \frac{\Delta\psi}{\frac{\Delta x}{v}} = \frac{\Delta\psi}{\Delta x}\, v. \tag{1.41}$$

In the limit as $\Delta x \to 0$ we recover equation (1.40).

A particularly interesting application of the concept of material derivative is provided by the calculation of the acceleration. Assume that the velocity field (of a blood flow, for

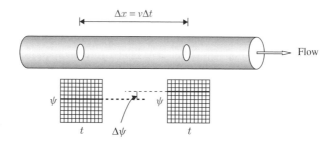

Figure 1.16 Meaning of the advection term

example) has been obtained experimentally as a *spatial* field, namely as

$$v = v(x, t). \tag{1.42}$$

This is the standard situation in fluids, where it is impractical, if not impossible, to follow the motion of particles over long periods of time. How do we obtain the value of the acceleration of the particle instantaneously located at a spatial position x? Since the acceleration is precisely the rate of change of the velocity *following the particle*, the answer to this question is given precisely by the material derivative of the velocity, namely,

$$a(x, t) = \frac{Dv}{Dt}. \tag{1.43}$$

Using equation (1.39), we obtain

$$a(x, t) = \frac{\partial v}{\partial t} + v \operatorname{grad} v. \tag{1.44}$$

Thus, the acceleration field in its Eulerian version contains an advection term which is fundamentally non-linear in the velocity.

Exercise 1.10.1   Obtain the material derivative of the temperature field (1.36) under the motion (1.29) in two different ways: by using the result of Exercise 1.9.2 and, independently, by using the result of Exercise 1.9.3 and equation (1.39). Verify that both results are in agreement with each other. Notice that the second method would be the only possibility available if the Eulerian velocity field alone were known, rather than the motion itself. The Eulerian velocity field may be available through experimental measurements, just like the given temperature field.

1.11   The Rate of Deformation

An important question that arises in physics is whether or not a quantity that can be measured has an intrinsic physical meaning independent of the observer making the measurement. Clearly, this question is also of great importance in continuum mechanics. There is, however, an additional problem of a somewhat similar nature peculiar to continuum mechanics, a problem that can be traced back to the introduction of the concept of reference configuration. Consider, for example, the fundamental notion of deformation gradient. By its very definition as a map from a reference configuration into physical space, it is clear that the deformation gradient depends on the reference configuration adopted. In fact, equation (1.16) shows explicitly how the deformation gradient is affected by a change of reference configuration. This dependence is transferred automatically to the various measures of strain. The velocity, on the other hand, although obviously dependent on the observer's state of motion, is independent of the reference configuration.

With these ideas in mind, we consider the material derivative of the deformation gradient and investigate how it is affected by a change of reference configuration. Starting from equation (1.16), and bearing in mind that a change of reference configuration is independent of time, we obtain

$$\frac{DF}{Dt} = \frac{D\hat{F}}{Dt} H. \tag{1.45}$$

We conclude that the material derivative of the deformation gradient still depends on the reference configuration, much in the same way as the deformation gradient itself. Consider, however, the combination

$$L = \frac{DF}{Dt}F^{-1}. \tag{1.46}$$

It is not difficult to verify that, by virtue of equations (1.16) and (1.45), this expression is independent of the reference configuration. We call this quantity L the *rate of deformation* or the *stretching*.[4]

In trying to justify this terminology, we observe that, if one were to adopt (as one certainly may) the configuration at time t as a reference configuration, the deformation gradient at any particle at time t would be equal to 1, and the deformation gradient at the same particle and at a nearby time $t + \Delta t$ would be very close to 1, say $1 + \Delta F$. Thus, in the limit as $\Delta t \to 0$ the ratio $\frac{\Delta F}{\Delta t}$ delivers L. It can be said, then, that the rate of deformation is the material derivative of the deformation gradient when the present configuration is adopted instantaneously as a reference configuration. Clearly, this is an intrinsic measure of the rate of change of the strain, regardless of which measure of stain is adopted, since all measures of strain coincide for deformation gradients very close to 1.

It is interesting that there is another way to arrive at the same concept. We have already encountered the spatial velocity gradient grad v as it appeared, for example, in equation (1.44), and later to appear in other contexts. Using equation (1.35), we obtain

$$\text{grad } v = (\text{Grad } v)F^{-1} = \frac{\partial}{\partial X}\frac{\partial x(X,t)}{\partial t}F^{-1}$$
$$= \frac{\partial}{\partial t}\frac{\partial x(X,t)}{\partial X}F^{-1} = \frac{\partial F(X,t)}{\partial t}F^{-1} = \frac{DF}{Dt}F^{-1} = L. \tag{1.47}$$

In this derivation, we use the fact that mixed *partial* derivatives are commutative. This is in general not true when the material derivative is expressed in terms of the spatial variable. It is for this reason that we were careful to indicate the independent variables involved at each of the steps of the derivation. We have arrived at the important kinematic result stating that the rate of deformation is equal to the spatial velocity gradient, which is usually readily amenable to direct measurement.

Exercise 1.11.1 Show that the rate of deformation L equals the material derivative of the logarithmic strain.

Exercise 1.11.2 Obtain the velocity gradient of the motion (1.29) by differentiation of the result of Exercise 1.9.3. Verify that this result is in agreement with equation (1.46) regardless of which reference configuration is used, whether it is the original one, the ones suggested in Exercises 1.8.3 of 1.8.4 or, for that matter, any other reference configuration.

1.12 The Cross Section

Always faithful to our policy of focusing on one-dimensional bodies, we must recognize that in many applications it may prove convenient to restore, at least in part,

---

[4] Although we try to preserve in this one-dimensional setting many of the notational features of the three-dimensional counterpart (such as carefully distinguishing between left and right multiplication and replacing the division by a scalar with multiplication by its reciprocal), there are some important differences attributable mainly to the fact that in the one-dimensional setting there are no rotations.

the three-dimensionality of the body by means of the notion of the *cross section*. Our one-dimensional bodies are essentially three-dimensional material entities in which all the 'action' takes place along a straight line, but the material situated in planes perpendicular to this line will also be affected by the process of deformation. We may say that the notion of the cross section is a price to be paid in exchange for the relative simplicity of a one-dimensional theory vis-à-vis the complexity of a fully-fledged three-dimensional setting. Denoting by $\hat{S}(X)$ the area of the cross section in the reference configuration, its counterpart in the current configuration will be some function

$$\hat{s} = \hat{s}(x,t). \tag{1.48}$$

This function will have to be prescribed by an *ad hoc* kinematic assumption representing the physical situation at hand.

We consider two important examples. The first example is, in fact, a very common occurrence in biomechanics whenever the tissue under study is, within experimental tolerance, *incompressible*. Incompressibility (or volume preservation) arises very often in applications because many biological tissues are made of a large proportion of water, which is nearly incompressible. Equating the referential and current volumes of an infinitesimally thin slice, we obtain

$$\hat{S}(X)dX = \hat{s}(x(X,t),t)\, dx, \tag{1.49}$$

which can also be written, with some abuse of notation, as

$$\hat{s} = \frac{\hat{S}}{F}. \tag{1.50}$$

We can render this condition independent of the reference configuration by taking the material derivative. The result is

$$\frac{D\hat{s}}{Dt} = -\hat{s}\, L. \tag{1.51}$$

Exercise 1.12.1   Prove equation (1.51) by differentiation of (1.50).

The second example corresponds to the flow of a fluid within a tapered rigid pipe or to a solid confined compression test. In this case, the cross sections at the spatial locations are prescribed by the shape of the pipe as some function $f(x)$ independent of time. We have, therefore,

$$\hat{s}(x,t) = f(x), \tag{1.52}$$

or, taking the material derivative,

$$\frac{D\hat{s}}{Dt} = \frac{df}{dx}\, v. \tag{1.53}$$

For a pipe of constant cross section, this condition reduces to

$$\frac{D\hat{s}}{Dt} = 0, \tag{1.54}$$

as expected.

Exercise 1.12.2 By invoking equations (1.51) and (1.53), prove that, if a fluid flowing within a rigid pipe is incompressible, the product $f(x)\,v(x,t)$ is independent of position x along the pipe. Justify this result on intuitive grounds. Notice that, due to the self-imposed limitations of a one-dimensional theory, the longitudinal velocity v is tacitly assumed to be constant throughout each cross section.

Exercise 1.12.3 A rigid pipe of known variable cross section $s = s(x,t)$ carries water between an inlet at $x = 0$ and an outlet at $x = L$. Express the acceleration of the water $a = a(x,t)$ in terms of the shape of the pipe and the velocity $v_L(t) = v(L,t)$ measured at the outlet.

2

Balance Laws

2.1 Introduction

Conservation laws or, more generally, *balance laws* are the result of a complete accounting
of the variation in time of the content of an *extensive* physical quantity in a given domain.
A useful analogy is the following. Suppose that we are looking at a big farming colony (the
domain of interest) and we want to focus attention on the produce (the physical quantity
of interest). As time goes on, there is a variation in the quantity of food contained in
the domain. At any given instant of time, we want to account for the rate of change of
this food content. There are some internal *sources* represented in this case by the rate at
which the land yields new produce (so many tons per week, say). There are also *sinks* (or
negative sources) represented by the internal consumption of food by workers and cattle,
damage caused by hail and pests, etc. We will call these sources and sinks the *production*
of the quantity in question. It is measured in units of the original quantity divided by the
unit of time.

In addition to these internal factors, there is also another type of factors that can cause
a change in content. We are referring to exchanges of food *through the boundary* of the
colony. These include the buying and selling of produce taking place at the gates, the
perhaps illegal activities of some members or visitors who personally take some food
away to other destinations, etc. At any given instant of time, we can estimate the rate
at which these exchanges take place at the boundary. We will call these transactions the
flux of the quantity in question. We may also have a flux arising from the fact that the
boundary of the domain of interest is changing (encroached upon by an enemy or by
natural causes, for example). Assuming that we have accounted for every one of these
causes, we may write the generic equation of balance in the following way:

$$\frac{d \text{ (content)}}{dt} = \text{production} + \text{flux}. \tag{2.1}$$

By convention, we consider the flux positive if it is an inflow and negative otherwise.

Equation (2.1) is a *global* statement, in the sense that it establishes the balance of a
physical quantity as it applies to the total domain of interest. In physically meaningful
examples (balance of energy, momentum, mass, electric charge, and so on), it is often the
case that the content, the production and the flux are somehow distributed or smeared out

The Elements of Continuum Biomechanics, First Edition. Marcelo Epstein.
© 2012 John Wiley & Sons, Ltd. Published 2012 by John Wiley & Sons, Ltd.

over the volume (in the case of the content and the production) or over the area of the boundary (in the case of the flux). In other words, these magnitudes are given in terms of *densities*, which vary from point to point and from one instant to the next. It is precisely this feature, whether real or assumed, that is responsible for the fact that we may be able to express the basic equation of balance *locally* in terms of differential equations. These differential equations are obtained by assuming that equation (2.1) applies to any subdomain, no matter how small, and then applying it to an infinitesimal volume.

The precise mathematical way to achieve the passage from the global to the local form of the balance equation requires the use of the *divergence theorem* of vector calculus. In the case of just one dominant spatial dimension, however, the divergence theorem mercifully reduces to the statement of the fundamental theorem of calculus of one variable (roughly speaking: 'differentiation is the inverse of integration'). For this reason, and also because the main ideas are preserved even in the case of just one spatial dimension, we will continue working for a while under the assumption that we have a one-dimensional spatial domain. As a consequence, we are left with only two independent variables, one to describe the spatial extent of the system (X or x) and one for the evolution of the system in time (t).

Corresponding to the two extreme points of view, Lagrangian and Eulerian, associated with the kinematics of a deformable material body, there exist two possible formulations of the laws of balance of an extensive property. Accordingly, we will deal with these two points of view in succession.

2.2   The Generic Lagrangian Balance Equation

2.2.1   Extensive Properties

As already pointed out, many of the quantities of interest in continuum mechanics represent *extensive properties*, such as mass, momentum and energy. An extensive property assigns a value to *each part of the body*. From the mathematical point of view, an extensive property can be regarded as a *set function*, in the sense that it assigns a value to each subset of a given set. Consider, for example, the case of the mass property. Given a material body, this property assigns to each sub-body its mass. Other examples of extensive properties are: volume, electric charge, internal energy, linear momentum. *Intensive properties*, on the other hand, are represented by *fields*, assigning to *each point of the body* a definite value. Examples of intensive properties are: temperature, displacement, strain.

As the example of mass clearly shows, very often the extensive properties of interest are *additive set functions*, namely, the value assigned to the union of two disjoint subsets is equal to the sum of the values assigned to each subset separately. Under suitable assumptions of continuity, it can be shown that an additive set function is expressible as the integral of a *density* function over the subset of interest. This density, measured in terms of property per unit size, is an ordinary pointwise function defined over the original set. In other words, the density associated with a continuous additive set function is an intensive property. Thus, for example, the mass density is a scalar field. The relation between an additive set function and its corresponding density is suggested in the following exercise.

Exercise 2.2.1   A scalar-valued set function is additive if its value on the union of any two disjoint subsets is equal to the sum of the values on the individual subsets. Let $\mathcal{G}$ be a set function defined on arbitrary (measurable) subsets $\mathcal{C}$ of $\mathbb{R}$ by means of the integral

$$\mathcal{G}[\mathcal{C}] = \int_{\mathcal{C}} G(X)\, dX, \tag{2.2}$$

where $G(X)$ is an integrable function. Verify that a set function $\mathcal{G}$ thus defined is automatically additive.

In the case of continuum mechanics, we will be interested not only in extensive properties defined over a body, but also in extensive properties defined over its boundary. The reason for our interest in these two kinds of extensive properties is implicit already in equation (2.1), where the content and the production are of the first kind, while the flux is of the second kind. Recall that we are idealizing our continuum as a one-dimensional entity. For specificity, you may want to think of $\mathcal{G}$ as the total elastic energy content in a tendon or a myofilament. It is important to realize that, since we will be considering only one spatial dimension, the lateral surface of this filament does not count as part of the boundary. On the contrary, the points on this small lateral surface are identified precisely with the interior points of the segment. What, then, is the boundary of the body? The answer is: the two end points, as shown in Figure 2.1. The density of an extensive property defined over a one-dimensional body is expressed in units of this property divided by units of length. The density of a property over the boundary of a one-dimensional body is expressed in units of the property itself.

Remark 2.2.2   If the spatial domain were two-dimensional, as in the case of a membrane, its boundary would be the perimeter curve, while the upper and lower faces of the membrane would be identified with the interior points. For a three-dimensional domain, the boundary is the whole bounding surface. In this case, the units of the density defined over the boundary of the body would be measured in units of this property divided by units of area.

2.2.2   *The Balance Equation*

Denoting the content, production and flux, respectively, by $\mathcal{G}$, $\mathcal{P}$ and $\mathcal{F}$, equation (2.1) reads:

$$\frac{d\mathcal{G}}{dt} = \mathcal{P} + \mathcal{F}. \tag{2.3}$$

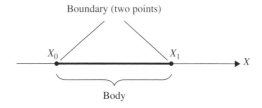

Figure 2.1   A one-dimensional body and its boundary

Window 2.1   The flux vector

The flux term is evidenced in the process of mentally cutting the body into two pieces, thus creating a new boundary point at each of the two sub-bodies. If we assume (following an idea similar to Newton's law of action and reaction) that whatever flow enters through the new boundary into one of the parts must necessarily be coming out of the new boundary of the other part, we realize that the flux is best represented by the dot product of a *flux vector* $\mathbf{H} = \mathbf{H}(X,t)$ with the unit vector (at the new boundary) *pointing in the outward direction of the part under study*. We call this last vector the *exterior unit normal*. We have two such vectors, $\mathbf{N}_L$ and $\mathbf{N}_R$, on the left and right sub-bodies, respectively, as shown in Figure 2.2. At the point X at time t, we obtain the following values for the flux entering the left and the right parts, respectively:

$$H_L = -\mathbf{H} \cdot \mathbf{N}_L, \quad H_R = -\mathbf{H} \cdot \mathbf{N}_R.$$

The reason for the minus sign in these expressions is that when the flux vector points to the right, the scalar flux through the cross section will be negative for the left part of the bar and positive (and of equal absolute value) for the right part of the bar. In other words, a right-pointing flux vector corresponds to an influx into the right part, according to our previously established sign convention. Notice that in our one-dimensional case, the flux vector has only one component, which we denote by $H = H(X,t)$. A positive value of this component corresponds to a flow of the physical quantity of interest to the right. Although in the one-dimensional case one can dispense with the vector approach, the concept of flux vector is very important and can be used directly in two- and three-dimensional spatial contexts.

New boundary pair

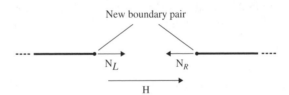

N_L N_R

H

Figure 2.2   The flux vector

If we assume that the quantity $\mathcal{G} = \mathcal{G}(t)$ is a continuous additive set function, we can express it in terms of a density $G = G(X,t)$ per unit length of the bar as

$$\mathcal{G}(t) = \int_{X_0}^{X_1} G(X,t)\,dX. \tag{2.4}$$

Similarly, the production $\mathcal{P} = \mathcal{P}(t)$ can be expressed in terms of a density $P = P(X,t)$ per unit length and per unit time as

$$\mathcal{P}(t) = \int_{X_0}^{X_1} P(X,t)\,dX. \tag{2.5}$$

We have assumed that a positive value of P corresponds to creation (source) and a negative value to annihilation (sink).

The flux term requires some further discussion. We imagine an amount of our physical quantity $H = H(X, t)$ flowing within the body and we assume it to be positive if flowing from left to right. In that case, the flux through the two-point boundary is given by

$$\mathcal{F}(t) = H(X_0, t) - H(X_1, t). \tag{2.6}$$

The signs are in agreement with our statement, after equation (2.1), to the effect that a boundary flux is positive if it measures an inflow.

Remark 2.2.3   The scalar flux H can also be expressed in terms of a flux vector **H**, as explained in detail in Window 2.1. This representation is particularly fruitful in the three-dimensional context, to be presented in Section 6.2.

Introducing our specific expressions for content, production and flux into the generic balance equation (2.1), we obtain

$$\frac{d}{dt} \int_{X_0}^{X_1} G(X, t)\, dX = \int_{X_0}^{X_1} P(X, t)\, dX + H(X_0, t) - H(X_1, t). \tag{2.7}$$

This is the equation of balance in its *global* or integral form. It is at this point that, if we wish to make the passage to the *local* or differential form of the equation, we need to invoke the fundamental theorem of calculus (or the divergence theorem in higher-dimensional contexts). Indeed, we can write

$$\int_{X_0}^{X_1} \frac{\partial H(X, t)}{\partial X}\, dX = H(X_1, t) - H(X_0, t). \tag{2.8}$$

Using equation (2.8), we rewrite equation (2.7) as

$$\frac{d}{dt} \int_{X_0}^{X_1} G(X, t)\, dX = \int_{X_0}^{X_1} P(X, t)\, dX - \int_{X_0}^{X_1} \frac{\partial H(X, t)}{\partial X}\, dX. \tag{2.9}$$

There is still a further assumption that needs to be made to obtain the local form. We must claim that equation (2.9) can equally well be applied to any sub-body of the original bar, without any change in the integrands. From the physical point of view, this means that we are disregarding any scale or boundary effects (such as surface tension and non-local interactions). Thus we claim that

$$\frac{d}{dt} \int_{a}^{b} G(X, t)\, dX = \int_{a}^{b} P(X, t)\, dX - \int_{a}^{b} \frac{\partial H(X, t)}{\partial X}\, dX, \tag{2.10}$$

for arbitrary values of a and b, with $X_0 \leq a < b \leq X_1$. We remark that these values, though arbitrary, are independent of time. This remark permits us to exchange in the first term of the equation the derivative with the integral, namely

$$\int_a^b \frac{\partial G(X,t)}{\partial t} \, dX \;=\; \int_a^b P(X,t) \, dX \;-\; \int_a^b \frac{\partial H(X,t)}{\partial X} \, dX, \qquad (2.11)$$

identically for all possible integration limits. We now claim that this identity is possible only if the integrands themselves are balanced, namely if

$$\frac{\partial G}{\partial t} = P - \frac{\partial H}{\partial X}. \qquad (2.12)$$

The truth of this claim can be verified by collecting all the integrands in equation (2.11) under a single integral and then arriving at a combined integrand whose integral must vanish no matter what limits of integration are used. Clearly, if the integrand is continuous and does not vanish at some point in the domain of integration it will also not vanish at any point in a small interval containing that point (by continuity). It will, therefore, be either strictly positive or strictly negative therein. Choosing, then, that small interval as a domain of integration, we would arrive at the conclusion that the integral does not vanish, which contradicts the assumption that the integration must vanish for all values of the limits. We conclude that equation (2.12) must hold true. It asserts that, under certain differentiability assumptions, the global Lagrangian equation of balance is equivalent to the local statement of equation (2.12) to the effect that the time derivative of the density of the balanced quantity is exactly compensated by the density of production minus the gradient of the flux.[1]

Remark 2.2.4   Using angle brackets to denote physical units, the following relations hold between the various quantities introduced:

$$\langle G \rangle = \langle \mathcal{G} \rangle \, \langle L \rangle^{-1}, \quad \langle P \rangle = \langle \mathcal{G} \rangle \, \langle L \rangle^{-1} \, \langle T \rangle^{-1}, \quad \langle H \rangle = \langle \mathcal{G} \rangle \, \langle T \rangle^{-1}, \qquad (2.13)$$

where L and T refer here to generic length and time. In many applications it is convenient to introduce into the one-dimensional treatment the three-dimensionality of the body via the idea of a *cross section* of area $\hat{S}(X)$, as discussed in Section 1.12, and to conceive of densities and fluxes per unit volume and per unit area, respectively. Specifically, if the balanced quantity per unit volume, the production per unit volume and per unit time, and the flux per unit area and per unit time are denoted, respectively, by $\hat{G}$, $\hat{P}$ and $\hat{H}$, we have

$$G = \hat{G}\hat{S}(X), \quad P = \hat{P}\hat{S}(X), \quad H = \hat{H}\hat{S}(X). \qquad (2.14)$$

For example, the mass per unit length (which would enter into the balance equation) is obtained by multiplying the mass density (mass per unit volume) by the cross sectional area.

Remark 2.2.5   The assumption that the flux is differentiable had to be made to be able to use the fundamental theorem of calculus. In this sense it can be said that the global (integral) balance equation (2.7) is more general than its local (differential) counterpart.

---

[1] In the three-dimensional context, the last term is represented by the divergence of the flux vector.

Exercise 2.2.6 Derive equation (2.12) by applying the integral equation of balance to a small (infinitesimal) slice of the bar, that is, to a slice contained between the cross sections at X and at $X + dX$.

A balance equation for which both the production and the flux vanish identically is called a *conservation equation*.

2.3 The Generic Eulerian Balance Equation

In the Eulerian formulation, we start by expressing the quantities of interest in terms of densities per unit *spatial* (rather than referential) length. We must, accordingly, replace equation (2.4) by

$$\mathcal{G}(t) = \int_{x(X_0,t)}^{x(X_1,t)} g(x,t)\, dx. \tag{2.15}$$

We observe that the integrand is now a function of x (rather than X) and t and, more importantly, that the limits of integration are no longer constant but functions of time. Indeed, these limits represent the spatial positions of the ends of the body as prescribed by the motion $x(X,t)$. The relation between the Lagrangian and Eulerian densities is given by

$$G(X,t)\, dX = g(x(X,t),t)\, dx \tag{2.16}$$

or, more briefly,

$$g = G/F. \tag{2.17}$$

Similarly, equation (2.5) is replaced by

$$\mathcal{P}(t) = \int_{x(X_0,t)}^{x(X_1,t)} p(x,t)\, dx, \tag{2.18}$$

with

$$p = P/F. \tag{2.19}$$

The flux, on the other hand, not being given by an integral (since the boundary consists of just two points), remains the same, namely

$$h(x(X,t),t) = H(X,t). \tag{2.20}$$

Accordingly, equation (2.7) is replaced by the global statement

$$\frac{d}{dt} \int_{x(X_0,t)}^{x(X_1,t)} g(x,t)\, dx = \int_{x(X_0,t)}^{x(X_1,t)} p(x,t)\, dx \;+\; h(x(X_0,t),t) - h(x(X_1,t),t). \tag{2.21}$$

As in the Lagrangian case, we can rewrite equation (2.21) as

$$\frac{d}{dt} \int_{x(X_0,t)}^{x(X_1,t)} g(x,t)\,dx = \int_{x(X_0,t)}^{x(X_1,t)} p(x,t)\,dx - \int_{x(X_0,t)}^{x(X_1,t)} \frac{\partial h(x,t)}{\partial x}\,dx. \tag{2.22}$$

On the left-hand side of this expression, however, we no longer claim that the time derivative can be interchanged with the integral. These two limiting processes are not interchangeable because the limits of the integral depend on the variable of differentiation. To overcome this obstacle, we may revert temporarily, by a change of variables (with $dx = F\,dX$), to the Lagrangian domain, as follows:

$$\frac{d}{dt} \int_{x(X_0,t)}^{x(X_1,t)} g(x,t)\,dx = \frac{d}{dt} \int_{X_0}^{X_1} g(x(X,t),t)\,F\,dX = \int_{X_0}^{X_1} \frac{D(g(x,t)\,F)}{Dt}\,dX. \tag{2.23}$$

Notice that, having transformed to the Lagrangian variables whereby the limits of integration are constant, it has become legitimate to exchange the derivative with the integral. Moreover, the time derivative is now to be interpreted as a material derivative (since the independent variables are X and t). The integrand on the right-hand side of equation (2.23) can be further expanded as

$$\frac{D(g(x,t)\,F)}{Dt} = \frac{Dg}{Dt}\,F + g\,\frac{DF}{Dt} = \left(\frac{Dg}{Dt} + gL\right)F, \tag{2.24}$$

where equation (1.46) has been used. Plugging this result in equation (2.23) yields

$$\frac{d}{dt} \int_{x(X_0,t)}^{x(X_1,t)} g(x,t)\,dx = \int_{X_0}^{X_1} \left(\frac{Dg}{Dt} + gL\right) F\,dX. \tag{2.25}$$

We are now in a position to revert to the original Eulerian variables, namely

$$\frac{d}{dt} \int_{x(X_0,t)}^{x(X_1,t)} g(x,t)\,dx = \int_{x(X_0,t)}^{x(X_1,t)} \left(\frac{Dg}{Dt} + gL\right) dx. \tag{2.26}$$

Exercise 2.3.1 Equation (2.26), which can be applied to any Eulerian function $g(x,t)$, is known as *Reynolds' transport theorem* in one dimension. Prove this theorem independently by explicitly writing the derivative as a limit. Equivalently, invoke the well-known *Leibniz integral rule* (6.13) of calculus for the derivative of a definite integral with variable limits.

Introducing equation (2.26) into (2.22), we obtain

$$\int_{x(X_0,t)}^{x(X_1,t)} \left(\frac{Dg}{Dt} + gL\right) dx = \int_{x(X_0,t)}^{x(X_1,t)} p(x,t)\,dx - \int_{x(X_0,t)}^{x(X_1,t)} \frac{\partial h(x,t)}{\partial x}\,dx. \tag{2.27}$$

Claiming now, as we did in the Lagrangian case, that this equation is valid for all possible subdomains of integration, we finally obtain the local Eulerian version of the generic equation of balance as

$$\left(\frac{Dg}{Dt} + gL\right) = p(x,t) - \frac{\partial h(x,t)}{\partial x}. \tag{2.28}$$

Exercise 2.3.2 The derivation of the local Eulerian form of the generic balance equation has been more laborious than expected. Use the following easier shortcut instead. Start with the local Lagrangian form (equation (2.12)) and implement equations (2.17), (2.19) and (2.20). The result (equation (2.28)) follows almost immediately.

Remark 2.3.3 An alternative derivation. Yet another conceptually fertile approach to deriving the Eulerian equation of balance involves considering a spatially fixed domain, rather than a material domain in motion with a fixed collection of particles. Let $x_0 \leq x \leq x_1$ be a fixed spatial interval. As far as this spatially fixed domain is concerned, the production term remains unchanged. The flux term, however, needs to be supplemented with an apparent flux stemming from the incoming and outgoing particles. This new contribution to the flux is, therefore, obtained by multiplying the density g by the particle velocities. It follows that the global statement of the balance equation is given by

$$\frac{d}{dt} \int_{x_0}^{x_1} g(x,t)\, dx = \int_{x_0}^{x_1} p(x,t)\, dx + h(x_0,t) - h(x_1,t)$$

$$+ g(x_0,t)v(x_0,t) - g(x_1,t)v(x_1,t). \tag{2.29}$$

Because the limits of integration are fixed, the integration and differentiation limiting processes commute, just as in the Lagrangian formulation. Moreover, the last two terms just introduced can be rewritten, by invoking the fundamental theorem of calculus, as

$$g(x_0,t)v(x_0,t) - g(x_1,t)v(x_1,t) = - \int_{x_0}^{x_1} \frac{\partial(gv)}{\partial x}\, dx. \tag{2.30}$$

The result,

$$\frac{\partial g}{\partial t} + \frac{\partial(gv)}{\partial x} = p - \frac{\partial h}{\partial x}, \tag{2.31}$$

follows suit and is clearly equivalent to equation (2.28).

2.4 Case Study: Blood Flow as a Traffic Problem

Not all problems relevant to continuum biomechanics necessarily involve the deformation of a body. Equations of balance of the type just presented arise in connection with a continuous background which is not necessarily a deformable material body. Here we present an example, originally developed within the context of vehicular traffic on a

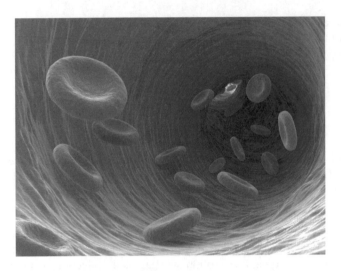

Figure 2.3   Blood flow as vehicular traffic

highway, which has been applied to the modelling of blood flow in arteries.[2] If we consider an artery, possibly with obstacles (such as plaque resulting from cholesterol build-up), the situation resembles that of a highway populated by travelling vehicles (red and white cells), as shown in Figure 2.3.

The quantity to be balanced is the number of cells. In the simplest model, we do not allow 'passing' (as on a single-lane road). In the language of continuum mechanics, this simply means that the flux vector vanishes.[3] Moreover, in a stretch of artery without tributaries or branches, the production term vanishes. Since the artery is fixed in space, the Eulerian formulation is most convenient. Interpreting the density $g(x, t)$ as the number of cells per unit length at the spatial location x along the artery at time t, equation (2.28) yields the conservation equation

$$\frac{Dg}{Dt} + g\frac{\partial v}{\partial x} = 0, \tag{2.32}$$

where equation (1.47) has been used. Expanding the material derivative, we obtain

$$\frac{\partial g}{\partial t} + \frac{\partial (gv)}{\partial x} = 0. \tag{2.33}$$

This is a single partial differential equation containing two unknown functions. What is missing, as we will see later in more generality, is a *constitutive equation* that establishes a further connection between these two variables. Indeed, the traffic model is based on the assumption that there exists a specific relation between the vehicle density and the speed of travel. In the case of vehicular traffic, the velocity of the cars may depend on a large number of factors, including the time of day, the weather and the traffic density. In the simplest model, the velocity will depend only on the traffic density, with larger densities giving rise to smaller speeds. In the case of blood flow, the constitutive relation

---

[2] Kachani et al. (2007).
[3] Clearly, as explained in Remark 2.3.3, there is flux through any spatially fixed section, but there is no flux through any *material* section.

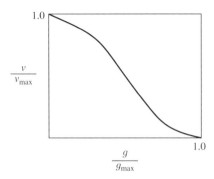

Figure 2.4   Generic velocity–density relation for blood flow or vehicular traffic

between cell density and velocity can be obtained approximately from experimental measurements using ultrasonography techniques in real-life blood vessels. In all cases, the constitutive function,

$$v = f(g),$$ (2.34)

will be monotonically decreasing within a range of practical interest. From practical considerations, since cars (and cells) have a finite length, there will be an upper bound g_{max} for the density, and it is sensible to assume that when this maximum is attained the traffic comes to a stop. It is also reasonable to consider an upper limit v_{max} for the speed, when the traffic density tends to zero, as shown in Figure 2.4. The simplest constitutive equation that one may adopt is the linear function

$$v = v_{max}\left(1 - \frac{g}{g_{max}}\right).$$ (2.35)

Introducing equation (2.35) into (2.33), we obtain

$$\frac{\partial g}{\partial t} + \frac{\partial g}{\partial x} v_{max}\left(1 - \frac{2g}{g_{max}}\right) = 0.$$ (2.36)

This is a quasi-linear first-order PDE. Its solutions, including the case of blockages and backward-travelling shock waves, can be studied by means of analytical or numerical methods.[4]

2.5   Case Study: Diffusion of a Pollutant

2.5.1   Derivation of the Diffusion Equation

Diffusive processes are prevalent in everyday life. They occur, for example, whenever a liquid or gaseous substance spreads within another (sneezing, pouring milk into a cup of coffee, intravenous injection, industrial pollution, etc.). The process of heat flow through a substance subjected to a temperature gradient is also a diffusive process. All these processes are characterized by thermodynamic irreversibility (the drop of milk poured into the coffee will never collect again into a drop).

---

[4] For an excellent elementary treatment of these and other related issues, see Knobel (2000).

Consider a tube filled with a liquid at rest in which another substance (the 'pollutant') is present with a variable concentration $g = g(x,t)$. Let $p = p(x,t)$ be the production of pollutant per unit length and per unit time. This production can be the result of perfusion through the lateral wall of the tube. The flux vector vanishes. If there is any input through the ends of the tube, it will have to be considered as part of the boundary conditions (which we have not yet discussed), rather than of the production term. The balance law, equation (2.28), can be rewritten as

$$\frac{\partial g}{\partial t} + \frac{\partial (gv)}{\partial x} = p. \tag{2.37}$$

Just as in the case of vehicular traffic, we need to supply a constitutive equation. In the case of diffusion of a pollutant (or, in general, a substance in small concentrations within another), it is possible to formulate a sensible, experimentally based, constitutive law directly in terms of the pollutant concentration. The most commonly used model, called *Fick's law*, states that

$$gv = -D \text{ grad } g, \tag{2.38}$$

where the positive constant D is a property that depends on the substances involved. The minus sign in equation (2.38) agrees with the fact that the pollutant tends to flow in the direction of smaller concentrations.[5]

Combining the last two equations, we obtain the following second-order linear PDE:

$$\frac{\partial g}{\partial t} - D \frac{\partial^2 g}{\partial x^2} = p. \tag{2.39}$$

In the absence of production we obtain the homogeneous equation

$$\frac{\partial g}{\partial t} - D \frac{\partial^2 g}{\partial x^2} = 0, \tag{2.40}$$

known as the *diffusion equation* and also as the *heat equation*.

Exercise 2.5.1 Spinal drug delivery. A drug has been injected into the spine so that at time $t = 0$ it is distributed according to the formula

$$g(x,0) = c + C \sin \frac{\pi x}{L}, \tag{2.41}$$

where the spine segment under study extends between $x = 0$ and $x = L$ and where c and C are constants. The drug concentration at the ends of the spine segment is artificially maintained at the value c for all subsequent times. If, in the absence of any production, T is the time elapsed until the difference between the concentration at the midpoint of the spine and c reaches one-half of its initial value, calculate the longitudinal diffusion coefficient D of the drug through the spinal meninges. For a rough order of magnitude, assume $L = 10$ mm and $T = 3$ hours. [Hint: verify that

$$g(x,t) = c + C \exp\left(-\frac{D\pi^2 t}{L^2}\right) \sin \frac{\pi x}{L}, \tag{2.42}$$

satisfies the PDE (2.39) and the initial and boundary conditions.]

---

[5] It is worth pointing out that the left-hand side of equation (2.38) is precisely the apparent flux discussed in Remark 2.3.3, where v is the velocity of the pollutant relative to the underlying substrate.

2.5.2 A Discrete Diffusion Model

A plausible *a priori* justification of Fick's law can be made by resorting to a statistical mechanics argument. This kind of argument can be very fruitful in biomechanics by providing a heuristic link between various levels of analysis.[6] Following a line of thought that can be regarded as a greatly simplified version of Einstein's celebrated 1905 explanation of Brownian motion, we postulate a discrete model of space and time, namely, we assume that all events take place at specific isolated sites and instants. The sites are assumed to be equally spaced along the real line according to the formula:

$$x_i = ih, \quad i = \ldots, -3, -2, -1, 0, 1, 2, 3, \ldots, \tag{2.43}$$

where $\Delta x = h$ is the distance between neighbouring sites. Similarly, the chosen instants of time are spaced at regular intervals according to the formula

$$t_j = jk, \quad j = 0, 1, 2, 3, \ldots, \tag{2.44}$$

with $\Delta t = k$ being the time interval between consecutive events. We assume, moreover, that at time $t = 0$ each site is occupied by a number N_i^0 of particles and we want to establish how this discrete mechanical system will evolve thereafter, namely, we want to predict the number N_i^j of particles at the site x_i at time t_j.

In the intended physical picture we imagine that there is an underlying ground substance (coffee, say) in equilibrium and that the particles of interest (a drop of milk, say) are being constantly bombarded by collisions with the molecules making up the ground substance. Because these collisions are random, we assume that each particle has an equal probability $\beta = 0.5$ of moving one space either to the right or to the left in the time interval $\Delta t = k$. Under this basic assumption, the rule of evolution of this system[7] is given by

$$N_i^{j+1} = 0.5 N_{i-1}^j + 0.5 N_{i+1}^j. \tag{2.45}$$

The link between the discrete model and the diffusion equation is obtained by formulating the latter as a finite-difference approximation on the assumed space-time grid. Setting $g(x_i, t_j) = N_i^j / h$ and using standard approximation formulae for first and second derivatives,[8] we obtain

$$\frac{N_i^{j+1} - N_i^j}{k} \approx D \frac{N_{i-1}^j - 2N_i^j + N_{i+1}^j}{h^2}. \tag{2.46}$$

Setting $D = h^2/2k$, we recover equation (2.45). The diffusion coefficient D is thus seen to be related directly to the average particle distance and the mean time between collisions.

Exercise 2.5.2 Modified model. Show that if the probability β has a value $0 < \beta < 0.5$, so that the particles have a positive probability $\alpha = 1 - 2\beta$ of staying put, the diffusion equation is still recovered, but with a different value for the diffusion coefficient D.

---

[6] In this regard, see also Remark 6.5.3 and Section 7.2.

[7] A discrete system governed by an evolution rule that determines the next state on the basis of the present state only is called a *cellular automaton* (see also Section 7.2.2).

[8] See, for example, Section 8.2.1.

Exercise 2.5.3 Biased diffusion. Let β^+ and $\beta^- = 1 - \beta^+$ represent, respectively, the generally different probabilities of a particle moving to the right or to the left. Obtain a PDE whose approximation matches the corresponding discrete model. Propose a physical interpretation.

Exercise 2.5.4 Finite domain. Modify the original discrete model so that it can accommodate a spatial domain of a finite extent. Consider two different kinds of boundary conditions, as follows: (1) The number of particles at each end of the domain remains constant. For this to be the case, new particles will have to be supplied or removed at the ends. (2) The total number of particles is preserved, with no flux of new particles through the end points of the domain. Implement the resulting model in a computer code and observe the time behaviour. What is the limit state of the system for large times under both kinds of boundary conditions?

2.6 The Thermomechanical Balance Laws

The generic balance law can be applied to formulate what presumably are to be considered fundamental universal laws governing all thermomechanical processes, regardless of the nature of the material medium involved. This sweeping statement may be too restrictive, particularly for the case of living tissue, in which more delicate issues than those typical of inert materials may arise. An example is provided by processes of biological growth, where the underlying system is open to exchanges of mass and to topological changes (such as the closing of a hole). That said, we will now formulate the laws of balance of classical continuum mechanics, leaving some of these more delicate questions for treatment in later chapters.

The fundamental balance laws of continuum mechanics can be summarized as follows:

1. Mechanical laws
   (a) Conservation of mass
   (b) Balance of linear momentum
   (c) Balance of angular momentum
2. Thermodynamical laws
   (a) Balance of energy
   (b) Dissipation inequality

The thermodynamical laws correspond to the classical first and second laws of thermodynamics. The balance of energy involves both thermodynamical fields (internal energy, heat flux, heat supply) and mechanical fields (stress, velocity). The dissipation inequality is, properly speaking, a law of imbalance, reflecting perhaps the inexorable unidirectionality of the passage of time.

2.6.1 Conservation of Mass

In the absence of sources and fluxes, the mass density ρ_0 per unit length in the reference configuration remains constant in time. Therefore, in accordance with the general prescription of equation (2.12),

$$\frac{\partial \rho_0(X,t)}{\partial t} = 0. \tag{2.47}$$

This is the Lagrangian equation of mass conservation.

The Eulerian counterpart, in terms of the mass density ρ per unit current length, is obtained, according to equation (2.28), as

$$\frac{D\rho}{Dt} + \rho L = 0. \tag{2.48}$$

Expanding the material derivative and using equation (1.47), equation (2.48) can also be written as

$$\frac{\partial \rho}{\partial t} + \frac{\partial(\rho v)}{\partial x} = 0. \tag{2.49}$$

This equation is known as the *continuity equation*. We have already used it in the example of traffic flow.

2.6.2 Balance of (Linear) Momentum

Although originally formulated for a single material particle and then extended to discrete systems, Newton's second law of motion is postulated to be applicable also to continuous systems. The quantity to be balanced is the total linear[9] momentum, whose Lagrangian density is $\rho_0 v$ (per unit length in the reference configuration). The sources of momentum are the external forces, which are assumed to be of two types only. The first type, giving rise to the production term in the balance equation, consists of the *body forces*, expressed in terms of a density B per unit reference length. A typical body force is obtained by placing a bar vertically,[10] in which case the weight of the bar is distributed axially along the bar.

The second type of external force is the *contact force* or *boundary traction*. When cutting the body, thereby creating two new boundary points, we must introduce the 'glueing' internal forces $N(X,t)$ that keep the uncut body together. In an intuitive picture, this is the direct result of the presence of intermolecular forces. The contact forces are of the 'flux' type, which we have already considered in the general context of balance laws. If we assume, by standard engineering convention, that a positive flux corresponds to tension (as opposed to compression), we are effectively reversing the sign convention adopted for all other fluxes. Equation (2.12) yields

$$\frac{\partial(\rho_0 v)}{\partial t} = B + \frac{\partial N}{\partial X}. \tag{2.50}$$

---

[9] The angular momentum does not play any role in the one-dimensional context, since rotations are excluded.

[10] Notice that our analysis would not be able to handle the weight of a horizontally placed bar, since the body force would then be transversal to the axis and would produce bending.

Combining this result with the statement of the conservation of mass provided by equation (2.47), we obtain

$$\rho_0 \frac{\partial v(X,t)}{\partial t} = B + \frac{\partial N}{\partial X}. \tag{2.51}$$

This is the Lagrangian version of the balance of linear momentum. The left-hand side consists of the expected 'mass times acceleration' expression.

In the Eulerian formulation, we obtain accordingly

$$\rho \frac{Dv}{Dt} = b + \frac{\partial n}{\partial x}, \tag{2.52}$$

with an obvious notation. Expanding the material derivative yields

$$\rho \frac{\partial v(x,t)}{\partial t} + \rho \frac{\partial v(x,t)}{\partial x} v(x,t) = b + \frac{\partial n}{\partial x}. \tag{2.53}$$

Remark 2.6.1 It is appropriate at this point to remind ourselves that our one-dimensional context severely restricts the applicability of the balance laws. Consider, for example, the flow of a viscous fluid within a pipe. The friction forces between the fluid and the conduit wall would have to be taken into consideration. If the pipe is tapered, the pressure between the fluid and the wall would have an axial component, even if the fluid were at rest. Moreover, the velocity profile of a fluid cross section is not constant, so that the velocity is not just a function of the longitudinal coordinate. In short, the problem cannot be handled in a one-dimensional context. Our discussion of blood flow as a traffic problem does not directly model the motion of a fluid. Instead, it ignores any cell interactions while replacing them with an empirical constitutive law relating velocity and density. Similar comments apply to our oversimplification of the diffusion problem, whereby Fick's law subsumes all the interactions. In short, those simplified (but extremely useful) treatments enforce the conservation of mass but ignore the balance of momentum. Similar approximations are also made in some applications involving the flow in porous media, where the balance of momentum is subsumed within Darcy's law, an empirical relation between pressure gradient and flow.

2.6.3 The Concept of Stress

In a strictly one-dimensional theory, the notion of internal force (Lagrangian $N(X,t)$ or Eulerian $n(x,t)$) is the legitimate representative of the notion of flux of momentum. Nevertheless, as we have already remarked, when trying to formulate constitutive equations (which derive from inherently three-dimensional experimental data) it is imperative to have fluxes defined in general per unit area of a boundary. In the particular case of the momentum flux, we can define the *Cauchy stress* σ as

$$\sigma(x,t) = \frac{n(x,t)}{\hat{s}(x,t)}, \tag{2.54}$$

namely, as the Eulerian internal force divided by the current area of the cross section. The three-dimensional counterpart of this concept is the *Cauchy stress tensor*, represented in each spatial coordinate system by a 3×3 matrix.

If instead of dividing by the current area $\hat{s}(x,t)$ we divide by the area in the reference configuration, we obtain the *Piola stress*, also known as the *first Piola–Kirchhoff stress*,

$$T(X,t) = \frac{n(x(X,t),t)}{\hat{S}(X)}. \tag{2.55}$$

From this definition, we can infer that the Piola stress is a hybrid quantity, since the numerator is a spatial quantity and the denominator a referential one. There is no compelling reason to prefer the Cauchy stress over the Piola stress, since ultimately the stress is a vehicle to express the constitutive equation rather than an end in itself.[11]

2.7 Case Study: Vibration of Air in the Ear Canal

Figure 2.5 is a schematic depiction of the human ear. Situated in the external ear between the external sources of sound and the tympanic membrane (or eardrum), the mean ear canal (or meatus) is an approximately straight cylindrical conduit about 25 mm in length and 7 mm in diameter. One of the functions of the meatus is to act as a passive pressure amplifier. In a crude first approximation, the ear canal can be represented as a relatively thin cylindrical tube open at one end to externally imposed vibrations and closed at the other end by a flexible membrane. Assuming air to be essentially non-viscous, and neglecting any turbulence effects, it can be reasonably assumed that the cross sections of air within the canal remain plane as they vibrate.

Having satisfied, at least to a reasonable approximation, the conditions of validity of our one-dimensional formulation, we proceed to formulate the equations governing the motion of air within the ear canal. We adopt an Eulerian point of view.

Starting with the conservation of mass, the density per unit length, $\rho(x,t)$, is obtained in terms of the physical density (mass per unit volume) $\hat{\rho}$ as

$$\rho(x,t) = \hat{\rho}(x,t)\hat{s}(x), \tag{2.56}$$

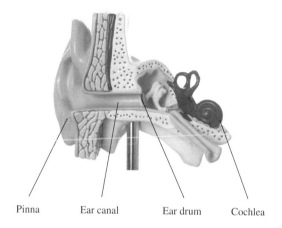

Pinna   Ear canal   Ear drum   Cochlea

Figure 2.5   Model of the human ear

---

[11] In the three-dimensional context it is customary to define yet another stress tensor known as the second Piola–Kirchhoff stress.

where $\hat{s}(x)$ is the cross sectional area of the canal. Substituting equation (2.56) into (2.49), we obtain

$$\frac{\partial \hat{\rho}}{\partial t} + \frac{\partial(\hat{\rho}v)}{\partial x} + \hat{\rho}v\frac{\partial \ln \hat{s}(x)}{\partial x} = 0. \tag{2.57}$$

If the cross section is constant the last term cancels out.

Similarly, the internal force $n(x,t)$, in the case of a gas, is obtained by multiplying the pressure $p(x,t)$ times the cross-sectional area $\hat{s}(x)$, namely,

$$n(x,t) = -p(x,t)\,\hat{s}(x). \tag{2.58}$$

The negative sign is placed for consistency, since a positive pressure corresponds to compression. In the absence of (axial) body forces, the equation of balance of momentum (equation (2.53)) yields

$$\hat{\rho}\frac{\partial v(x,t)}{\partial t} + \hat{\rho}\frac{\partial v(x,t)}{\partial x}v(x,t) = -\frac{\partial p}{\partial x} - p(x,t)\frac{\partial \ln \hat{s}(x)}{\partial x}. \tag{2.59}$$

Once again, if the cross section is constant (which we will henceforth assume) the last term cancels out.

Equations (2.56) and (2.59) constitute a system of two first-order PDEs for the three unknown fields $\hat{\rho}(x,t)$, $\hat{p}(x,t)$ and $v(x,t)$. To complete the system we need to provide a constitutive equation or *equation of state* for the air within the canal, assumed to be rigid. As we know from elementary physics, the equation of state of a gas involves not just the density and the pressure but also the temperature. Since at this stage we are not particularly interested in the thermal contribution, an assumption may be made as to the nature of the process taking place. One possibility is to assume that the air vibrations take place isothermically, that is, at constant temperature. Yet another possibility is to assume that the process is adiabatic, namely, without transfer of heat. The second choice is closer to reality. For our purposes, however, all we need to declare at this point is that the temperature influence has been factored out one way or another and that, therefore, we have at our disposal a constitutive equation of the form

$$\hat{\rho} = k(\hat{p}), \tag{2.60}$$

where $k(\cdot)$ is some monotonically increasing function.

A problem involving differential equations cannot be said to have been well posed until suitable boundary and/or initial conditions have been specified. In the case of systems of PDEs, such as those governing fluid motion, the determination of appropriate conditions is far from trivial and depends on the type of equations under consideration. The situation at hand can be shown to belong to the case of *hyperbolic systems*. From the physical point of view, these are time-dependent phenomena that admit the propagation of waves at finite speeds. Conditions have to be specified on the three sides of the space-time domain indicated in Figure 2.6 with thick lines. On the horizontal segment $0 \leq x \leq L_c$ (where L_c is the length of the canal) *initial values* for v and $\hat{p}$ (or $\hat{\rho}$) need to be specified. As far as the boundary conditions are concerned (on the vertical lines $x = 0$ and $x = L_c$), it can be shown[12] that a well-posed problem is obtained by specifying the value of the velocity only. Specifically, we assume that at the open end ($x = 0$) of the canal there exists a known externally imposed air velocity $v(0,t) = v_0(t)$ arising from the sound

---

[12] See, for example, Courant and Hilbert (1962, p. 475).

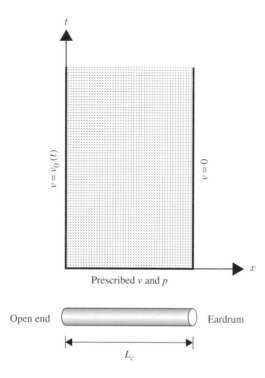

Figure 2.6   Space-time solution domain

source, such as a musical instrument. At the closed end ($x = L_c$), if we assume, at least provisionally, that the eardrum is rigid, we obtain that the velocity vanishes, namely $v(L_c, t) = 0$.

A useful way to obtain some approximate information about the solution of a system of PDEs is to resort to the technique of *linearization* around a known solution, usually an equilibrium state. We have already hinted at such a technique when introducing, in Section 1.6, the notion of small strain. Denoting by $\hat{p}_s$ and $\hat{\rho}_s$, respectively, the pressure and density of static equilibrium (with zero velocity throughout the canal), we write

$$\hat{p} = \hat{p}_s + \Delta\hat{p}, \quad \hat{\rho} = \hat{\rho}_s + \Delta\hat{\rho}. \tag{2.61}$$

We now assume that the increments $\Delta\hat{p}$ and $\Delta\hat{\rho}$ are small in relation to $\hat{p}$ and $\hat{\rho}$, respectively, so that their mutual products and squares can be neglected. In particular, by virtue of the equation of state (2.60), we can relate the increments of density and pressure approximately by

$$\Delta\hat{\rho} = \left.\frac{dk}{d\hat{p}}\right|_{\hat{p}_s} \Delta\hat{p} = \frac{1}{c^2}\Delta\hat{p}. \tag{2.62}$$

The constant c is the *speed of sound* in air at the state of static equilibrium assumed. Its value, under normal conditions, is about 340 metres per second.

Finally, we assume that the advection term in the acceleration, namely the term

$$a_{\text{adv}}(x, t) = v \text{ grad } v, \tag{2.63}$$

can be neglected.

Implementing these smallness assumptions in equations (2.57) and (2.59), we obtain the following linear system of PDEs for the two unknown fields $\Delta\hat{p}(x,t)$ and $v(x,t)$:

$$\frac{1}{c^2}\frac{\partial\Delta\hat{p}}{\partial t} + \hat{\rho}_s\frac{\partial v}{\partial x} = 0 \tag{2.64}$$

and

$$\hat{\rho}_s\frac{\partial v(x,t)}{\partial t} = -\frac{\partial\Delta\hat{p}}{\partial x}. \tag{2.65}$$

It is now possible to explicitly eliminate one of the fields so as to obtain a single second-order equation for the other field. Differentiating the first equation with respect to t and the second with respect to x, for example, we obtain the following linear second-order PDE for the excess pressure:

$$\frac{\partial^2\Delta\hat{p}}{\partial t^2} = c^2\frac{\partial^2\Delta\hat{p}}{\partial x^2}, \tag{2.66}$$

which can be recognized as the *one-dimensional wave equation*. The same equation governs the excess density $\Delta\hat{p}(x,t)$ and the velocity $v(x,t)$.

Exercise 2.7.1   Show that the excess density $\Delta\hat{p}(x,t)$ and the velocity $v(x,t)$ independently satisfy the one-dimensional wave equation.

Assume that an oscillatory motion emanating from some sound source, such as a musical instrument, produces an excess pressure of the form

$$\Delta\hat{p}(0,t) = A\sin(\omega t), \tag{2.67}$$

at the entrance of the ear canal, where A is the amplitude and ω is the circular frequency of the sound. This takes care of the boundary condition at that end. At the eardrum end, where the velocity has been assumed to vanish, equation (2.65) delivers the boundary condition

$$\frac{\partial\Delta\hat{p}}{\partial x}(L_c,t) = 0. \tag{2.68}$$

We are interested in finding a particular solution of equation (2.66) with boundary conditions (2.67) and (2.68), regardless of any specific initial conditions. In physical terms, we are looking for a stabilized solution when the effect of the initial conditions has presumably been dissipated away by energy losses.[13] We try a solution of the form

$$\Delta\hat{p}(x,t) = B(x)\sin(\omega t). \tag{2.69}$$

In other words, we investigate the possibility of a response of our system that moves, so to speak, in unison with the excitation at the entrance of the canal, the amplitude $B(x)$

---

[13] Theoretically, since our model does not include dissipative terms, the effects of the initial conditions will persist indefinitely. Even then, the solution of the equation will consist of the sum of two parts, one representing the effect of the initial conditions and the other a particular solution of the type we seek. This additive decomposition is peculiar to linear systems only.

depending on position along the canal. Introducing the assumed form of the solution into equation (2.66), we obtain the following ODE for the amplitude function:

$$\frac{d^2B}{dx^2} + \left(\frac{\omega}{c}\right)^2 B = 0, \tag{2.70}$$

whose general solution is

$$B(x) = C \cos\left(\frac{\omega}{c}x\right) + D \sin\left(\frac{\omega}{c}x\right), \tag{2.71}$$

where C and D are constants to be determined by the boundary conditions. Note that we have found a particular solution of the form

$$\Delta\hat{p}(x,t) = \left(C \cos\left(\frac{\omega}{c}x\right) + D \sin\left(\frac{\omega}{c}x\right)\right) \sin(\omega t). \tag{2.72}$$

At the open end ($x = 0$), the boundary condition (2.67) yields

$$C = A. \tag{2.73}$$

At the eardrum end ($x = L_c$), equation (2.68) implies

$$D = A \tan\left(\frac{\omega L_c}{c}\right), \tag{2.74}$$

where the result (2.73) was also used. Combining all these partial results, we can write our particular solution (2.72) in the final form:

$$\Delta\hat{p}(x,t) = A \left(\cos\left(\frac{\omega}{c}x\right) + \tan\left(\frac{\omega L_c}{c}\right) \sin\left(\frac{\omega}{c}x\right)\right) \sin(\omega t). \tag{2.75}$$

Because of the presence of the tan function, this formula has the potential of blowing up to infinity for particular values of the exciting frequency ω. This will happen for the following spectrum of frequency values:

$$\omega = (2n + 1)\frac{\pi c}{2L_c}, \quad n = 0, 1, 2, \ldots. \tag{2.76}$$

The lowest value, corresponding to $n = 0$, is called the (first) *resonant frequency* of the system. The resonant frequency of the ear canal is, therefore,

$$\omega_r = \frac{\pi c}{2L_c}. \tag{2.77}$$

For the average length of the canal in adults ($L_c = 25\,\text{mm}$) and the standard speed of sound ($c = 340\,\text{m s}^{-1}$), we obtain the resonant frequency:

$$\omega_r = 21\,400\,\frac{1}{\text{s}}, \tag{2.78}$$

or, in cycles per second (Hz),

$$\omega_r = 3400\,\text{Hz}. \tag{2.79}$$

In practice, the amplitude of the response is attenuated by frictional losses (as well as other factors, such as the elastic compliance of the membrane and the fact that our linearization required that amplitudes be small in the first place), but the amplification at frequencies near the theoretical resonance can be severe enough to damage the tympanic membrane. If we consider that a frequency of 3400 Hz corresponds roughly to the musical note a'''' (namely, to the note A7, which is the A in the uppermost octave of a modern piano), we can understand why many orchestra musicians suffer from various hearing impairments, including hearing loss and tinnitus (constant ringing noise in the ear). In children, as the ear canal is shorter, the resonant frequency is higher. The resonant frequency of the ear canal plays a role in the design of hearing aids.

It is interesting to point out that, since the normal hearing range of humans can reach frequencies as high as 20 kHz, equation (2.76) provides two other values of n ($n = 1$ and $n = 2$), that is, two higher resonant frequencies, for which the amplitude grows without bound, well within the auditory range. These (second and third) resonant frequencies are approximately 10 200 Hz and 17 000 Hz.

2.8 Kinetic Energy

Although elementary, it may be a good idea to review the concept of kinetic energy and its consequences in the case of a single particle of mass m moving along the x-line with instantaneous velocity v. In this case, the kinetic energy K is defined as

$$K = \frac{1}{2}mv^2. \tag{2.80}$$

As the particle moves, its kinetic energy will, in general, change in time. Its rate of variation is obtained directly as

$$\frac{dK}{dt} = mv\frac{dv}{dt}. \tag{2.81}$$

So far, this is just a formal expression arising directly from the definition. If we now take into consideration Newton's law of motion,

$$F = \frac{d(mv)}{dt} = m\frac{dv}{dt}, \tag{2.82}$$

where F is the force acting on the particle along the x-axis, we conclude that

$$\frac{dK}{dt} = Fv. \tag{2.83}$$

The product of the force times the velocity is the *power* of the force, which we denote by W:

$$W = Fv. \tag{2.84}$$

Introducing this definition into equation (2.83), we obtain

$$\frac{dK}{dt} = W. \tag{2.85}$$

Expressed in words: the rate of change of the kinetic energy of a particle is equal to the power of the force acting on it.

The force acting on the particle, on the other hand, may be of a completely arbitrary nature. The particle may represent, say, a small rail car in an amusement park and the force may be applied to the car in response to the instantaneous intensity of the laughter of the children riding on it. In this most general situation, therefore, the force is some arbitrary function $F = F(x, v, t)$. On the other hand, in many applications, the force acting on the particle may be the result of a pre-existing *external force field*, independent of time, namely, $F = F(x)$. Such is the case of the force of gravity, or rather its tangential component, acting on a rail car in a frictionless roller coaster, or a toboggan accelerating downhill. We define the *potential*, or *potential energy*, of a force field as the integral

$$V(x) = -\int_{x_0}^{x} F(\xi)d\xi. \tag{2.86}$$

We can, therefore, write

$$F(x) = -\frac{\partial V}{\partial x}. \tag{2.87}$$

The following observations are appropriate:

1. The arbitrary choice x_0 for the lower limit of the integral will affect the value of the potential only by an additive constant. Equation (2.87) remains unaffected by this choice.
2. In the two- or three-dimensional case, there is no natural way to replace equation (2.86). In fact, given an external force field (namely, a force depending on the spatial coordinates alone), there may not exist a scalar potential function of the coordinates such that the components of the force are obtained as partial derivatives of the potential. If such a function does exist, the force field is said to be *conservative*. In one dimension, therefore, every force field is automatically conservative (assuming always that it depends on position alone).
3. The minus sign in the definition of potential energy is a matter of physical convenience. We wish to be able to assert, for example, that the potential energy of a stone is larger the higher the stone is. Since the force of gravity points downwards while the height increases upwards, the minus sign is required. The word 'potential' refers precisely to the ability of the stone to produce work on arrival at ground level. This ability is clearly enhanced with height.

Assuming that the force acting on the particle is conservative, we can define the *total energy* of the particle as the sum:

$$E_T = K + V. \tag{2.88}$$

The total energy depends, of course, on the reference point x_0 adopted in the definition of the potential. A different choice of reference point affects the total energy only by an additive constant.

We now investigate the rate of change of the total energy. Using equation (2.85), we obtain

$$\frac{dE_T}{dt} = \frac{dK}{dt} + \frac{dV}{dt} = W + \frac{dV}{dx}\frac{dx}{dt} = W - Fv = 0. \tag{2.89}$$

We conclude that, for a particle moving in a conservative force field, the total energy remains constant throughout the motion. The value of this constant is determined by the initial conditions. This is, in a nutshell, the *principle of conservation of mechanical energy* for a particle in a conservative force field. It can be shown that the same principle applies for a rigid body. We stress the fact that the principle of conservation of mechanical energy is not an independent statement, but rather a consequence of the law of motion and of the assumption of the existence of a potential for the force.

We have reviewed the elementary case of a single particle in perhaps excessive detail because we want to show in what sense it may or may not apply to a deformable continuum. As a preliminary example in this direction, we will consider a much simpler case: two particles, of masses m_1 and m_2, interacting with each other. The interaction force may be of gravitational origin (such as the force between the Earth and the Moon), or the result of mechanical actuators (springs, dampers, and so on) connecting the two masses, or any other interaction, which we assume to be governed by Newton's third law (action and reaction). Always working, for simplicity, in a one-dimensional spatial context, let the (internal) force exerted by m_2 on m_1 be denoted by (the scalar) f_{12}. The force exerted by m_1 on m_2 is then $f_{21} = -f_{12}$, by the assumed validity of Newton's third law. For definiteness, we introduce the notation N for the internal force consisting of the *pair* f_{12} and f_{21}, and we declare that N is positive if the forces are attractive, and negative if they are repulsive. In addition to these internal forces, we consider external forces F_1 and F_2 applied to the masses m_1 and m_2, respectively. The total kinetic energy is calculated as

$$K = \frac{1}{2}m_1 v_1^2 + \frac{1}{2}m_2 v_2^2, \tag{2.90}$$

where v_1 and v_2 are the velocities of m_1 and m_2, respectively. Taking its time derivative, we obtain

$$\frac{dK}{dt} = m_1 v_1 \frac{dv_1}{dt} + m_2 v_2 \frac{dv_2}{dt}. \tag{2.91}$$

According to Newton's second of law motion, however, each particle independently must satisfy it in terms of the *total* force acting upon it, regardless of its origin. In other words, we must have

$$m_1 \frac{dv_1}{dt} = F_1 + f_{12}, \tag{2.92}$$

$$m_2 \frac{dv_2}{dt} = F_2 + f_{21} = F_2 - f_{12}. \tag{2.93}$$

Returning with these two formulas to equation (2.91), we can write

$$\frac{dK}{dt} = F_1 v_1 + F_2 v_2 + f_{12}(v_1 - v_2). \tag{2.94}$$

The expression

$$W_{\text{ext}} = F_1 v_1 + F_2 v_2 \tag{2.95}$$

is clearly identified as the *power of the external forces*. As for the last term,

$$W_{\text{int}} = f_{12}(v_1 - v_2), \tag{2.96}$$

on the right-hand side of equation (2.94), we call it the *power of the internal forces*. If we denote by e the distance between the particles, namely

$$e = |x_2 - x_1| \tag{2.97}$$

(x_1 and x_2 being the coordinates of the particles), we can write the power of the internal forces in terms of the internal force N as

$$W_{int} = -N \frac{de}{dt}, \tag{2.98}$$

since attraction leads to positive power if the particles are getting closer, and vice versa. Our final result reads:

$$\frac{dK}{dt} = W_{ext} + W_{int}. \tag{2.99}$$

In short, for a system of interacting particles, the rate of change of the kinetic energy is equal to the sum of the powers of the external and the internal forces. Notice that in the case of a rigid system, since the distances between particles are fixed by definition, the power of the internal forces vanishes and we recover the result of a single particle.

We are now in a position to tackle the case of a continuous body. Working in the Lagrangian formulation, the kinetic energy is expressed as

$$K = \int_{X_0}^{X_1} \frac{1}{2}\rho_0 \left(v(X,t)\right)^2 \, dX. \tag{2.100}$$

Taking the time derivative, we obtain

$$\frac{dK}{dt} = \int_{X_0}^{X_1} \rho_0 v(X,t) \frac{\partial v(X,t)}{\partial t} \, dX, \tag{2.101}$$

where we have cavalierly exchanged the order of differentiation and integration (as can be done in the Lagrangian formulation) and where we have also exploited the Lagrangian form of the conservation of mass, equation (2.47). We still have at our disposal the Lagrangian version of the balance of momentum, as given by equation (2.51). Introducing this result into equation (2.101), we obtain

$$\frac{dK}{dt} = \int_{X_0}^{X_1} v(X,t) \left(B + \frac{\partial N(X,t)}{\partial X} \right) \, dX, \tag{2.102}$$

The first term on the right-hand side, namely the integral

$$W_{ext}^{body} = \int_{X_0}^{X_1} v(X,t) \, B \, dX, \tag{2.103}$$

is easily interpreted as the *power of the body forces*. It is the interpretation of the second term that needs more work. The crucial step consists of using integration by parts as follows:

$$\int_{X_0}^{X_1} v(X,t) \left(\frac{\partial N}{\partial X} \right) dX = v(X_1,t)N(X_1,t) - v(X_0,t)N(X_0,t)$$

$$- \int_{X_0}^{X_1} N(X,t) \left(\frac{\partial v(X,t)}{\partial X} \right) dX. \tag{2.104}$$

Recall that, from the treatment of flux in previous sections, $N(X_0,t)$ and $N(X_1,t)$ are nothing but the boundary tractions, positive if in tension, according to our convention. Thus, the expression

$$W_{\text{ext}}^{\text{boundary}} = v(X_1,t)N(X_1,t) - v(X_0,t)N(X_0,t) \tag{2.105}$$

represents the power of the external forces (if any) applied at the boundaries. The total external power is given by

$$W_{\text{ext}} = W_{\text{ext}}^{\text{body}} + W_{\text{ext}}^{\text{boundary}}. \tag{2.106}$$

Finally, the term

$$W_{\text{int}} = - \int_{X_0}^{X_1} N(X,t) \left(\frac{\partial v(X,t)}{\partial X} \right) dX \tag{2.107}$$

is defined as the *power of the internal forces* (in the sense of fluxes introduced in the continuum when dealing with the equation of balance of momentum). Notice that, significantly, the power of the internal forces can also be written as

$$W_{\text{int}} = - \int_{X_0}^{X_1} N(X,t) \left(\frac{\partial e(X,t)}{\partial t} \right) dX, \tag{2.108}$$

where $e(X,t)$ is the engineering strain. Putting all these results together, we obtain the final formula,

$$\frac{dK}{dt} = W_{\text{ext}} + W_{\text{int}}. \tag{2.109}$$

This result, identical to that of the simpler two-particle counterpart, was obviously derived independently. Nevertheless, the analogy with the two-particle system is useful because of its clearer physical appeal. In the discrete case, there is no obvious analogue to the step involving integration by parts.[14] It is worth stressing, as we did for the discrete case, that equation (2.109) governing the balance of kinetic energy, is not an independent law, but was derived entirely on the basis of the conservation of mass and the balance

---

[14] In a three-dimensional continuum context, the divergence theorem of vector calculus would need to be invoked.

of momentum. The same final result is obtained if using the Eulerian formulation. In particular, the power of the internal forces is given by

$$W_{\text{int}} = - \int_{x(X_0,t)}^{x(X_1,t)} n(x,t) L(x,t) \, dx, \tag{2.110}$$

where L is the velocity gradient.

Remark 2.8.1 A question we have not asked, either in the two-particle case or in the continuum context, is whether or not some principle of conservation of energy might apply. Assuming that the external forces are conservative, the question still remains as to whether such a concept can be extended to internal forces. In the discrete case, it can be shown that this is the case provided the internal forces depend on the inter-particle distance alone. We will come back to this issue and its continuum counterpart when dealing with the notion of *elasticity*.

2.9   The Thermodynamical Balance Laws

2.9.1   Introduction

The mechanical equations of balance, as we have already found out in various case studies, need to be supplemented with constitutive equations that relate the internal forces to the kinematical variables. In the most general case, the constitutive equation may involve a dependence on the present configuration as well as on all the configurations attained by the body in the past. This history dependence can be observed in many everyday materials that exhibit one form of material memory or another. But even in the most extreme case, one has to deal with a *purely mechanical system* whose response can be analysed by solving a system of differential or, perhaps, integro-differential equations.

Example 2.9.1   Elasticity. The most widespread purely mechanical theory is that of an *elastic material*. An elastic constitutive law stipulates a dependence of the internal force at each point of the body on the present value of the strain at that point, namely,

$$N(X,t) = f(E(X,t)). \tag{2.111}$$

It is not difficult to verify that, with such a constitutive equation and with appropriate initial and boundary conditions, the laws of conservation of mass and of balance of momentum provide a sufficient number of equations to solve for the motion of the body.

It would appear, therefore, that the mechanics of deformable bodies is a self-contained subject and that its domain of applicability comprises all observed material phenomena, at least at the macroscopic scale. The everyday experience of twisting a paper clip until it breaks, however, immediately reveals that there may be a problem: the clip becomes hotter. From the microscopical point of view, there is a massive movement of crystal dislocations producing internal friction and thus dissipating heat, all of which manifests itself macroscopically in the phenomenon known as metal plasticity. In a more biologically relevant experiment, which has been replicated for over a hundred years in the study of skeletal muscle, a muscle is placed in a calorimeter and activated electrically so as to

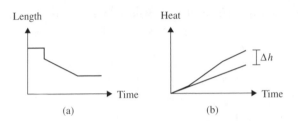

Figure 2.7   (b) Heat released (a) by muscle during contraction

produce maximal contraction of the fibres. The muscle is first kept for a while at constant length (isometric contraction) and the heat released (sometimes called maintenance heat) is measured. The muscle is then allowed to shorten at constant force (isotonic contraction) and the heat released is measured again. The observation reveals an excess heat Δh, as depicted schematically in Figure 2.7.

The details of this and similar experiments are not important at this point, but only the awareness that, either because of passive phenomena (such as the case of metal plasticity) or because of the existence of exothermal chemical reactions (such as the hydrolization of ATP during muscle contraction), mechanics and thermodynamics are inextricably linked. This linkage necessitates the introduction of a new variable, the *temperature*, which, in the continuum context, is regarded as a field variable varying with position and time. The internal forces become functions not only of the kinematic variables (such as the strain) and their history, but also of the temperature field and its history. Moreover, new thermodynamic quantities, such as internal energy and entropy, depending constitutively on both the kinematics and the temperature, enter the picture. The mechanical and thermodynamical fields are irretrievably coupled. The universal formulation of this coupling, regardless of the particular material constitution, is the content of the law of balance of energy, also known as the *first law of thermodynamics*.

2.9.2   Balance of Energy

The first law of thermodynamics starts by postulating the existence of an extensive physical quantity called the *internal energy*, U, and by asserting that this quantity is constitutively determined.[15] The *total energy* of the body under consideration is defined as the sum of its kinetic and internal energies. This sum, $K + U$, is the quantity whose rate of change is to be balanced.

The first law establishes that the rate of change of the total energy is accounted for by the sum of a mechanical contribution and a non-mechanical (or thermal) contribution. The mechanical contribution is given by the power of the external forces, which is exactly

---

[15] In classical thermodynamics, this last assertion is expressed by saying that the internal energy is a *function of state*. One should, however, be mindful of the fact that the internal energy, as much as all other constitutive quantities, may depend not just on the present state of the system (deformation and temperature), but also on its history. This discrepancy can, of course, easily be bridged by a more comprehensive definition of the notion of state.

what we called W_{ext} in Section 2.8. The non-mechanical contribution, which may also be called *thermal power*, we denote by $W_{thermal}$. The balance of energy, therefore, reads as follows:

$$\frac{d(K + U)}{dt} = W_{ext} + W_{thermal}. \tag{2.112}$$

Invoking the kinetic energy identity (equation (2.109)), we can write the balance of energy in the reduced form

$$\frac{dU}{dt} = W_{thermal} - W_{int}. \tag{2.113}$$

We can say, therefore, that the first law of thermodynamics establishes that the rate of change of the internal energy is accounted for by the thermal power minus the mechanical power of the internal forces. Notice that in the case of a rigid body, where the internal mechanical power vanishes identically, the thermal power input goes directly into causing an increase of the internal energy.

To obtain a local (differential) statement of this law in the Lagrangian formulation, we start by expressing the internal energy as the integral

$$U = \int_{X_0}^{X_1} \rho_0 \bar{u} \, dX. \tag{2.114}$$

We have followed here the common practice of defining the internal energy in terms of an internal energy density *per unit mass*, which we have denoted by $\bar{u}$. This explains the presence of the referential density under the integral. Naturally, it would have been perfectly acceptable to express the internal energy in terms of an energy per unit length.[16]

As far as the thermal power is concerned, we follow the general policy of resolving it into the sum of production and flux terms. In a three-dimensional context, for example, the production term consists of sources and sinks of heat per unit volume and per unit time, while the flux terms consist of inflow or outflow of thermal energy through the boundary per unit area and per unit time. In the one-dimensional case, as we have discussed at length in Section 2.2, the production terms are specified per unit length and per unit time, while the boundary consists of just two points and the flux is expressed as energy per unit time. Accordingly, emulating the treatment of Section 2.2 for the Lagrangian description, we write

$$W_{thermal} = \int_{X_0}^{X_1} \rho_0 r \, dX + Q(X_0, t) - Q(X_1, t). \tag{2.115}$$

In this equation, $Q(X, t)$ is (the only component of) the *heat flux vector*. In physical terms, the heat flux corresponds to the transfer of thermal energy by *conduction*. Just like

---

[16] Note that in defining the body forces B and b, we could have used instead a force, β say, per unit mass. The relation between these quantities is $B = \rho_0 \beta$ and $b = \rho \beta$.

the internal energy, the production term has been expressed in terms of a heat production r per unit mass, rather than per unit length. The heat production r is sometimes called *radiation*, a terminology that may lead to some confusion.

Collecting the previous expressions and using equation (2.108), we can write equation (2.113) in the form

$$\frac{d}{dt} \int_{X_0}^{X_1} \rho_0 \bar{u} \, dX = \int_{X_0}^{X_1} \rho_0 r \, dX + Q(X_0, t) - Q(X_1, t) + \int_{X_0}^{X_1} N(X, t) \left(\frac{\partial e(X, t)}{\partial t} \right) dX.$$

(2.116)

Finally, expressing the boundary terms as an integral and invoking the assumption that the law is valid for all sub-bodies, as explained in detail in Section 2.2, we obtain the Lagrangian local form of the equation of balance of energy as

$$\rho_0 \frac{\partial \bar{u}(X, t)}{\partial t} = \rho_0 r - \frac{\partial Q(X, t)}{\partial X} + N \frac{\partial e(X, y)}{\partial t}.$$

(2.117)

The Eulerian counterpart of this equation is

$$\rho \frac{D\bar{u}}{Dt} = \rho r - \frac{\partial q(x, t)}{\partial x} + nL.$$

(2.118)

Recall that in the one-dimensional context, the Eulerian flux q is numerically equal to the Lagrangian flux Q.

2.9.3 The Entropy Inequality

The mechanical balance equations (mass and momentum) and the first law of thermo-dynamics (energy balance), when supplemented with appropriate boundary and initial conditions and with constitutive equations that characterize the material properties, are in themselves sufficient to determine the evolution of the motion and the temperature of a continuous body. The imposition of any additional physical principle appears, therefore, to be superfluous and to open the door to potential inconsistencies. The fact, however, that certain natural processes do not seem to occur spontaneously (such as the molecules of a gas spontaneously collecting in a small region of a large container) and that some material properties never occur in reality (such as a negative fluid viscosity that would produce, rather than dissipate, energy), seems to indicate the existence of an underlying principle of directionality in nature, formally known as the *second law of thermodynamics*.

This law does not act as an additional balance equation that might possibly clash with the complete system mentioned above. Rather, the second law of thermodynamics takes the form of an inequality not to be violated by any conceivable process that the material at hand might undergo. This inequality is expressed in terms of a new function of state called the *entropy*. As we will see, this *entropy inequality* results often in restrictive conditions to be satisfied by the material properties so that all offending processes are automatically ruled out. In living organisms, the application of the second law of thermodynamics tends to be complicated by the fact that biological systems are usually open to flow of mass and to many external mechanisms, such as subtle chemical reactions and electrochemical activation controlled by the central nervous system. As a result, it is often impossible to account for all sources and sinks of entropy, thus weakening the reliability of the inequality in the form to be presented here. Nevertheless, there is no doubt that the entropy inequality should be brought to bear in one form or another in studying the response of biological materials.

The second law of thermodynamics postulates the existence of a new constitutive quantity (or 'function of state'), S, called the *entropy*. In the case of a continuous medium, it is assumed that this is an extensive property expressible as an integral of an entropy s per unit mass that varies from point to point. In trying to balance the total entropy, that is, in attempting to explain its rate of change, there is a universally identifiable production $\mathcal{P}$ and a universally identifiable flux $\mathcal{F}$, both of thermal origin. By 'universally identifiable' we mean that these contributions to balancing the entropy are present in all processes. The second law asserts that the rate of change of the entropy is never less than those universal contributions, namely

$$\frac{dS}{dt} \geq \mathcal{P} + \mathcal{F}. \tag{2.119}$$

A process for which the equality holds is called *reversible*. Otherwise, the process is called *irreversible*. Given an irreversible process, it may be possible to identify and even quantify the sources of entropy production (such as friction, microcracking, and so on). But, according to the second law, all these contributions cannot be universally assessed *a priori* and, more importantly, they only contribute to a combined *increase* of entropy.

The specific forms of the two contributions $\mathcal{P}$ and $\mathcal{F}$ necessitate the introduction of the concept of *absolute temperature*,[17] which we will denote by θ. Although any empirical measure of 'hotness' appears to be legitimate, the notion of absolute temperature goes deep to the core of natural phenomena. The absolute temperature (which, historically, appeared first in the context of the behaviour of ideal gases) has two salient properties. The first is that it is a strictly positive number ($\theta > 0$). The second feature is a particular *scale*, which can be inferred from the laws of ideal gases. Notice the difference between a scale and a particular unit of measurement, which can be quite arbitrary. By the absolute temperature scale we mean that the *ratio* between two absolute temperatures must be independent of the units being used. In terms of the absolute temperature, the two contributions alluded to above are given by the following expressions:

$$\mathcal{P} = \int_{X_0}^{X_1} \frac{\rho_0 r}{\theta} \, dX = \int_{x(X_0,t)}^{x(X_1,t)} \frac{\rho r}{\theta} \, dx \tag{2.120}$$

and

$$\mathcal{F} = \frac{Q(X_0,t)}{\theta(X_0,t)} - \frac{Q(X_1,t)}{\theta(X_1,t)} = \frac{q(x(X_0,t),t)}{\theta(x(X_0,t),t)} - \frac{q(x(X_1,t),t)}{\theta(x(X_1,t),t)}. \tag{2.121}$$

With these expressions in place, the entropy inequality (2.119) attains the form

$$\frac{d}{dt} \int_{X_0}^{X_1} \rho_0 s \, dX \geq \int_{X_0}^{X_1} \frac{\rho_0 r}{\theta} \, dX + \frac{Q(X_0,t)}{\theta(X_0,t)} - \frac{Q(X_1,t)}{\theta(X_1,t)}, \tag{2.122}$$

---

[17] Also called *thermodynamic temperature*.

in the Lagrangian formulation. The Eulerian version is

$$\frac{d}{dt} \int\limits_{x(X_0,t)}^{x(X_1,t)} \rho s \, dx \geq \int\limits_{x(X_0,t)}^{x(X_1,t)} \frac{\rho r}{\theta} \, dx + \frac{q(x(X_0,t),t)}{\theta(x(X_0,t),t)} - \frac{q(x(X_1,t),t)}{\theta(x(X_1,t),t)}. \tag{2.123}$$

Either version is known as the *Clausius–Duhem inequality*.

The local versions are obtained by following the standard procedure that we have used for all other balance laws, a procedure that applies equally well to inequalities. The results are

$$\rho_0 \frac{\partial s(X,t)}{\partial t} \geq \frac{\rho_0 r}{\theta} - \frac{\partial}{\partial X}\left(\frac{Q}{\theta}\right) \tag{2.124}$$

and

$$\rho \frac{Ds}{Dt} \geq \frac{\rho r}{\theta} - \frac{\partial}{\partial x}\left(\frac{q}{\theta}\right). \tag{2.125}$$

Further treatment and interpretation of the Clausius–Duhem inequality will be pursued in Section 3.5.

2.10   Summary of Balance Equations

1. Lagrangian formulation
   (a) Conservation of mass

$$\frac{\partial \rho_0(X,t)}{\partial t} = 0. \tag{2.126}$$

   (b) Balance of momentum

$$\rho_0 \frac{\partial v(X,t)}{\partial t} = B + \frac{\partial N}{\partial X}. \tag{2.127}$$

   (c) Balance of energy

$$\rho_0 \frac{\partial \bar{u}(X,t)}{\partial t} = \rho_0 r - \frac{\partial Q(X,t)}{\partial X} + N \frac{\partial e(X,y)}{\partial t}. \tag{2.128}$$

   (d) Entropy inequality

$$\rho_0 \frac{\partial s(X,t)}{\partial t} \geq \frac{\rho_0 r}{\theta} - \frac{\partial}{\partial X}\left(\frac{Q}{\theta}\right), \tag{2.129}$$

2. Eulerian formulation
   (a) Conservation of mass

$$\frac{\partial \rho}{\partial t} + \frac{\partial(\rho v)}{\partial x} = 0. \tag{2.130}$$

   (b) Balance of momentum

$$\rho \frac{\partial v(x,t)}{\partial t} + \rho \frac{\partial v(x,t)}{\partial x} v(x,t) = b + \frac{\partial n}{\partial x}. \tag{2.131}$$

(c) Balance of energy

$$\rho \frac{D\bar{u}}{Dt} = \rho r - \frac{\partial q(x,t)}{\partial x} + nL. \tag{2.132}$$

(d) Entropy inequality

$$\rho \frac{Ds}{Dt} \geq \frac{\rho r}{\theta} - \frac{\partial}{\partial x}\left(\frac{q}{\theta}\right). \tag{2.133}$$

At this point, a simple head count of unknowns and equations is not out of order. Considering, for example, the Eulerian formulation, we have a total of three independent equations of balance (mass, momentum, energy). As already remarked, the entropy inequality plays a different role and does not directly participate in the solution of a problem, but acts as a control so that no possible solution can violate the second law of thermodynamics. The unknown independent fields explicitly appearing in these three balance equations are: the density ρ, the velocity v, the internal force n, the internal energy density $\bar{u}$ and the heat flux q, for a total of five unknown fields. We note that the velocity gradient L is the spatial derivative of v, so it does not count as an independent field. Moreover, the body force b and the body heat source r are externally specified for any given problem. From this elementary count, it appears that we have a defect of two equations. This gap should be filled by the *constitutive equations* representing the specific material response of the body at hand. In principle, the constitutive equations must specify the internal force n, the internal energy density $\bar{u}$ and the heat flux q as functions of the (history of the) deformation and the temperature θ. The deformation and the velocity are related by differentiation, but the temperature is a new field to be solved for. Thus, we have a total of six equations (three balance equations and three constitutive equations) for a total of six fields $(\rho, v, n, \bar{u}, q, \theta)$.

2.11   Case Study: Bioheat Transfer and Malignant Hyperthermia

Just as there are many practical problems that can be modelled as purely mechanical (ignoring thermal effects), so too there are many important situations in which the deformations of the underlying continuum are insignificant, while the transfer of thermal energy is paramount. A discipline that is almost entirely based on this last assumption is *bioheat transfer*. One of the applications of this discipline is to *ablative surgical procedures* in general and *malignant hyperthermia* in particular, consisting on a number of techniques for the destruction of cancerous cells by targeted heating. A problem with these treatments is how to raise the temperature of the cancerous tissue to lethal levels while maintaining the temperature of the surrounding healthy tissue at acceptable levels so that it is not destroyed. The success of the technique depends on the enhanced sensitivity of cancerous cells to heat, due in part to the fact that blood perfusion, with its attendant cooling effect, is less abundant in cancerous cells than in healthy tissue.

Assuming, for practical purposes, the substratum to be rigid and at rest, the equations of mechanical balance (mass and momentum) are trivially satisfied. Moreover, there is no difference between the Lagrangian and Eulerian formulations. The energy balance reduces to

$$\rho_0 \frac{\partial \bar{u}(X,t)}{\partial t} = \rho_0 r - \frac{\partial Q(X,t)}{\partial X}. \tag{2.134}$$

For many materials, a good empirical constitutive law for increments $\Delta \bar{u}$ in internal energy density due to corresponding increments $\Delta \theta$ in temperature is given by the linear relation

$$\Delta \bar{u} = c \Delta \theta, \tag{2.135}$$

where c is a constant known as the *specific heat (capacity)*.[18] The internal energy depends in general also on the deformation gradient, which in our case has been ignored.

As far as the heat flux is concerned, *Fourier's law of heat conduction* is an empirical relation valid for most materials within limited temperature ranges. It establishes that the heat flux per unit area is proportional to the gradient of the temperature. In our notation, this law is expressed as

$$Q = -k\hat{S} \frac{\partial \theta}{\partial X}, \tag{2.136}$$

where $\hat{S}$ is the cross-sectional area and k is the *thermal conductivity* of the material, a positive constant. The minus sign expresses the fact that heat flows spontaneously from higher to lower temperatures.[19]

Introducing the constitutive equations (2.135) and (2.136) into the energy balance equation (2.134) and assuming, for definiteness, a constant cross-sectional area, we obtain

$$\hat{\rho}_0 c \frac{\partial \theta}{\partial t} - k \frac{\partial^2 \theta}{\partial X^2} = \hat{\rho}_0 r, \tag{2.137}$$

where $\hat{\rho}_0$ is the mass density per unit volume, as explained in Remark 2.2.4. This equation is identical in form to equation (2.39) governing the diffusion of one substance through another, as we studied in Section 2.5. For this reason, this equation is known both as the (non-homogeneous) *diffusion equation* and as the *heat equation*. The adjective 'non-homogeneous' refers here to the fact that there are body sources. Thus, the equation would be called homogeneous if the right-hand side were zero. On the other hand, the material itself may have properties, such as the specific heat or the thermal conductivity, varying from point to point, in which case it is the body (rather than the equation) which would be called inhomogeneous. In deriving equation (2.137), in fact, it was assumed that the coefficient of thermal conductivity k was constant throughout the domain of interest. If, instead, k, $\hat{\rho}_0$ and/or c are functions of position (i.e., if the material is inhomogeneous), equation (2.137) should be replaced by

$$\hat{\rho}_0(X) \, c(X) \, \frac{\partial \theta}{\partial t} - \frac{\partial}{\partial X} \left(k(X) \frac{\partial \theta}{\partial X} \right) = \hat{\rho}_0(X) \, r. \tag{2.138}$$

So far, equation (2.137) (or 2.138) has been derived without particular reference to the problem of bioheat transfer. It is the right-hand side of equation (2.137), containing the various sources of heat r, that will reflect the bioheat context. It is precisely the complexity of the transport of thermal energy in the human body that makes it difficult

---

[18] This positive 'constant' is obtained through calorimetry by measuring the amount of heat necessary to raise a unit of mass by one degree. For many solids under normal conditions, the distinction between specific heats at constant volume (or deformation) and at constant pressure (or stress) does not need to be made.

[19] That this is the case can be shown to be consistent with the dissipation inequality, as will be discussed in Chapter 3.

to give an explicit expression to the variety of mechanisms involved, most particularly the perfusive heat transfer (cooling effect) r_b between tissue and blood, the metabolic production r_m and the radiation of the heating device r_r:

$$r = r_b + r_m + r_r. \tag{2.139}$$

The first explicit model for the quantification of the blood perfusion phenomenon was proposed by Pennes (1948), and *Pennes' equation* is still widely used. Specifically, the term due to blood perfusion has the form

$$\hat{\rho}_0 r_b = w_b c_b (\theta_a - \theta). \tag{2.140}$$

In this equation, w_b is the *blood perfusion rate*, namely, the mass of blood per unit volume of tissue that traverses this volume per unit time; c_b is the specific heat of blood; and θ_a is the temperature of the arterial blood. It is interesting to remark that this heat sink (whenever $\theta_a < \theta$) is not given in a completely explicit numerical manner, but it depends on the solution (θ) itself.[20]

The metabolic production r_m, though not necessarily negligible, will be neglected here in comparison with the radiation of the heating device. In our numerical example we are following the work of Kouremenos and Antonopoulos (1988), in which this device consists of an antenna developed by Uzunoglu et al. (1987) producing electromagnetic radiation and heating the tissue according to the following empirical formula:

$$r_r = sP \exp(a(x - 0.01\text{m})), \tag{2.141}$$

where s and a are antenna constants, x is the tissue depth and P is the transmitted power. One of the rare merits of the work of Kouremenos and Antonopoulos (1988) is that, in addition to parametric studies and critical discussion of the results, they provide a complete list of material constants, so that their results can in principle be reproduced in every detail. Here, we will content ourselves with a one-dimensional analysis along a line perpendicular to the skin at the centre of the antenna applied at the skin surface (line AB in Figure 2.8). The antenna constants are: $s = 12.5\,\text{kg}^{-1}$, $a = -127\,\text{m}^{-1}$, $P = 20\,\text{W}$, $\hat{\rho}_0 = 1000\,\text{kg m}^{-3}$, $c = 4180\,\text{J kg}^{-1}\,{}^\circ\text{C}^{-1}$, $k = 0.5016\,\text{W m}^{-1}\,{}^\circ\text{C}^{-1}$, $w_b = 8\,\text{kg m}^{-3}\,\text{s}^{-1}$, $c_b = 3344\,\text{J kg}^{-1}\,{}^\circ\text{C}^{-1}$. The arterial blood temperature θ_a will be taken as the normal body temperature of 37°C, while the outer surface of the skin will be assumed to be kept at the room temperature of $\theta_s = 20^\circ$C. Before the application of the heating device, the transition between this temperature and the normal body temperature is assumed (in our calculations) to take place gradually over a thickness of 10 mm according to the formula

$$\theta_{\text{transition}}(x) = 37^\circ\text{C} - (37^\circ\text{C} - \theta_s)\,((x - 0.01\,\text{m})/0.01\,\text{m})^2, \quad 0 \le x \le 0.01\,\text{m}. \tag{2.142}$$

The initial condition for our heat transfer problem is given, accordingly, by

$$\theta(x,0) = \theta_{\text{initial}}(x) = \begin{cases} \theta_{\text{transition}}(x) & \text{if } 0 \le x \le 0.01\,\text{m}, \\ 37^\circ\text{C} & \text{if } x > 0.01\,\text{m}. \end{cases} \tag{2.143}$$

The boundary conditions are

$$\theta(0,t) = \theta_s \tag{2.144}$$

----

[20] In this respect, the blood perfusion term functions very much like a *follower force*, just as when inflating a balloon the direction of the applied external force is determined by the unknown normal direction to the current balloon shape.

and

$$\theta(0.05\,\mathrm{m}, t) = 37^\circ\mathrm{C}. \tag{2.145}$$

In other words, it has been assumed that at a depth of 5 cm within the affected tissue, the temperature is the usual body temperature. This situation would approximately correspond to the depth of soft tissue in the human thigh.

Remark 2.11.1   The heat equation (2.138) is a second-order linear PDE of the *parabolic* type. The wave equation (2.66), on the other hand, which is of the *hyperbolic* type, admits at each point of its domain two different *characteristic directions*, representing two different finite speeds of propagation of signals. In the case of the heat equation, the two characteristic directions coincide and the PDE is of the *parabolic* type, with the common speed of propagation being infinite. In the space-time domain of Figure 2.8, the characteristic direction is everywhere horizontal, namely, a line with $t = $ constant. Since, in particular, we want to prescribe initial conditions at time $t = 0$, we find ourselves in the presence of a so-called *characteristic problem*. It so happens that in this case, to obtain a properly formulated initial value problem, it is sufficient to specify on that line the value of the unknown function alone (and not its time derivative too, as one might have

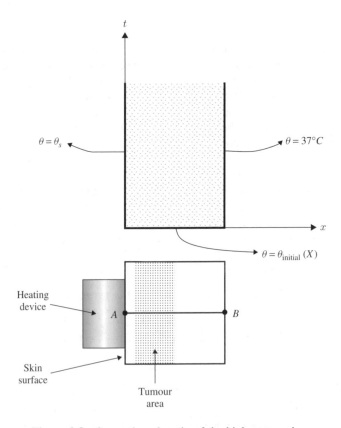

Figure 2.8   Space-time domain of the bioheat equation

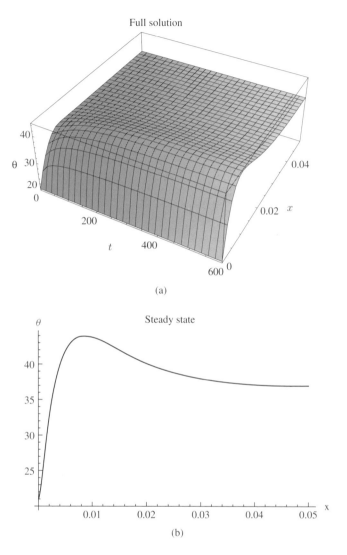

Figure 2.9 Hyperthermia reaction in normal tissue

expected because the equation is of second order). Indeed, the in-line (spatial) second derivative of the initial condition determines, via the PDE itself, the time derivative at the initial time.

Two simulations were carried out using Mathematica®. In the first one, it is assumed that normal tissue alone is being heated. The results are shown in Figure 2.9. The full transient solution (a) is shown for the time period $0 \le t \le 600$ s. The steady state (b) is achieved with an accuracy of $0.3°$C at $t = 400$ s. It will be observed that the maximum temperature at the steady state is about $44.1°$C at a depth of 8.6 mm.

In the second simulation, it is assumed that a tumour occupies the zone 0.003 m $\le x \le 0.024$ m. The tumour tissue has a reduced thermal conductivity of

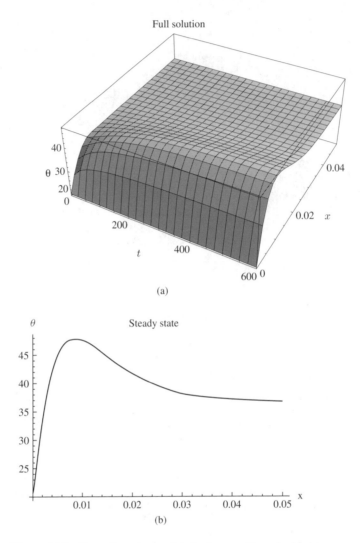

Figure 2.10 Hyperthermia reaction in tissue with embedded tumour

$k_{\text{tumour}} = 0.4\,\text{W}\,\text{m}^{-1}\,{}^{\circ}\text{C}^{-1}$ and a reduced perfusion rate of $w_{b_{\text{tumour}}} = 5\,\text{kg}\,\text{m}^{-3}\,\text{s}^{-1}$, while the other properties remain unchanged. Equation (2.138) holds in this case, since the material properties vary with position. Figure 2.10 shows the solution. The steady state is achieved with an accuracy of 0.3°C at $t = 500\,s$. The maximum temperature in the tumour reaches a value of 47.95°C at a depth of $8.6\,\text{mm}$. The difference of almost 4°C between the healthy tissue simulation and the tumour is significant. In fact, if these values are reliable, the power of the antenna should be slightly reduced so as not to exceed temperatures in the range of $42-46^{\circ}\text{C}$ in the tumorous tissue.

```
rho[x_]:=1000
c[x_]:=4180
k[x_]:=Whch[x<0.003,0.5016,x>0.024,0.5016,True,0.4]
p=20
s=12.5
wab[x_]:= Whch[x<0.003,8,x>0.024,8,True,5]
cb=3344
ts=20
init[x_]:=if[x<0.01,37-(37-ts)*((x-0.01)/0.01)^2,37]
plot[init[x],{x,0,0.05},plotRange→All]
sol=
 NDSolve[{rho[x]*c[x]*D[u[t,x],t]-D[k[x]*D[u[t,x],x],x]==
    wab[x]*cb*{37-u[t,x]}+rho[x]*s*p*Exp[a*(x-0.010)],u[0,x]==init[x],u[t,0]==ts,
    u[t,0.050]==37},u,{t,0,1000},{x,0,0.050}]
Plot3D[Evaluate[u[t,x]/.First[sol]],{t,0,600},{x,0,0.050},PlotPoints→50,
 PlotRange→All,AxesLabel→{"t","x","θ"},PlotLabel→"full solution"]
Plot[Evaluate[u[1000,x]/.First[sol]],{x,0,0.050},PlotRange→All,AxesLabel→{"x","θ"},
 PlotLabel→"Steady state"]
```

Figure 2.11   Mathematica® code corresponding to Figure 2.10

The Mathematica® code corresponding to the second simulation is shown in Figure 2.11.

References

Courant, R. and Hilbert, D. (1962) *Methods of Mathematical Physics*, Vol. II. New York: Interscience Publishers.

Kachani, S., Shmatov, K. and Weinberger, J. (2007) Applications of vehicular traffic theory to blood rheology. In *Proceedings of TRISTAN VI, The Sixth Triennial Symposium on Transportation Analysis*, Phuket, Thailand, 10–15 June.

Knobel, R (2000) *An Introduction to the Mathematical Theory of Waves*. Providence, RI: American Mathematical Society.

Kouremenos, D.A. and Antonopoulos, K.A. (1988) Heat transfer in tissues radiated by a 432 MHz directional antenna. *International Journal of Heat and Mass Transfer* **31**(10), 2005–2012.

Pennes, H.H. (1948) Analysis of tissue and arterial blood temperature in the resting human forearm. *Journal of Applied Physiology* **1**, 93–122.

Uzunoglu, N.K., Angelikas, E.A. and Cosmidis, P.A. (1987) A 432-MHz local hyperthermia system using an indirectly cooled, water-loaded waveguide applicator. *IEEE Transactions on Microwave Theory Techniques* **35**, 106–111.

3

Constitutive Equations

3.1 Introduction

We have already discussed briefly the need for constitutive equations, a need that is both mathematically driven (so as to render the governing system of equations complete, as suggested in Section 2.10) and physically inspired (so as to incorporate a description of the material properties in the phenomenological model, as already mentioned in Section 1.1). And, indeed, in all of our case studies we had to take into consideration the constitutive factor on a case by case basis, whether as an empirical velocity–density relation in the blood/traffic problem, or Fick's law in the diffusion of a pollutant, or the equation of state of air in the ear canal, or Fourier's law of heat conduction in the bioheat equation.

In view of these examples, one may adduce a purely experimental method for the discovery of constitutive equations. All one has to do, one may argue, is perform a sufficient number of laboratory tests and then fit a formula to the data. There are, however, some serious objections that may be raised against this epistemological point of view. To have asked a pre-Newtonian experimentalist for the value of the gravitational constant would have been certainly incomprehensible, if not impossible. Similarly and less glamorously, to ask someone to measure the viscosity of a fluid presumes a previous common understanding of what a viscous fluid is and how the viscosity might enter into some pre-existing formula pertinent to the mathematical model of a viscous fluid. In other words, in most cases, experiments are carried out either to put to test the correctness of a model (the Michelson–Morley experiment comes to mind), or, having established its validity, to determine the precise values of its parameters (Millikan's oil-drop experiment, for example).

There are still other, more practical reasons, to have a theoretical framework at hand before venturing into the laboratory. If, either by malfunctioning of a measuring device or by misinterpretation of the data, one should record, for example, a negative viscosity or a negative thermal conductivity, a fundamental law of nature, in this case the second law of thermodynamics, would appear to have been violated. So, unless there is a very good reason to believe in this anomalous behaviour, one should become suspicious of the experimental results and repeat the experiment with redoubled care.[1]

---

[1] A historically important biomechanical example of the interaction between experiment and theory is Archibald Vivian Hill's (1886–1970) life-long involvement with the modelling of skeletal muscle. The experiments reported

These introductory comments, which are by no means incontrovertible or complete, may serve as a justification for what we are about to do. Our programme involves establishing certain general guidelines or principles for the formulation of constitutive equations. These guidelines are not written in proverbial stone, however, and are subject to enlightened criticism from both within and without the discipline. The presentation will follow the more or less accepted tenets of continuum mechanics, even though, as befits any scientific discipline worthy of its name, these principles have long worn and still wear the garb of work in progress.

3.2 The Principle of Determinism

The *principle of determinism* establishes that all the constitutive quantities of a body B at time t are completely determined by the histories of the motion and of the temperature of the body up to and including the present time t. By 'constitutive quantity' we mean any of the following fields: internal force, heat flux, internal energy density and entropy density.[2]

Several remarks are in order. The first one has to do with the irrelevance of the future in the determination of the constitutive fields. This can be interpreted physically as a *principle of causality*, to which we shall strictly adhere.

The second remark brings into focus the fact that, as formulated above, the principle of determinism asserts that the past and the present of the motion and the temperature *alone* completely determine the present constitutive fields everywhere in the body, and that nothing else is required. In this regard, however, we must mention that there exist legitimate material theories of great practical relevance where more information is needed. Examples of such theories that are of interest in biomechanical applications comprise models of materials with added *internal structure* and theories with *internal state variables*. The former involve extra kinematic degrees of freedom, as could be the case of a material background made of extra-cellular matrix in which a large number of cells, with different properties, are embedded. One may then think of attributing internal deformations to the cells in some sense independent of those of the matrix. Among models with internal state variables, we mention the rich variety of biological phenomena of *remodelling*, *growth* and *ageing*. These phenomena are particular cases of theories of *material evolution*, which usually require the introduction of a measure of the extent of evolution (such as a damage parameter, a plastic strain, and so on) governed by additional evolution equations. For theories with extra independent variables, the principle of determinism should be extended, always adhering to the principle of causality.

The third important remark is designed to emphasize and distinguish two aspects of the principle of determinism: a temporal aspect and a spatial one. The temporal aspect refers to the possible influence of the whole history of the body, from $t = -\infty$ to the present time. In physical terms, we speak of *materials with memory*. Constitutive equations of materials

---

in Hill's celebrated 1938 article supported the force–velocity relation in terms of thermodynamic arguments. Hill's paper concerned itself mainly with measurements of heat production, and it was through hypothesizing certain relations between measured heat rates and velocities that Hill's essentially correct mechanical model was originally derived. A candid account of the crystallization of these ideas is provided by Hill himself in his delightful book (Hill, 1970).

[2] In some of our case studies we have used constitutive equations that relate other quantities, such as velocity with density. These cases, however, should be understood as shortcuts that take the place of balance equations altogether. For example, in Section 2.5 we used mass conservation together with Fick's law, but we did not enforce the balance of momentum.

with memory are often expressed in terms of integrals over the time variable extending over the whole time interval from $t = -\infty$ to the present time. This infinite extent of the memory range is seldom enforced in an arbitrary way. Rather, many materials remember much more strongly the more immediate than the more distant past. Such materials are said to have *fading memory* and are more manageable mathematically, although the theory of materials with fading memory is notoriously difficult and seldom applied in its full generality. It is to be noted, however, that many biological materials are among the most likely candidates to be treated with this delicate tool. One only has to think of skeletal muscle, in which the process of activation (an internal variable, by the way) triggers complex chemical reactions whose effects wane in time. But even without the waning effect, it is clear that the history of the activation of a muscle, on the one hand, and of the stretch, on the other hand, plays a decisive role in the resulting internal force, so that two identical fibres may have equal present values of activation and stretch and yet sustain different internal forces.

Many materials do not exhibit any such memory effects. By this we mean that only the present state of affairs of the motion and the temperature completely determine the present constitutive fields. By 'present state of affairs', however, we do not necessarily mean just the deformation, but also any number of time derivatives of the motion evaluated at the present time. A common example is that of a viscous fluid, in which the present value of the velocity gradient appears in the constitutive laws. If no further time derivatives are involved, we speak of a *first-order material*, reserving the name of *higher-order materials* for those that involve higher-order time derivatives. These materials do not possess memory, although the higher the order of the material the more one may reasonably claim that the past makes itself felt in the present in the following intuitive sense: there are more histories with the same first time derivative than histories with both the same first and second time derivatives evaluated at the current time.

The spatial aspect of the principle of determinism is almost the exact spatial counterpart of the phenomenon of memory, except for the fact that there is no intrinsic spatial directionality, like past and future. If the motion of the whole body is involved in the determination of the constitutive quantities, we speak of a *non-local material*.[3] It is common to express this non-locality by means of integrals that extend over the whole body. Nevertheless, since the strength of the interaction between distant body particles usually diminishes rapidly with their mutual distance, it is not uncommon to have those integrals extend only over a finite range around the particle on which the quantity is evaluated. This is clearly the spatial counterpart of fading memory. If the value of a constitutive variable is affected only by the state of the deformation (and of the temperature) at the point, we speak of *local materials*. The state of deformation at the point may involve any number of spatial derivatives of the deformation, the first one being what we have called the deformation gradient. A local material for which just the deformation gradient appears in the constitutive equations is said to be a *first-grade* or *simple material*. Higher-grade materials are similarly defined, with second-grade materials having found a number of interesting practical applications, including the modelling of certain aspects of muscle tissue. Notice that one should not make the mistake of coupling the definitions of memory and non-locality. A material may be simple (hence, local) and yet possess memory. Conversely, a material may have no memory and yet be non-local.

---

[3] This curious negative definition (reminiscent of Stanislaw Ulam's *bon mot* on the 'physiology of non-elephants') raises the question of why not to call these materials *global*. However, we reserve this terminology for yet more general materials, for which even the concept of flux itself cannot be localized.

Needless to say, the vast majority of successful material models used in classical engineering applications are local and without memory. One of the challenges of biomechanics, however, is that most biological materials, particularly soft tissues, exhibit some kind of memory effects. For this reason, we devote the last two sections of this chapter to an introductory treatment of memory effects. For now, however, we turn to the other guiding principles of the constitutive theory.

3.3   The Principle of Equipresence

When studying *classes of materials*, the starting point is a statement as to which arguments are to be used in the constitutive functionals. We have already established that these functionals relate the internal force, the heat flux, the internal energy density and the entropy density to the histories of the deformation and the temperature. Which aspects of the latter two are to be retained is precisely what determines the class of materials under study and allows us to develop the particular theories of elasticity, thermoelasticity, fading memory, and so on. One theory may consider as the causes of all constitutive effects the present value of the deformation gradient, the temperature and the temperature gradient, thus leading to the theory of simple thermoelastic heat conductors. Another theory may include also the time derivative of the deformation gradient so as to account for viscous effects.

What the *principle of equipresence* frowns upon is an artificial *a priori* separation of the various aspects of the material response and the attribution of different causes to each. Thus, to quote from Truesdell (1966, p. 42),[4] the 'separation of temperature gradient as the "cause" of heat flux but not of stress, and velocity gradient as the "cause" of stress but not of heat flux, is unnatural and unjustified by physical principle. Resulting only from the gradual discovery of individual phenomena, it reflects old opinions that break physics up into compartments. Theorists should not propose constitutive equations which artificially divert theories into disjoint channels.' In Truesdell's words, the principle of equipresence establishes that 'a quantity present as an independent variable in one constitutive equation is so present in all, to the extent that its appearance is not forbidden by the general laws of physics or rules of invariance.' It is expected, however, that these laws of physics and principles of invariance will precisely restrict the way in which a particular variable may occur in a particular constitutive equation. The separation of effects, rather than being postulated *a priori*, should emerge as an *a posteriori* consequence of the correct application of these laws, such as the second law of thermodynamics.

3.4   The Principle of Material Frame Indifference

The notion of *observer* in classical mechanics is not without its difficulties. It presupposes the existence of a peculiar geometric structure that can support the following concepts: (1) the *simultaneity* of physical events is observer independent; (2) all observers agree on the relative size of time intervals and, as a consequence, can agree on a common time unit; (3) each space of simultaneous events has an observer-independent *affine and metric structure*, whereby the ideas and theorems of Euclidean geometry hold true. This geometric structure permits us to define an observer (or a *frame*) as a sentient entity

---

[4] This opuscule, already almost half a century old, still stands as the most concise and beautifully written presentation of the aims, scope, method and style of continuum mechanics.

attached to a rigid triad of mutually perpendicular directions and carrying a standard ruler and a clock. The most general relative motion between different observers is a translation between their origins and a rotation between their rigid triads, both arbitrarily dependent on time.

The tenets of classical mechanics, however, recognize that among all observers there exists a privileged class, known as the class of *inertial observers*, for which the laws attain a particularly simple form (force equals mass times acceleration, say). All observers of this class are in an arbitrary state of mutual relative translation at constant speed, the relative orientation of their respective triads being frozen in time. Knowing just one of these inertial observers is, therefore, sufficient to determine them all. The balance laws of continuum mechanics, as generalizations of the laws for classical systems of particles, are also limited in this way. In other words, their validity is restricted to inertial observers. For non-inertial observers, additional terms need to be incorporated in the equations. One might think, accordingly, that the non-inertial observers have no role to play in continuum mechanics and that the material properties of the body may also be affected by the non-inertiality of its carrier. That this is actually not the case is the essential content of the *principle of material frame indifference*.

To quote again from Truesdell (1966, p. 6), whose eloquence in these matters is beyond compare: 'Take a spring and on one end hang a weight of one pound. The spring lengthens, say, by one inch. Now lay the spring on a horizontal table, fastening one end to the center, and leaving the weight attached to the other end. Spin the table, and adjust the angular speed until the spring again stretches exactly one inch. On seeing this demonstration in the laboratory, the freshmen, happy to participate in the experimental foundation of science, take it as obvious that the force exerted by the spring is again one poundal'. Continues Truesdell: 'What has been assumed, tacitly, is that the elastic law or constitutive equation of the spring is invariant under rotation. For an observer standing on the floor as well as for an observer seated upon the table, or, for that matter, for an observer watching the experiment in a plane mirror as he is shot from the mouth of a cannon, one inch of extension corresponds to one poundal of force.'[5]

The principle of material frame indifference states that *the material functionals are invariant under arbitrary observer transformations*. The motion itself, the velocity, the deformation gradient, the heat flux vector and other constitutive quantities are not themselves necessarily invariant under observer transformations. But the constitutive functionals relating these quantities are. Since in our restricted one-dimensional context rotations do not exist, we are left merely with arbitrary translations and possibly reflections (just like Truesdell's strange observer shot from the mouth of a cannon and watching the events in a non-distorting mirror). The applications of the principle in the one-dimensional context are quite trivial when compared with its consequences in the full three-dimensional case, which will be dealt with in Chapter 6.

Denoting with a star quantities measured by a second observer, the spatial and temporal variables are related by the formulas

$$x^* = x + c(t) \tag{3.1}$$

and

$$t^* = t + b, \tag{3.2}$$

---

[5] It is possible to object that some materials may be constituted of nanogyroscopes which would be sensitive to experiments in rotating frames. If this is indeed the case, the principle of material frame indifference can be construed as excluding such materials.

where $c(t)$ is an arbitrary function of time and b is a constant. This is the most general observer transformation (or change of frame) in a one-dimensional world. Assuming that both observers agree on the notion of material particle, the observed velocities and accelerations of a given particle are related, respectively, by

$$v^* = v + \frac{dc}{dt} \tag{3.3}$$

and

$$a^* = a + \frac{d^2c}{dt^2}. \tag{3.4}$$

The deformation gradients, on the other hand, are equal,

$$F^* = F, \tag{3.5}$$

and so are the internal forces,

$$N^* = N, \tag{3.6}$$

and the temperatures,

$$\theta^* = \theta. \tag{3.7}$$

Example 3.4.1 Assume that a constitutive equation is proposed by the first observer of the form

$$N = g(F, \theta, v, a, t). \tag{3.8}$$

According to the principle of material frame indifference, the second observer will necessarily have to adopt the law

$$N^* = g(F^*, \theta^*, v^*, a^*, t^*), \tag{3.9}$$

with *the same constitutive function* $g(\cdot)$. By equations (3.2)–(3.7), we can write (3.9) as

$$N = g\left(F, \theta, v + \frac{dc}{dt}, a + \frac{d^2c}{dt^2}, t + b\right). \tag{3.10}$$

Comparing with equation (3.7), we obtain the identity

$$g(F, \theta, v, a, t) = g\left(F, \theta, v + \frac{dc}{dt}, a + \frac{d^2c}{dt^2}, t + b\right), \tag{3.11}$$

to be satisfied by the constitutive function $g(\cdot)$ for all functions $c(t)$ and all constants b. It is this kind of identity that allows us to *reduce* the form of the constitutive functional. In our particular example, we first adopt $c(t) \equiv 0$ and obtain

$$g(F, \theta, v, a, t) = g(F, \theta, v, a, t + b). \tag{3.12}$$

Taking on both sides the derivative with respect to b, we immediately conclude that

$$\frac{dg}{dt} = 0. \tag{3.13}$$

The principle of frame indifference has yielded as a first result the conclusion that the constitutive functional cannot depend explicitly on the time variable. It might depend, though, on time intervals. For example, if a defining fabrication event has triggered a process of material ageing, then it would not go against the principle of material frame indifference to impose a dependence of the constitutive law on the time elapsed from that event, since both observers would certainly agree on the measure of time differences. We are left with

$$g(F,\theta,v,a) = g\left(F,\theta,v + \frac{dc}{dt}, a + \frac{d^2c}{dt^2}\right). \tag{3.14}$$

Consider now a relative motion at constant speed, such as

$$c(t) = kt, \tag{3.15}$$

where k is an arbitrary constant. We obtain

$$g(F,\theta,v,a) = g(F,\theta,v + k,a). \tag{3.16}$$

Since k is arbitrary, we differentiate with respect to it and get

$$\frac{dg}{dv} = 0. \tag{3.17}$$

We remark again that the constitutive law might still depend on a relative velocity (as is the case in theories of mixtures, where relative diffusive velocities are at play). Finally, considering a uniformly accelerated relative motion, we eliminate the acceleration argument as well:

$$\frac{dg}{da} = 0. \tag{3.18}$$

No extra independent conditions can be imposed. The final reduced form of the constitutive law is

$$N = g(F,\theta). \tag{3.19}$$

As demonstrated by this simple example, the principle of material frame indifference saves the experimentalist from attempting to measure constitutive properties that cannot exist. It would be futile to attempt to measure a putative dependence of the internal force on the velocity, for example, and the experimental effort should concentrate exclusively on the remaining variables (the deformation gradient and the temperature).

3.5   The Principle of Dissipation

The *principle of dissipation* or of *thermodynamic consistency* establishes the consistency of any proposed constitutive law with the Clausius–Duhem inequality. No material can exist which, under any conceivable circumstances, will be able to undergo any process that contravenes this fundamental inequality.

Recall the (Eulerian) local version of the Clausius–Duhem inequality, equation (2.125):

$$\rho \frac{Ds}{Dt} \geq \frac{\rho r}{\theta} - \frac{\partial}{\partial x}\left(\frac{q}{\theta}\right). \tag{3.20}$$

A cursory look at this inequality suggests a potential difficulty in enforcing it for arbitrary processes. This has to do with the fact that the body source r can be arbitrarily controlled from outside the system and, therefore, whatever process might be proposed the inequality could be violated by an appropriate choice of the external heat source. Nevertheless, it has to be remembered that processes must also satisfy the first law of thermodynamics, namely equation (2.118):

$$\rho \frac{D\bar{u}}{Dt} = \rho r - \frac{\partial q(x,t)}{\partial x} + nL. \tag{3.21}$$

Combining the two laws, we can eliminate the heat source and obtain the following reduced form of the Clausius–Duhem inequality:

$$\frac{\rho}{\theta} \frac{D\bar{u}}{Dt} - \rho \frac{Ds}{Dt} - \frac{1}{\theta} nL + \frac{1}{\theta^2} q \frac{\partial \theta}{\partial x} \leq 0. \tag{3.22}$$

In this form, the inequality contains only constitutive quantities and can be exploited to impose restrictions on constitutive equations.

At this point, it proves convenient, although not strictly necessary, to define a new constitutive quantity (or function of state) called the *Helmholtz free-energy density* per unit mass, denoted by ψ, as

$$\psi = \bar{u} - \theta s. \tag{3.23}$$

We can now replace the internal energy in favour of the new variable and write the reduced Clausius–Duhem inequality as follows:

$$\rho \frac{D\psi}{Dt} + \rho s \frac{D\theta}{Dt} - nL + \frac{1}{\theta} q \frac{\partial \theta}{\partial x} \leq 0, \tag{3.24}$$

where we have used the fact that $\theta > 0$.

The physical meaning of the Helmholtz free energy can be gathered by considering a process at constant and uniform temperature, so that both the temporal and spatial derivatives of the temperature vanish identically. Under these circumstances, equation (3.24) stipulates that

$$\rho \frac{D\psi}{Dt} - nL \leq 0. \tag{3.25}$$

This result can be interpreted as follows. The term nL represents the power of the internal force, which is positive when work is being performed *against* the system (rather than by it).[6] According to equation (3.25), the free energy can grow, but its rate of growth is limited by the product nL. If we integrate over an interval of time, it follows that the growth in internal energy is less than (or, at most, equal to) the work entering the system. The difference $nL - \rho \frac{D\psi}{Dt}$, integrated over time, represents energy (per unit volume) 'lost' or *dissipated*. This loss is irretrievable, except in the reversible case, where the equal sign holds. Indeed, as the system tries to do work against the environment ($nL < 0$), the absolute value of this work is limited precisely by the previous increase of free energy, so that when the free energy returns to its original value, more work has been done against the system than recovered from it. It follows that the free energy of Helmholtz can be regarded as the maximum amount of mechanical work that can be extracted from the

---

[6] Recall that n is positive in tension and L is positive during extension.

system under isothermal conditions. This maximum is attained if and only if the process involved in the extraction of work is reversible.

A convenient way to restate the Clausius–Duhem inequality consists of first defining the *dissipation* per unit (spatial) volume as

$$\delta = nL - \rho\frac{D\psi}{Dt} - \rho s\frac{D\theta}{Dt} - \frac{1}{\theta}q\frac{\partial\theta}{\partial x}. \tag{3.26}$$

The Clausius–Duhem inequality states that the dissipation is always non-negative,

$$\delta \geq 0. \tag{3.27}$$

Exercise 3.5.1   The flow of heat. Assume that a rigid bar has attained a steady-state temperature field. By a careful consideration of the sign convention for the heat flux vector, show that the Clausius–Duhem inequality implies that heat flows from hotter to colder places.

We have worked in the Eulerian formulation only, but similar results hold, *mutatis mutandis*, in the Lagrangian formulation.

Example 3.5.2   A thermoelastic heat conductor. As an example of the application of the Clausius–Duhem inequality to derive restrictions on any proposed set of constitutive equations, we consider the case in which the independent constitutive variables are: the deformation gradient F, the temperature θ and its gradient $\operatorname{grad}\theta = \frac{\partial\theta}{\partial x}$. According to the principle of equipresence, we must assume that *all* constitutive equations have these variables as arguments, namely:

$$n = n(F, \theta, \operatorname{grad}\theta), \tag{3.28}$$

$$q = q(F, \theta, \operatorname{grad}\theta), \tag{3.29}$$

$$\psi = \psi(F, \theta, \operatorname{grad}\theta) \tag{3.30}$$

and

$$s = s(F, \theta, \operatorname{grad}\theta), \tag{3.31}$$

where we have substituted the free energy ψ for the internal energy $\bar{u}$. The heuristic value of the principle of equipresence is that we are not presupposing the possible independence of the free energy, for example, on the temperature gradient (as could be argued on legitimate physical grounds). Rather, we are letting the Clausius–Duhem inequality tell us whether or not that prejudice is justified. The fact that, as it will turn out, it is, is one of the merits of the inequality.

Having established the general form of the constitutive equations for this particular class of materials (known as *thermoelastic heat conductors*), we proceed to substitute them into the Clausius–Duhem inequality (3.24) and use the chain rule of differentiation. The result is

$$\rho\left(\frac{\partial\psi}{\partial F}\frac{DF}{Dt} + \frac{\partial\psi}{\partial\theta}\frac{D\theta}{Dt} + \frac{\partial\psi}{\partial(\operatorname{grad}\theta)}\frac{D(\operatorname{grad}\theta)}{Dt}\right) + \rho s\frac{D\theta}{Dt} - nL + \frac{1}{\theta}q\frac{\partial\theta}{\partial x} \leq 0, \tag{3.32}$$

or, collecting some terms,

$$\left(\rho\frac{\partial\psi}{\partial F}-nF^{-1}\right)\frac{DF}{Dt}+\rho\left(\frac{\partial\psi}{\partial\theta}+s\right)\frac{D\theta}{Dt}+\left(\frac{\partial\psi}{\partial(\mathrm{grad}\,\theta)}\right)\frac{D(\mathrm{grad}\,\theta)}{Dt}+\frac{1}{\theta}q\frac{\partial\theta}{\partial x}\le 0,$$

(3.33)

where we have used equation (1.46). This inequality has to be satisfied identically for all possible processes. This means that we can choose any values for DF/Dt, $D\theta/Dt$, $D(\mathrm{grad}\,\theta)/Dt$ and $\partial\theta/\partial x$ and still the inequality has to be satisfied. As a clarifying digression, consider a linear expression of the form

$$a_1 x + a_2 y + a_3 z + a_4 \le 0,$$

(3.34)

where a_1, a_2, a_3, a_4 are constants or functions of variables other than x, y, z. If this inequality has to be satisfied identically for all values of the variables x, y, z, then a_1, a_2, a_3 must necessarily vanish. For suppose that $a_1 > 0$. We can choose a value $x > -a_4/a_1$ and $y = z = 0$, thereby violating the inequality. We conclude that $a_1 = a_2 = a_3 = 0$ and we are left with the *residual inequality* $a_4 \le 0$.

Returning now to inequality (3.33), we observe that it is indeed linear in the variables DF/Dt, $D\theta/Dt$ and $D(\mathrm{grad}\,\theta)/Dt$, since these expressions, according to our constitutive assumptions, are not among the list of independent variables. The situation is different in regard to $\partial\theta/\partial x$, which is included in that list, so that the coefficient of $\partial\theta/\partial x$ is itself a function of $\partial\theta/\partial x$. From the previous reasoning we necessarily conclude that

$$\rho\frac{\partial\psi}{\partial F}=nF^{-1},$$

(3.35)

$$\frac{\partial\psi}{\partial\theta}=-s,$$

(3.36)

$$\frac{\partial\psi}{\partial(\mathrm{grad}\,\theta)}=0,$$

(3.37)

and the residual inequality

$$q\frac{\partial\theta}{\partial x}\le 0.$$

(3.38)

According to equation (3.37), the free energy ψ turns out to be independent of the temperature gradient, without having assumed it beforehand. Moreover, as equations (3.35) and (3.36) clearly show, the internal force n and the entropy s are completely determined by the Helmholtz free energy, which acts as some kind of potential for these quantities. In particular, both n and s are independent of the temperature gradient! Finally, the residual inequality (3.38) imposes a restriction on the constitutive function for the heat flux. Assume that the following generalized form of Fourier's law (already encountered in Section 2.11) is suggested:

$$q = -k(F,\theta)\,\mathrm{grad}\,\theta,$$

(3.39)

where $k(\cdot)$ is a function of the deformation gradient and the temperature. Substituting this expression in equation (3.38) yields

$$-k(F,\theta)\,(\mathrm{grad}\,\theta)^2 \le 0.$$

(3.40)

We arrive at the conclusion that the function $k(\cdot)$ (the heat conductivity of the material) must be non-negative.

To sum up, the Clausius–Duhem inequality has imposed the following restrictions on the proposed constitutive equations:

1. The material response is completely determined by just two functions: the Helmholtz free energy ψ and the heat flux q.
2. The free energy is independent of the temperature gradient. Since it acts as a potential for the internal force n and the entropy s, they too are independent of the temperature gradient.
3. The conduction of heat proceeds from hot to cold (see Exercise 3.5.1). Moreover, in a Fourier-like conduction law the conductivity must be non-negative.

In closing this example, it may prove useful to evaluate the dissipation δ. This can only be done *a posteriori*, not as an assumption of the theory. Using equations (3.35)–(3.37) in conjunction with equation (3.26), we obtain

$$\delta = -\frac{1}{\theta} q \frac{\partial \theta}{\partial x}. \tag{3.41}$$

By equation (3.38), it follows that $\delta \geq 0$, as expected. The conclusion is that in a thermoelastic heat conductor all the dissipation is due to the irreversibility of the conduction of heat. Processes at uniform temperature throughout the body are reversible, even if the uniform temperature changes in time.

3.6   Case Study: Memory Aspects of Striated Muscle

Muscles are the force-producing units in animals. They are divided into two types: striated and smooth. This terminology derives from the appearance of striated muscle under the microscope as an alternation of light and dark bands, unlike smooth muscle where such a pattern is not to be discerned. The underlying reason for the striation is the existence of a basic unit in striated muscle called a *sarcomere*, whose length is of the order of a few micrometres. Striated muscle can be subdivided into skeletal muscle and cardiac muscle, the latter being involuntary (controlled, as smooth muscle, by the autonomic nervous system). In the case of striated muscle, it is of great interest to relate the macroscopic properties of muscle fibres or even entire muscles to those of the basic sarcomeric unit. Without entering into any degree of detail of this complex issue, we will content ourselves here with accepting the following facts for the purpose of analysis:

1. Sarcomeres (and longer fibres) can be electrically *activated*. If unconstrained, activated sarcomeres would shorten (by a mechanism known as the *sliding filament* model), thus explaining the production of force when the ends are fixed. The activation can attain a maximum value that produces the maximal contraction or *tetanized state*.[7]

---

[7] In living muscle, the activation is the result of repeated pulses sent by *motor neurons*. Each isolated pulse produces a *twitch*, and it is the (non-linear) superposition of repeated pulses that causes the saturated tetanized state.

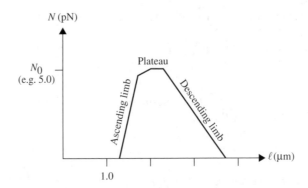

Figure 3.1   Force–length relation for a single sarcomere

2. In the absence of activation, the sarcomere is in a *passive state*, and can be elongated (within reasonable bounds) without the application of any significant force. Consider the following force–length protocol: (a) A passive elongation is applied up to some length ℓ. (b) The sarcomere is held between fixed supports at length ℓ and is then tetanized. (c) The resulting tensile internal force N is recorded. (d) The activation is removed. Step (b) of this protocol is known as *isometric contraction*. If this protocol is repeated for a large number of lengths ℓ and a graph of force vs. length is drawn, the result obtained is always of the shape shown in Figure 3.1. This graph, or its mathematical expression $N = N_e(\ell)$, is known as the *force–length relation*. The graph consists usually of three parts, known as the *ascending limb*, the *plateau* and the *descending limb*. The approximate middle of the plateau, where the sarcomere produces the maximum force N_0, is sometimes called the *optimal sarcomere length*.

3. Consider now the following protocol: (a′) The sarcomere is first fully activated isometrically at the optimal length, but then one of the supports is removed and replaced with a weight (connected perhaps by means of a nanopulley)[8] exactly equal to the force provided by the force–length relation. (b′) The weight is suddenly reduced to some smaller amount. (c′) The ensuing contraction is plotted against time until motion ceases. (d′) The average speed of contraction is recorded (or, more elegantly, it is observed that after a sudden partial contraction the speed stabilizes for a while, and this value v_c is recorded). Step (b') is known as an *isotonic contraction*. If this protocol is repeated for many values of the weight reduction and the remaining weight is plotted against the speed of contraction, as shown in Figure 3.2, the resulting graph, or its mathematical expression $N = N_v(v_c)$, is known as the *force–velocity relation*.

Once in possession of these two functions, N_e and N_v, it might appear that we should be able to construct a plausible complete constitutive equation by means of the formula

$$N = N(\ell, v_c) = \frac{N_e(\ell)\, N_v(v_c)}{N_0}. \tag{3.42}$$

One may even venture into the question of partial activation by introducing an extra internal variable α which takes values in the interval $[0, 1]$, with $\alpha = 0$ corresponding to

---

[8] Historically, these experiments were first carried out in whole muscles.

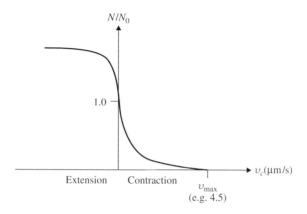

Figure 3.2   Force–velocity relation for a single sarcomere

passivity and $\alpha = 1$ to the tetanized state. We would then have

$$N = \hat{N}(\ell, v_c, \alpha) = \alpha\, N(\ell, v_c) = \alpha\, \frac{N_e(\ell)\, N_v(v_c)}{N_0}. \tag{3.43}$$

These formulas, however, fail to represent the observed facts. A suspicious point is already evident in the existence of a descending limb in the force–length relationship. Assume that we have arrived at the end of step (a′) of the second protocol described above, but at a length above optimal. In other words, we have first brought the sarcomere passively to a length beyond the plateau and we have tetanized it isometrically and, finally, replaced one of the supports with a pulley and a weight equal exactly to the force provided by the force–length relation. The system is, by construction, in equilibrium. If we now increase the load slightly (even if we do this quasi-statically), we find ourselves with the strange situation that, according to formula (3.42), the length of the sarcomere should decrease! And if we were to reduce the weight slightly (just as we do in the isotonic experiment) we would obtain an elongation, rather than a shortening. This is a manifestation of the *insta-bility* of the equilibrium for points on the descending limb of the force–length relation. This instability is not observed experimentally. On the contrary, experiments show that the equilibrium is stable. Moreover, an increase in the force leads to an increase in the length. The force increase can be carried out as slowly as desired (i.e., quasi-statically), always leading to an increase in length. We may say, therefore, that the force–length relation, even in quasi-static conditions, is only part of the story. The material does not follow the force–length curve, which was obtained as the result of many separate and independent experiments, each following the prescription of the first protocol described above: elongate passively, fix the length, tetanize isometrically, remove the activation and start over with a different length. The material, instead, *remembers* the length at which it was first activated and builds upon this memory until it is deactivated.

From the preceding description, it should be clear that striated muscle calls for a constitutive law of the memory type. But the scheme just described, with the need for keeping track of the last length of activation, does not easily fit the fading memory paradigm. Nor is this scheme commonly enforced in computations,[9] in spite of the fact

---

[9] A model that incorporates this aspect was proposed by Epstein and Herzog (1998), where an intuitive description in terms of the mechanics of two brushes in mesh is suggested. A further numerical implementation into a complete muscle model can be found in Lemos et al. (2004).

that any serious implementation of muscle movement (such as in sports applications) requires a fibre recruitment policy to be coupled with the constantly varying length of the fibres, some of which may be operating below or above the optimal length. There is, however, another aspect of striated muscle memory that is more commonly enforced, namely, the velocity dependence of the force produced. The force–velocity experiment, important as it is as a cornerstone of the understanding of muscle response, does not provide enough detail as to the variation of the force in time. Typical viscous-like effects such as creep and relaxation (i.e., elongation at constant load and force reduction at constant length), are just two manifestations of the presence of fading memory. In fact, the isotonic experiment is a partial measure of creep. These phenomena were originally erroneously attributed to the viscosity of a putative fluid flowing within muscular tissue and much was made of this issue (in a way curiously reminiscent of the theory of aether). Although it is known today that such fluid does not exist and that the observed phenomena are rooted in the underlying chemistry, the constitutive equation must be able to account phenomenologically for the observed time-dependent behaviour.

Figure 3.3 shows typical results of experiments on an entire muscle subjected to a variety of loading protocols. The fading memory effects are evident in the exponential-like

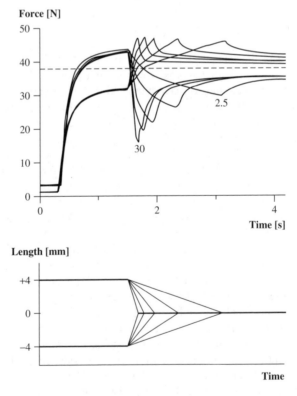

Figure 3.3 Force and length histories of isometric–shortening–isometric and isometric–stretch–isometric contractions for a cat soleus muscle. The horizontal dashed line indicates the isometric force value of the muscle at the final length. The final length, indicated as 0 mm, corresponds to a soleus length at an included ankle angle of 80°. Force enhancement and depression are evident in the fact that the force does not converge to the corresponding isometric value. From Herzog and Leonard (1997), with permission from Pergamon

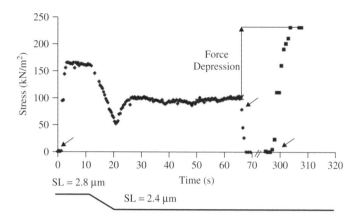

Figure 3.4 Isolated myofibril response when activated at an average sarcomere length of 2.8 μm, shortened to 2.4 μm, deactivated and activated again. These sarcomere lengths (SL) fall within the descending limb of the force–length relation. Note that the stress in the actively shortened condition is considerably smaller than in the purely isometric contraction at either the initial or final lengths. The arrows, from left to right, indicate the times of activation (with SL = 2.8 μm), deactivation (SL = 2.4 μm) and reactivation (SL = 2.4 μm). From Herzog et al. (2010), with permission from Springer

decay of the force over time. More interesting, though, are the effects of *force depression* and *force enhancement*, reflecting the fact that the muscle remembers the length at which it was activated and, upon shortening or elongation, does not attain the same force as it would if it were activated at the new length, as already noted above. It is interesting to point out that recent experiments[10] reveal that these effects are present even at the level of isolated sarcomeres. The importance of this result, shown in Figure 3.4, cannot be overemphasized. It is crucial for the explanation of muscle stability.

Of all these phenomena (and others not mentioned above) the theory of fading memory, to which we turn in Section 3.8, will be able to account only for those related with linear or non-linear viscoelastic effects. Other effects necessitate the inclusion of extra internal state variables. Nevertheless, the theory of fading memory is an important part of the picture, not only for striated muscle but also for many other soft biological tissues.

3.7 Case Study: The Thermo(visco)elastic Effect in Skeletal Muscle

The 1922 Nobel Prize in Medicine or Physiology was shared by Archibald Vivian Hill (1886–1977) and Otto Fritz Meyerhof (1884–1951). Hill's citation was 'for his discovery relating to the production of heat in the muscle', while Meyerhof's was 'for his discovery of the fixed relationship between the consumption of oxygen and the metabolism of lactic acid in the muscle'. This would be the only Nobel Prize awarded for muscle physiology to this day. Both citations highlight the basic fact that force production in muscle is intimately related with heat dissipation and chemical reactions. This should not be surprising, since muscles are essentially motors, just like the internal combustion engines of cars. They transform chemical into mechanical energy, releasing heat in the process.

---

[10] See Leonard et al. (2010) and Herzog et al. (2010).

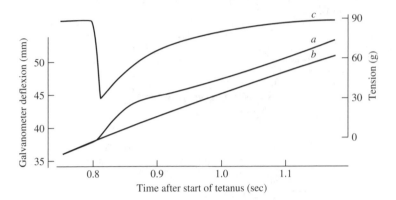

Figure 3.5 Heat produced ((a), (b)) and tension (c) in tetanic contractions of toad sartorius muscle at 0°C. Muscle mass = 91 mg. Optimal length = 25.5 mm. Galvanometer scale 2.75×10^{-5} J mm^{-1}. Graph (b) corresponds to isometric conditions. For graphs (a) and (c), sudden shortening = 0.5 mm. From Woledge (1961), with permission from The Physiological Society

Hill used calorimetric measurements to get at the heart of the mechanical phenomena. One of the crucial experiments involved measuring the (accumulated) heat released by an isometrically tetanized muscle subjected to a sudden shortening. Figure 3.5 shows the results of one such experiment. The extra heat is represented by the difference between the two lower graphs, which is shown in Figure 3.6. Notice that upon shortening an increase in the heat released is registered, but as the tension redevelops at the new shorter length, heat is absorbed (negative slope of the difference curve).

The explanation of these phenomena must surely be rooted in two causes or, most likely, a combination thereof. The first explanation attributes the heat production to chemical reactions triggered by the process of length relaxation. The second, which is our concern here, attributes at least part of the heat production to the so-called *thermoelastic effect*.

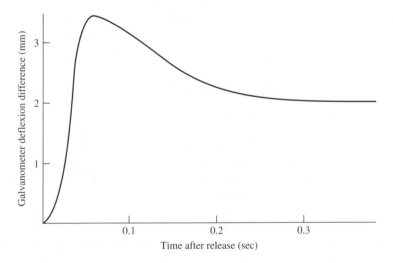

Figure 3.6 Difference between graphs (a) and (b) in Figure 3.5. Adapted from Woledge (1961), with permission from The Physiological Society

This effect can intuitively be regarded as the counterpart of the everyday phenomenon of thermal expansion: if, on heating an unloaded rod, the length increases (or, in some rubber-like materials, decreases), it is 'natural' to expect that a (rapid) change in length could be accompanied by heat. We propose to investigate this expectation somewhat more rigorously and to extend the analysis of the thermoelastic effect so as to include a velocity-dependent material response, which is the case in muscle.

We start from the velocity-independent case, which is easier to grasp, and only thereafter will we proceed to its generalization to the velocity-dependent case. In all cases, however, we will neglect the conduction of heat. Although it is true that to a certain extent the response functions may depend on the temperature gradient (which is usually very small within the muscle under controlled experimental conditions), we assume that the constitutive equations are of the form

$$n = n(F, \theta), \tag{3.44}$$

$$q = q(F, \theta), \tag{3.45}$$

$$\psi = \psi(F, \theta) \tag{3.46}$$

and

$$s = s(F, \theta). \tag{3.47}$$

Since this is clearly a particular case of Example 3.5.2, we conclude from the residual inequality (3.38) that $q = 0$. We are, in other words, considering a thermoelastic insulator. The remaining constitutive restrictions (3.35) and (3.36) are still valid. Moreover, the dissipation δ vanishes identically, so that all our processes are necessarily reversible. An interesting result, known as *Maxwell's law* (and applicable also to the conducting case), is the following:

$$\rho_0 \frac{\partial s}{\partial F} = -\frac{\partial n}{\partial \theta}, \tag{3.48}$$

where we have used equations (2.17), (3.35) and (3.36) and exploited the symmetry of mixed partial derivatives. Most materials have a tendency to elongate when heated, so that preventing the elongation while raising the temperature will result in compression, namely,

$$\alpha(F, \theta) := -\frac{1}{\rho} \frac{\partial n}{\partial \theta} > 0. \tag{3.49}$$

Although we are not adducing direct experimental evidence, it is surmised that tetanized muscle behaves in this way (while, possibly, passive muscle may exhibit the contrary effect, just like rubber). The material property α can be regarded as the *thermoelastic coefficient*,[11] which depends, in general, on the deformation and the temperature. On the other hand, the entropy inequality in its original form (2.125) reduces to

$$\rho \frac{Ds}{Dt} = \frac{\rho r}{\theta}, \tag{3.50}$$

---

[11] To relate this coefficient to the usual coefficient of thermal expansion of strength of materials, we need to multiply the former by Young's modulus and divide by the density.

since $q = 0$ and the process is reversible, as observed above. By the chain rule, we have

$$\rho \left(\frac{\partial s}{\partial F} \frac{DF}{Dt} + \frac{\partial s}{\partial \theta} \frac{D\theta}{Dt} \right) = \frac{\rho r}{\theta}. \tag{3.51}$$

If the shortening is performed under (approximately) isothermal conditions, we have $D\theta/Dt = 0$ It follows, then, from equations (3.48), (3.49) and (3.51) that

$$r = \theta \alpha L, \tag{3.52}$$

where L is the rate of deformation (or velocity gradient). Equation (3.52) embodies quantitatively the *thermoelastic effect*. Since $\alpha > 0$, a positive value of L (namely, a lengthening) results in a positive r (namely, the muscle absorbs heat), and, conversely, a shortening is accompanied by emission of heat. In the case of tetanized muscle, these effects are observed upon instantaneous shortening or stretching. As the tension returns to its original value (ignoring here the force depression or enhancement phenomena), the opposite effect takes place, because of the nature of the sliding filament origin of muscle contraction (if the fibre has been shortened rapidly, the force recovery entails a stretching of the filaments).

The preceding discussion seems to be quite satisfactory in general terms. There are, however, at least three points that may call for further consideration. The first is that experiments have shown that the thermoelastic effect is not symmetric. If a small stretch is applied at exactly the same speed and of the same magnitude of a small shortening, the amount of heat absorbed is smaller in the first case than the amount of heat emitted in the second case. The second point to consider is that, since the experiments are performed at relatively high speed, one expects the force–velocity relation to play some role. Finally, the force depression or enhancement effects are certainly present in the experiments. All these three issues are clearly interconnected and they seem to point to the need to include the rate of deformation in the list of arguments of the constitutive equations. In so doing, we will discover the existence of a *thermoviscoelastic effect* that may help explain some of the issues just raised.

We start with constitutive laws of the form

$$n = n(F, \theta, L), \tag{3.53}$$

$$q = q(F, \theta, L), \tag{3.54}$$

$$\psi = \psi(F, \theta, L) \tag{3.55}$$

and

$$s = s(F, \theta, L). \tag{3.56}$$

Using these constitutive equations in conjunction with the Clausius–Duhem inequality (3.24) and collecting terms yields

$$\left(\rho \frac{\partial \psi}{\partial F} - nF^{-1} \right) \frac{DF}{Dt} + \rho \left(\frac{\partial \psi}{\partial \theta} + s \right) \frac{D\theta}{Dt} + \frac{\partial \psi}{\partial L} \frac{DL}{Dt} + \frac{1}{\theta} q \frac{\partial \theta}{\partial x} \leq 0. \tag{3.57}$$

By a reasoning similar to that in Example 3.5.2, we obtain

$$\frac{\partial \psi}{\partial \theta} = -s, \tag{3.58}$$

$$\frac{\partial \psi}{\partial L} = 0, \tag{3.59}$$

$$q = 0, \tag{3.60}$$

and the residual inequality

$$\left(\rho \frac{\partial \psi}{\partial F} F - n \right) L \leq 0. \tag{3.61}$$

With the heat flux vanishing, the dissipation is now entirely due to the viscous effect, namely, the dependence of n on L. Note that equation (3.59), on the other hand, implies, via equation (3.58), that the entropy is independent of the rate of deformation L.

Since the internal force n no longer derives from the free energy as a potential, Maxwell's law (3.48), which was used in quantifying the thermoelastic effect, is no longer valid. Nevertheless it is fortunately still possible to evaluate a *thermoviscoelastic effect* by a somewhat more indirect and elegant reasoning. Indeed, notice that, for fixed F and θ, the function of L given by

$$\left(\rho \frac{\partial \psi}{\partial F} F - n \right) L \tag{3.62}$$

has a maximum at $L = 0$. This is so because it vanishes at $L = 0$ and, according to the residual inequality (3.61), it is non-positive elsewhere. Assuming differentiability, it follows that

$$\rho \frac{\partial \psi}{\partial F} F = n_0, \tag{3.63}$$

where we have denoted

$$n_0 = n_0(F, \theta) = n(F, \theta, 0). \tag{3.64}$$

Physically, n_0 represents the internal force arising in very slow (quasi-static) processes. We have, therefore, the following modified form of Maxwell's law:

$$\rho_0 \frac{\partial s}{\partial F} = -\frac{\partial n_0}{\partial \theta}, \tag{3.65}$$

where we have used equations (3.63) and (3.58) and the symmetry of mixed partial derivatives. The material property

$$\alpha = -\frac{1}{\rho} \frac{\partial n_0}{\partial \theta} \tag{3.66}$$

has the same meaning as before. It should be measured in quasi-static processes.

To obtain the heat rate associated with an isothermal process ($D\theta/Dt = 0$), we can use the first law of thermodynamics, equation (2.118). The result is

$$\rho r = \rho \frac{D\bar{u}}{Dt} - nL = \rho \left(\frac{D\psi}{Dt} + \frac{Ds}{Dt} \right) - nL. \tag{3.67}$$

Using the chain rule of differentiation and always enforcing isothermal conditions, we obtain

$$\rho r = \left(\rho \frac{\partial \psi}{\partial F} F + \rho \frac{\partial s}{\partial F} F\theta - n \right) L. \tag{3.68}$$

Invoking Maxwell's equation in the form (3.65), definition (3.66) and equation (3.63), we finally obtain

$$\rho r = \big(\rho\alpha\theta - (n - n_0)\big) L. \tag{3.69}$$

We now compare this result (the thermoviscoelastic effect) with its thermoelastic counterpart, equation (3.52). For a force–velocity relation which is monotonically increasing (as is the case in muscle), the extra term $(n - n_0)L$ is always non-negative, so that, upon shortening, more heat is emitted than before, but less heat than before is absorbed in stretching. The dependence on the rate of deformation results, therefore, in an asymmetry of behaviour that can be traced back to the dissipative nature of the processes involved. This asymmetry is qualitatively consistent with experimental observations.

   We have remarked that the force–velocity relation is monotonically increasing. It can be shown that this result can be *derived* (at least in the vicinity of $L = 0$) from the residual inequality and, therefore, that it is a natural consequence of the second law of thermodynamics. To obtain this result, we observe that (always assuming differentiability) at a maximum the second derivative must be non-positive. Applying this observation to the function (3.62) at $L = 0$, we obtain

$$\left.\frac{\partial n}{\partial L}\right|_{L=0} \geq 0. \tag{3.70}$$

Exercise 3.7.1   Thermodynamic consistency of Hill's force–velocity relation. Prove that inequality (3.70) is true.

   In closing this case study, it is worth pointing out that we have tacitly assumed that no chemical reactions are triggered by the shortening. For this to be the case, it is probable that the change of length has to be fast enough so that the energy released is that stored by the elasticity of the cross bridges in the sliding filament model. That the heat released in slower processes turns out, in experiments, to be qualitatively consistent with the thermoelastic effect in a non-reacting velocity-dependent material may have been a contributing factor to some reluctance to discard Hill's original hypothesis that muscle is essentially viscoelastic.

3.8   The Theory of Materials with Fading Memory

3.8.1   Groundwork

The theory of materials with fading memory makes use of the language and theorems of functional analysis, thus placing itself without the realm of usability of many people interested in those applications where fading memory matters the most. It is possible, however, to gain a basic working knowledge of the theory without necessarily mastering the totality of its rather specialized mathematical apparatus. The following presentation may serve as an introduction to the subject and as a motivation to study the pertinent literature. Although it is possible to produce viable models of fading memory by resorting to collections of springs and dampers, as is often done in muscle mechanics, our presentation aims at a greater generality. For concreteness, we concentrate on the purely mechanical case and, more particularly, on simple (or first-grade) materials as defined in Section 3.2.

The first building block in the theory of materials with memory is the notion of *history*. In a purely mechanical theory, we need to concern ourselves just with the strain history, leaving aside the temperature history. Before defining the strain history, we will make the assumption that the material under consideration has been in existence for all past times up to, and including, the present, that is, for all times τ such that $-\infty < \tau \le t$, where t is the present time. We further assume that we have access to all past and present information pertaining to the material point of interest and, most particularly, to its strain $e = e(\tau)$ with respect to a given reference configuration. The *total history* of the strain is defined as the function $e(\tau)$. We emphasize the word *function*, because the history is not a specific value of the strain nor even the collection of all its values. It is the function itself which is the object of our attention. Pictorially, you may want to identify the function with its graph.

If we want to consider a collection of possible histories, we immediately notice that different histories may have different domains of definition in the sense that they all start at $-\infty$, but may run up to different times t. For this and other reasons, it is convenient to introduce a physically meaningful change of variable, namely

$$s = t - \tau. \tag{3.71}$$

The physical meaning of the new time variable s is a backward-running time, starting at the present and going toward the past. In other words, s is a measure of the time elapsed between past events and the present. We can, therefore, define for each history $e(\tau)$ a new function,

$$e_t(s) = e(t - s), \tag{3.72}$$

which we will also call the total history of the strain, since the notation will make clear which function is being used. All histories e_t have exactly the same domain of definition, namely, the half line $0 \le s < \infty$. It is sometimes convenient to distinguish between a *total history* and the corresponding *past history*, which is simply the restriction of the total history to the interval $0 < s < \infty$.

The next step towards a theory of materials with memory consists of considering the collection of all *admissible* histories. By 'admissible' we mean that not every real-valued function e_t will qualify as a strain history, whether for physical or mathematical reasons. In plain words, we want to make sure that we have eliminated from the picture all kinds of 'weird' functions. In disregard of mathematical rigour, we do not pursue this point and assume that we have at our disposal a *function space* $\mathcal{E}$ of nice functions e_t, each of which maps the half line $[0, \infty)$ into the real line.

So far, no constitutive statements have been made, since we were only laying down the groundwork. We want to define now the idea of the *material response functional* $\mathcal{N}$. As a matter of terminology, we start by agreeing to use the term *functional* for a function whose argument is itself a function. The response functional $\mathcal{N}$ is, therefore, a real-valued functional, that is, it assigns to every strain history e_t a value of the internal force $N = \mathcal{N}[e_t]$. More formally:

$$\mathcal{N} \colon \mathcal{E} \to \mathbb{R}$$
$$e_t \mapsto \mathcal{N}[e_t]. \tag{3.73}$$

We use square brackets as a reminder that the argument of the functional $\mathcal{N}$ is a whole function rather than just a real variable.

Remark 3.8.1 **Examples of functionals.** It is not difficult to produce examples of real-valued functionals. It is important in each case to realize that the argument of the functional is a whole function.

1. Let $\mathcal{E}$ be the space of continuous real-valued functions on $\mathbb{R}$ and let x denote the running variable. The *Dirac* functional,

$$\delta : \mathcal{E} \to \mathbb{R}, \tag{3.74}$$

   assigns to each function $f \in \mathcal{E}$ its value at $x = 0$, that is,

$$\delta[f(x)] = f(0). \tag{3.75}$$

2. Let $\mathcal{E}$ be the same as in the previous example. We define the functional

$$\mathcal{A} : \mathcal{E} \to \mathbb{R} \tag{3.76}$$

   by

$$f(x) \mapsto \mathcal{A}[f(x)] = \int_0^1 f(x)dx. \tag{3.77}$$

   This functional assigns to each function $f(x)$ the area under its graph contained between 0 and 1.

3. A functional may have more than one argument. Consider, for instance, the *inner product*

$$\mathcal{I} : \mathcal{E} \times \mathcal{E} \to \mathbb{R}, \tag{3.78}$$

   defined by

$$\mathcal{I}[f(x), g(x)] = \langle f, g \rangle = \int_0^1 f(x)g(x)dx. \tag{3.79}$$

Before going on, it is important to comment on the fact that, by adopting the generic form of equation (3.73), we have already made an implicit assumption: that two histories that are identical except for a time translation give rise to exactly the same value for the internal force N. This is a consequence of our change of variables, in which $s = 0$ always points at the present time. An immediate consequence of this hidden assumption is that if a defining external event were to occur, our response functional would not reflect its effect. A good example of what we are leaving out is precisely such things as the activation of a muscle. As we have already remarked, live biological tissue is, in most cases, so complex, that one needs to include extra *internal state variables* in the picture, variables that are usually non-mechanical but do affect the mechanical response (by, for example, changing the stiffness by several orders of magnitude in going from the passive to the active state). In other words, the history of the strain alone is not sufficient to determine the internal force, but more information is needed. In the case of striated muscle, therefore, the theory of fading memory will account only for effects that happen at constant activation. These phenomena include creep and relaxation, and some aspects of the intriguing observations of *force depression* following a contraction and *force enhancement* following a stretch.

Consider a *constant history*, namely

$$e(\tau) = \text{constant}, \quad \text{for all } \tau \leq t. \tag{3.80}$$

An immediate consequence of the assumed invariance of the response functional under time translations is that constant histories give rise to constant internal force for all (terminal) times t. In particular, if $e(\tau) = 0$ we obtain the *residual stress* with respect to the chosen reference configuration. When there is no room for confusion we will denote by e the constant history $e(\tau) = e(t)$. Thus, e represents the constant history with strain equal to the present value of the strain. When evaluated on constant histories, the response functional $\mathcal{N}$ becomes an ordinary function $N_\infty(e)$ called the *equilibrium response function*.

A useful device consists of resolving any given strain history $e_t(s)$ into the sum of the constant history e (just defined) and the *difference history* $e_t^d(s)$ defined as

$$e_t^d(s) = e_t(s) - e. \tag{3.81}$$

The constitutive equation (3.73) can, therefore, be regarded as a function of two functions,

$$N(t) = \mathcal{N}[e, e_t^d]. \tag{3.82}$$

With some abuse of notation, we have used the same symbol ($\mathcal{N}$) as in equation (3.73) to denote the new constitutive functional in equation (3.82).

3.8.2 Fading Memory

Roughly speaking, the *principle of fading memory* establishes that 'deformations that occurred in the distant past should have less influence in determining the present value of the stress than those that occurred in the recent past'.[12] To express this general statement in terms amenable to mathematical treatment is not a straightforward task.

We may start by observing that if two histories differ very little in the recent past, even if they may differ a lot in the distant past, the results of applying a response functional $\mathcal{N}$ to these two histories should differ by a small amount. To quantify what we mean by the expression 'two histories differ little', we need to introduce some idea of distance in the function space $\mathcal{E}$ of admissible histories. Window 3.1 summarizes some of the technical concepts and terminology pertaining to the definition of a distance between functions.

For the particular application of fading memory, we would like to define a notion of distance that is heavily biased toward the recent past and, at the same time, more indifferent to, or oblivious of, the distant past. With this idea in mind, we define a weight function $h = h(s)$, called an *obliviator of order r*, satisfying the following conditions:

1. $h(s) > 0$ in its domain of definition $0 \leq s < \infty$.
2. $h(0) = 1$.
3. $\lim_{s \to \infty} s^r h(s) = 0$ for some positive $r \in \mathbb{R}$.

---

[12] Quoted from Truesdell and Noll (1965), where a concise and rigorous treatment of the theory can be found. See also Dill (1975).

Window 3.1   Functions as vectors

Real-valued functions of a real variable defined over the same domain can be added (pointwise) and multiplied (pointwise) by real numbers to produce new functions. We assume that the proposed function space $\mathcal{E}$ is closed under these operations. For example, the space of continuous functions over $\mathbb{R}$ satisfies this condition, since any finite linear combination of continuous functions is continuous. Technically, the space $\mathcal{E}$ is an (infinite-dimensional) *real vector space*, as defined in more detail in Section 5.2.2. In a vector space, the notion of size of a vector does not exist in general. To be able to speak of the size of a vector (in our case, the size of a function) we need to introduce a norm. A *norm* in the vector space $\mathcal{E}$ is any real-valued functional G, also denoted by $\| \cdot \|$, with the following properties:

1. $\|f + g\| \leq \|f\| + \|g\|$,   for all $f, g \in \mathcal{E}$.
2. $\|\alpha f\| = |\alpha|\,\|f\|$,   for all $\alpha \in \mathbb{R}, f \in \mathcal{E}$.
3. $\|f\| = 0$ if and only if $f = 0$.

It is easy to show, using Properties 1 and 2, that a norm is necessarily non-negative. A vector space endowed with a norm is said to be a *normed vector space*.[a]

The *distance* between two functions $f, g \in \mathcal{E}$ is defined as the norm of their difference. A sequence of functions $f_1, f_2, \ldots$ in a normed vector space $\mathcal{E}$ is called a *Cauchy sequence* if, for sufficiently large n, the distance between any two functions f_i, f_j (with $i, j > n$) is as small as desired. If every Cauchy sequence in a normed vector space $\mathcal{E}$ converges to a function in $\mathcal{E}$, $\mathcal{E}$ is said to be *complete*. A complete normed vector space is also called a *Banach space*. By a standard procedure, every normed vector space can be *completed* (by formally including all Cauchy sequences) to become a Banach space.

---

[a]The defining properties of a norm are abstracted from the properties of Euclidean geometry: (i) a side of a triangle is smaller than the sum of the other two; (ii) multiplying a vector by a number changes its magnitude accordingly; (iii) the only vector with zero length is the zero vector.

The *standard obliviator* is the exponential

$$h(s) = \exp(-ks), \quad k > 0. \tag{3.83}$$

Checking Condition 3 for this standard obliviator, it follows that it is satisfied for all $r > 0$. We say that the exponential is an obliviator of arbitrary order.

Exercise 3.8.2   Prove that Condition 3 is satisfied for the standard obliviator (3.83). In intuitive terms, the standard obliviator is faster in its decay than any polynomial in its growth. As a pictorial exercise, plot the function $s^2 \exp(-s)$ and observe how, beyond a value of about 8, the exponential factor overcomes the quadratic growth and the product decays exponentially. The coefficient k can be used to adjust the memory range.

In possession of an obliviator $h(s)$, we define the *h-norm*[13] of a function $f(s)$ in the space $\mathcal{E}$ by

$$\|f(s)\|_h = \sqrt{f^2(0) + \int_0^\infty f^2(s)h^2(s)\, ds}. \tag{3.84}$$

The *h-distance* between two functions in $\mathcal{E}$ is defined as the *h*-norm of their difference. We will assume that our space $\mathcal{E}$ of admissible functions with the norm induced by a given obliviator h is a Banach space. What this means is that, although we may have started from a space of admissible histories (continuous, say), the completion of this space with the *h*-norm brings into the picture other histories obtained as limits of Cauchy sequences (which may be, for example, discontinuous functions).

An even stronger structure can be conferred upon the function space $\mathcal{E}$. Indeed, we can define the *h-inner product* of two functions f and g in $\mathcal{E}$ as

$$\langle f, g \rangle_h = f(0)g(0) + \int_0^\infty f(s)g(s)h^2(s)\, ds. \tag{3.85}$$

By (3.84) we have the following relation between the inner product and the norm:[14]

$$\|f\|_h = \sqrt{\langle f, f \rangle_h}. \tag{3.86}$$

With all this metric structure in place, it is just a matter of using the traditional '$\delta - \varepsilon$' definition (or its more modern counterpart) to define the *continuity* of a functional. Intuitively, we want that a small change in the functions (as measured by our norm) be reflected in a small change in the functional (as measured by the usual absolute value of the difference of real numbers). The *(weak) principle of fading memory* states that there exists an obliviator $h(s)$ of order $r > 1/2$ such that the response functional $\mathcal{N}$ is continuous with respect to the corresponding *h*-norm. In simple terms, it establishes that if the distance between two histories is very small, so is the difference between the corresponding internal forces.[15]

3.8.3 Stress Relaxation

As an example of the application of the principle of fading memory, we will now show that a material satisfying this principle necessarily exhibits the phenomenon of *stress relaxation*, in a precise sense explained below. Intuitively speaking, stress relaxation is made evident by applying some protocol of deformation and then suddenly freezing it at a point of time t_0, leaving it thereafter at a fixed length (under isometric conditions, in the language of muscle mechanics). One then measures the internal force as time goes on

---

[13] Since the obliviator $h(s)$ is positive definite, some authors define the norm using h itself instead of h^2.

[14] Thus, an inner product induces a norm, but not vice versa. A Banach space whose norm is induced by an inner product is called a *Hilbert space*. We assume that our space $\mathcal{E}$ is a Hilbert space under the *h*-inner product.

[15] It is also possible to define the notion of differentiability of a functional. A stronger principle of fading memory can be established by requiring the response functional to be differentiable.

and observes that its magnitude decreases steadily. The strain history that corresponds to such a protocol is the following:

$$e(\tau) = \begin{cases} f(\tau) & \text{for } \tau < t_0 \\ f(t_0) & \text{for } \tau \geq t_0, \end{cases} \tag{3.87}$$

where $f(\tau)$ is a history defined up to at least t_0. Technically, the history $e(\tau)$ is called the *static continuation* of $f(\tau)$ after t_0. In terms of the notation introduced in equation (3.72), the static continuation reads:

$$e_t(s) = \begin{cases} f_t(t-t_0) & \text{for } s \leq t - t_0 \\ f_t(s) & \text{for } s > t - t_0. \end{cases} \tag{3.88}$$

Denoting by f_0 the constant history with value $f(t_0)$, we have, using (3.88),

$$\lim_{t \to \infty} \|f_0 - e_t\|_h = \lim_{t \to \infty} \sqrt{\int_{t-t_0}^{\infty} (f(t_0) - f_t(s))^2 \, h^2(s) ds}. \tag{3.89}$$

If we assume that the history e_t is bounded and we denote by M an upper bound of $|f_0 - e_t(s)|$, we obtain

$$\lim_{t \to \infty} \|f_0 - e_t\|_h = \lim_{t \to \infty} \|f_0 - e_t\|_h \leq M \lim_{t \to \infty} \sqrt{\int_{t-t_0}^{\infty} h^2(s) ds}. \tag{3.90}$$

Since we have assumed that the order of the obliviator is larger than $1/2$, the integral on the right-hand side tends to zero. Hence,

$$\lim_{t \to \infty} \|f_0 - e_t\|_h = 0. \tag{3.91}$$

By the principle of fading memory, the response functional $\mathcal{N}$ is continuous with respect to the h-norm. Therefore, having just shown that the distance between the static continuation and the state of constant strain is arbitrarily small (for sufficiently large times t), we conclude that

$$\lim_{t \to \infty} \mathcal{N}[e_t] = \mathcal{N}[f_0]. \tag{3.92}$$

This is precisely the statement of the stress relaxation phenomenon.

3.8.4 Finite Linear Viscoelasticity

A functional $\mathcal{L} : \mathcal{E} \to \mathbb{R}$ is *linear* if it satisfies the equation

$$\mathcal{L}[\alpha f + \beta g] = \alpha \mathcal{L}[f] + \beta \mathcal{L}[g], \quad \text{for all } \alpha, \beta \in \mathbb{R}, \, f, g \in \mathcal{E}. \tag{3.93}$$

In other words, a linear functional maps linear combinations into corresponding linear combinations or, in physical terms, a linear functional satisfies the *principle of superposition*.

In a space endowed with an inner product, such as the inner product (3.85) induced on $\mathcal{E}$ by $h(s)$, to every element $f \in \mathcal{E}$ one can assign a continuous linear functional $\mathcal{L}_f$ by the prescription

$$\mathcal{L}_f[g] = \langle f, g \rangle_h. \tag{3.94}$$

Under certain conditions, that we assume to be satisfied by $\mathcal{E}$, the converse is also true,[16] that is, every continuous linear functional can be represented as the inner product with a fixed element of $\mathcal{E}$.

In Section 3.8.1 we introduced the idea of resolving any given strain history $e_t(s)$ into the sum of the the the constant history $e = e_t(0)$ and the difference history $e_t^d(s) = e_t(s) - e$. The constitutive equation can be considered as a function of two functions,

$$N(t) = \mathcal{N}[e, e_t^d]. \tag{3.95}$$

If we assume that, within the recent past ($t - s$, with s not too large) the difference history is relatively small, it makes sense to consider that the excess internal force over and above the equilibrium value $N_\infty(e)$ depends linearly on the difference history e_t^d, while the dependence on the first argument (the present value of the strain) can be any non-linear function. In this way, we obtain the theory of finite linear viscoelasticity: finite, because there is in principle no limitation placed on the magnitude of the present strain; linear, because the dependence on the strain difference is linear. The general form of the constitutive functional that arises from these considerations is, therefore,

$$N(t) = N_\infty(e) + \mathcal{L}_e[e_t^d], \tag{3.96}$$

where the linear functional $\mathcal{L}_e$ is a different functional for each value of e. By the representation theorem of linear functionals as inner products, we conclude that, for each e there exists a function (with finite h-norm) $K_e(s) \in \mathcal{E}$ such that

$$\mathcal{L}_e[e_t^d] = \int_0^\infty K_e(s) e_t^d(s) h^2(s)\, ds. \tag{3.97}$$

What this means is that, in ultimate analysis, the constitutive equation of a finite linear viscoelastic material is completely specified by means of two ordinary functions: (i) the *equilibrium response function* $N_\infty(e)$; (ii) the product of the *viscoelastic kernel* $K_e(s)$, a function of two variables, e and s, times the obliviator $h(s)$. Thus,

$$N(t) = N_\infty(e) + \int_0^\infty K_e(s) e_t^d(s) h^2(s)\, ds. \tag{3.98}$$

It is customary to define, for each value of e, a primitive (integral) function $G_e(s)$ such that

$$\frac{dG_e(s)}{ds} = K_e(s) h^2(s). \tag{3.99}$$

Clearly, this function is determined only up to an arbitrary additive function of e. A convenient choice is obtained by setting $G_e(\infty) = 0$. Using the function $G_e(s)$ thus defined in the integrand of equation (3.98), we obtain

$$N(t) = N_\infty(e) + \int_0^\infty \frac{dG_e(s)}{ds} e_t^d(s)\, ds. \tag{3.100}$$

---

[16] The space must be a Hilbert space, namely, a complete inner-product space. The truth of this statement is known as the *Riesz–Fréchet representation theorem*.

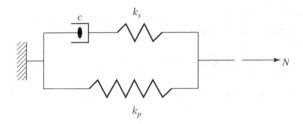

Figure 3.7 The Kelvin rheological model

This expression can be further transformed by reverting to the original time variable τ and by integration by parts, for example, as shown in Exercise 3.8.3.

The first formulation of fading memory in muscle along these lines is due to Bergel and Hunter (1979).[17] In this work, agreement with Hill's force–velocity relation is achieved by the ingenious device of equating the right-hand side of (3.98) not directly to the internal force but rather to a function of the internal force suggestive of Hill's formula.

Exercise 3.8.3 Consider histories $e(\tau)$ that vanish for all $\tau < 0$ (so that we may say that they start at $\tau = 0$). If $e(\tau)$ is differentiable for all $\tau > 0$, show that

$$N(t) = N_\infty(e) + G_e(t)e(0) + \int_0^t G_e(t - \tau)\frac{de(\tau)}{d\tau}\, d\tau. \qquad (3.101)$$

This form of the constitutive equation is most commonly found in textbooks.

Exercise 3.8.4 Rheological toys. A *Kelvin* rheological model consists of two linear springs and a dashpot arranged as shown in Figure 3.7. In each spring the force is proportional to the elongation (with proportionality constants k_s and k_p) and in the dashpot the force is proportional to the elongation rate (with proportionality constant c).

1. Show that the total force applied $N(t)$ and the total elongation $e(t)$ satisfy the following differential equation:

$$k_s N + c\frac{dN}{dt} = k_s k_p e + c(k_s + k_p)\frac{de}{dt}. \qquad (3.102)$$

2. Assume that the function $G_e(t)$ in equation (3.101) is of the exponential form:

$$G_e(t) = G \exp(-\lambda t), \qquad (3.103)$$

where G and λ are positive constants. Assume, moreover, that the equilibrium response function is linear, namely

$$N_\infty(e) = Ae, \qquad (3.104)$$

where A is a constant. Under these conditions, show that equation (3.101) provides the solution of equation (3.102) for any given history of the type considered in

[17] Hunter et al. (1998).

```
ks=6; kp=2; c=1; e[t_]:=If[t<1.6,-2.5*t,-4]:ep[t_]:=If[t<1.6,-2.5,0]
NDSolve[{c*n'[t]+ks*n[t]==c*(ks+kp)*ep[t]+ks*kp*e[t],n[0]==0},n[t],{t,0,4}]
Plot[Evaluate[n[t]/.%],{t,0,3}, AspectRatio®0.3,PlotRange®All]
```

Figure 3.8 Sample Mathematica® code for Exercise 3.8.5

Exercise 3.8.3. Identify the values of the constants A, λ, G in terms of the constants k_s, k_p, c. Check that the initial conditions are satisfactory.

Exercise 3.8.5 Force depression. In the Kelvin model, choose the following numerical values: $k_s = 6$, $k_p = 2$, $c = 1$. Referring to Figure 3.3, choose various deformation histories consisting of a constant velocity shortening at various speeds (from 2.5 to 30 mm s^{-1}) up to a shortening of 4 mm, with the length remaining constant thereafter. Solve the differential equation (3.102) and plot the results. A Mathematica® code of the type shown in Figure 3.8 can do this for you. Compare with the tail curves in Figure 3.3. Repeat the simulation with various values of k_p and compare the results in terms of the values of the force depression effect. Notice that, for fixed k_p the final value of the force depression is independent of the shortening speed. Experiments show that this is not the case. Investigate this aspect in the literature. Propose a way to modify the Kelvin model to make the value of the force depression dependent on speed.

References

Bergel, D.H. and Hunter, P.J. (1979) The mechanics of the heart. In H.H.C. Hwang, D.R. Gross and D.J. Patel (eds), *Quantitative Cardiovascular Studies, Clinical and Research Applications of Engineering Principles*, pp. 151–213. Baltimore, MD: University Park Press.

Dill, E.H. (1975) Simple materials with fading memory. In A.C. Eringen (ed.), *Continuum Physics*, Vol. II, pp. 284–403. New York: Academic Press.

Epstein M and Herzog W (1998) *Theoretical Models of Skeletal Muscle*, Chichester: John Wiley & Sons, Ltd.

Herzog, W. and Leonard, T.R. (1997) Depression of cat soleus forces following isokinetic shortening. *Journal of Biomechanics* **30**(9), 865–872.

Herzog, W., Joumaa, V. and Leonard, T.R. (2010) The force-length relationship of mechanically isolated sarcomeres. *Advances in Experimental Medicine and Biology (Muscle Biophysics)* **682**, 141–161.

Hill, A.V. (1970) *First and Last Experiments in Muscle Mechanics*. Cambridge: Cambridge University Press.

Hunter, P.J., McCulloch, A.D. and ter Keurs, H.E.D.J. (1998) Modelling the mechanical properties of cardiac muscle. *Progress in Biophysics and Molecular Biology* **69**, 289–331.

Lemos, R.R., Epstein, M., Herzog, W. and Wyvill, B. (2004) A framework for structured modeling of skeletal muscle. *Computer Methods in Biomechanics and Biomedical Engineering* **7**(6), 305–317.

Leonard, T.R., DuVall, M. and Herzog, W. (2010) Force enhancement following stretch in a single sarcomere, *American Journal of Physiology, Cell Physiology* **299**(6), C1398–C1401.

Truesdell, C. (1966) *Six Lectures on Modern Natural Philosophy*. Berlin: Springer-Verlag.

Truesdell, C. and Noll W. (1965) *The Non-linear Field Theories of Mechanics*. Volume III/3 of *Handbuch der Physik* (S. Flügge, ed.). Berlin: Springer-Verlag.

Woledge R.C. (1961) The thermoelastic effect of change of tension in active muscle. *Journal of Physiology* **155**, 187–208.

4

Mixture Theory

4.1 Introduction

Everything is mixture. But the need for a continuum *theory of mixtures*[1] arises whenever: (i) there is a relative motion between the components of the mixture, and/or (ii) the components exchange mass. The first situation is a case of *diffusion*, while the second is encompassed under the notion of *chemically reacting mixtures*. Living organisms tend to exhibit both phenomena, hence the enormous potential of mixture theory for biomechanics. On the other hand, the use of mixture theory in biomechanics is not as widespread as one would expect, a fact that may be due both to the intrinsic difficulties of the subject and to the lack or, in some cases, ironically, the excess of specific qualitative and quantitative knowledge of the underlying biological phenomena. The example of force production in muscle serves as an obvious illustration. It is never easy to transform discrete information into a viable continuum counterpart. If, for instance, one ATP molecule is required per stroke cycle according to some versions of the sliding filament model of force production in striated muscle, what is the continuum counterpart of this fact? The challenges and opportunities for the use of mixture models are wide open, from the development of specific theories to the implementation of computational tools.

4.2 The Basic Tenets of Mixture Theory

A *mixture* is made of a finite number[2] M of identifiable *components* (also called *species* or *constituents*). The fundamental tenets of mixture theory are as follows:

1. At each instant of time, every point of the spatial domain of the mixture is simultaneously occupied by all the components of the mixture.
2. With suitable choices of variables, the mixture as a whole abides by the standard balance equations of classical continuum mechanics.

---

[1] See Truesdell and Toupin (1960), Bowen (1976) and Rajagopal and Tao (1995).
[2] In principle, the number of components may be infinite, the presence of each component being measured by a probability-like distribution defined over a continuous parameter space, such as the real interval $[0, 1]$.

Both of these tenets call for some reflection. The first one can be regarded as a self-imposed limitation on the scope of the theory. Indeed, if, for example, a high temperature triggered a new product of a chemical reaction that would otherwise not be present, the first tenet would be contradicted. A similar observation can be made with regard to a possible change of phase. The second principle is a formal requirement which may be challenged on physical grounds. It is an implicit recognition that whatever happens inside a material, even one as complex as a mixture, is an internal matter only. Another way to express this assumption is by affirming the existence of *mixture particles*, with identifiable kinematic properties.

Each of the M components of a mixture follows its own individual motion, namely

$$x = x_\alpha(X_\alpha, t), \quad \alpha = 1, \ldots, M. \tag{4.1}$$

We use Greek indices to denote properties pertaining to one of the components of the mixture. These properties may be referred to as *individual*, *peculiar*, or *specific*. The individual velocity of the αth species is given by

$$v_\alpha(X_\alpha, t) = \frac{\partial x_\alpha(X_\alpha, t)}{\partial t}, \tag{4.2}$$

where the notation leaves no room for confusion as to which is the variable that has been kept fixed (i.e., X_α). The individual acceleration is

$$a(X_\alpha, t) = \frac{\partial v_\alpha(X_\alpha, t)}{\partial t}. \tag{4.3}$$

Clearly, in our one-dimensional context, the velocity and the acceleration can be regarded as scalar fields.

If the velocity is expressed (via the inverse motion) as a function of x and t, the acceleration can be written as

$$a(x, t) = \frac{D_\alpha v(x, t)}{Dt}, \tag{4.4}$$

where the specific material derivative of a field $\psi(x, t)$ *along the motion of the component number* α is defined as

$$\frac{D_\alpha \psi(x, t)}{Dt} = \frac{\partial \psi(x, t)}{\partial t} + \frac{\partial \psi(x, t)}{\partial x} v_\alpha(x, t). \tag{4.5}$$

We have, so far, described the motion of each component of the mixture. It is time now to introduce the idea of the motion of the mixture as a whole. Let $\rho_\alpha = \rho_\alpha(x, t)$, $\alpha = 1, \ldots, M$, denote the spatial mass densities of the individual components. By the first tenet of the theory, these are well-defined quantities at each spatial place occupied by the mixture. The density of the mixture at point x and at time t is, therefore, given by

$$\rho(x, t) = \sum_{\alpha=1}^{M} \rho_\alpha(x, t). \tag{4.6}$$

The *mass concentration* (or, simply, the *concentration*) of the component α is defined as

$$c_\alpha = \frac{\rho_\alpha}{\rho}. \tag{4.7}$$

It follows from this definition and from equation (4.6) that

$$\sum_{\alpha=1}^{M} c_\alpha = 1. \tag{4.8}$$

With these definitions in place, we can define the *velocity of the mixture* as

$$v(x,t) = \sum_{\alpha=1}^{M} c_\alpha v_\alpha. \tag{4.9}$$

By virtue of equation (4.7), the velocity of the mixture can also be expressed as

$$v(x,t) = \frac{1}{\rho} \sum_{\alpha=1}^{M} \rho_\alpha v_\alpha. \tag{4.10}$$

This expression suggests that a more appropriate designation of the velocity of the mixture should be *barycentric velocity*.[3]

The *diffusion velocity* of the constituent α is defined as

$$u_\alpha(x,t) = v_\alpha(x,t) - v(x,t). \tag{4.11}$$

The name appropriately describes the deviation of the velocity of each component from the mean velocity of the mixture. Diffusion velocities, being relative velocities, are objective, that is, observer independent, quantities. The diffusion velocities satisfy the identity

$$\sum_{\alpha=1}^{M} \rho_\alpha u_\alpha = 0, \tag{4.12}$$

or equivalently,

$$\sum_{\alpha=1}^{M} c_\alpha u_\alpha = 0. \tag{4.13}$$

These identities, which follow directly from the definitions, express the fact that the total mass flux associated with the diffusive motions vanishes.

We have introduced in equation (4.5) the material derivative associated with the motion of each constituent. The material derivative following the motion of the mixture as a whole is given by

$$\frac{D\psi(x,t)}{Dt} = \frac{\partial \psi(x,t)}{\partial t} + \frac{\partial \psi(x,t)}{\partial x} v(x,t). \tag{4.14}$$

The relationship between these material derivatives is given by

$$\frac{D_\alpha \psi(x,t)}{Dt} = \frac{D\psi(x,t)}{Dt} + \frac{\partial \psi(x,t)}{\partial x} u_\alpha(x,t). \tag{4.15}$$

---

[3] There are other physically meaningful ways of defining a mean velocity of the mixture (by using weights other than the mass concentrations).

4.3   Mass Balance

Because of the possible presence of chemical reactions, the mass of a constituent of the mixture may increase at the expense of the masses of the other constituents. If we denote by π_α the *mass supply* of the constituent α per unit time and per unit spatial volume of the mixture, the integral form of the equation of mass balance for this constituent is

$$\frac{d}{dt} \int_{x(X_0,t)}^{x(X_1,t)} \rho_\alpha(x,t)\, dx = \int_{x(X_0,t)}^{x(X_1,t)} \pi_\alpha(x,t)\, dx, \tag{4.16}$$

where we have followed the notation of Section 2.3. The local form of this equation is

$$\frac{\partial \rho_\alpha}{\partial t} + \frac{\partial (\rho_\alpha v_\alpha)}{\partial x} = \pi_\alpha. \tag{4.17}$$

According to the second tenet of mixture theory, as discussed in Section 4.2, we should be able to recover the usual form of the equation of mass conservation (equation (2.49)) for the mixture as a whole. To this effect, we add the mass balances of all the constituents and, invoking equations (4.6) and (4.10), we obtain

$$\frac{\partial \rho}{\partial t} + \frac{\partial (\rho v)}{\partial x} = \sum_{\alpha=1}^{M} \pi_\alpha. \tag{4.18}$$

Comparing this result with equation (2.49), we conclude that the mixture abides by the usual conservation of mass if, and only if, the sum of the mass supplies of all the constituents vanishes:

$$\sum_{\alpha=1}^{M} \pi_\alpha = 0. \tag{4.19}$$

In other words, we are explicitly excluding processes of mass growth at this stage. With this condition, the balance of mass of the mixture attains the standard form,

$$\frac{\partial \rho}{\partial t} + \frac{\partial (\rho v)}{\partial x} = 0. \tag{4.20}$$

Exercise 4.3.1   Constant concentrations. Show that

$$\rho \frac{Dc_\alpha}{Dt} = \pi_\alpha - \frac{\partial (\rho_\alpha u_\alpha)}{\partial x}. \tag{4.21}$$

Conclude that if there is no diffusion and no chemical reactions (i.e., if for all α, $u_\alpha = \pi_\alpha = 0$), the concentrations c_α are constant (following the motion of the mixture).

Remark 4.3.2   Chemical reactions. When definite chemical reactions are at play, further limitations are imposed on the mass exchanges between the various species. We will discuss these conditions below in Section 4.8.

4.4   Balance of Linear Momentum

4.4.1   Constituent Balances

Following the constituent α in its motion, the integral form of the balance of momentum is

$$
\frac{d}{dt} \int_{x(X_0,t)}^{x(X_1,t)} \rho_\alpha v_\alpha \, dx = \int_{x(X_0,t)}^{x(X_1,t)} b_\alpha \, dx + \int_{x(X_0,t)}^{x(X_1,t)} \left(\pi_\alpha v_\alpha + p_\alpha \right) dx + n_\alpha(x(X_1,t)) - n_\alpha(x(X_0,t)).
$$

$$(4.22)$$

In this equation, b_α is the body force associated with the constituent α and n_α is the *partial internal force* of this constituent. The volume term $\pi_\alpha v_\alpha$ is the momentum contributed by the entrant mass, provided this mass (π_α) is entering at the same velocity as the substratum (i.e., the velocity v_α). Any other contribution to the momentum is included in the additional term p_α. Specifically, this term (called the *momentum supply*) can be associated, for instance, with the fact that the entering mass (the product, perhaps, of a chemical reaction) may enter at a different velocity v'_α. In this picture, the α-momentum supply would read

$$
p_\alpha = \pi_\alpha(v'_\alpha - v_\alpha).
$$

$$(4.23)$$

There are, of course, many other possible sources of momentum supply, including a *viscous drag* proportional to the relative velocity between the components and a so-called *virtual mass effect* arising from their relative acceleration. The local form of the balance of momentum for the species α is obtained following the standard procedure of Section 2.3 and exploiting the balance of mass, equation (4.17). The result is

$$
\rho_\alpha \frac{D_\alpha v_\alpha}{Dt} = b_\alpha + \frac{\partial n_\alpha}{\partial x} + p_\alpha.
$$

$$(4.24)$$

Intuitively speaking, the molecules entering at the pre-existent velocity do not affect the ensuing motion. For this reason, it is not surprising that only the extra momentum supply p_α intervenes in the equation of motion.

4.4.2   Mixture Balance

We now proceed to show, as demanded by the second tenet of mixture theory, that the mixture as a whole abides by the usual form of the equation of balance of momentum for a single substance. This result should emerge essentially by adding the balance equations of the constituents and interpreting accordingly the sums of the individual contributions (such as the internal forces). The balance of momentum for each constituent has been formulated by following the motion of that individual constituent. Because of the presence of diffusion, however, each constituent travels at a different instantaneous velocity. Before proceeding to add the individual contributions, therefore, it proves useful to change the perspective somewhat and to ask: what is the rate of change of the momentum associated with a given species *within a fixed region of space*? Beyond the forces acting on this region and the momenta contributed by the mass source π_α and the momentum supply p_α, we must take into consideration the fact that for a fixed region of space (even without the contributions just mentioned) there are particles flowing in and out and carrying momentum. We could, therefore, adopt this position and carefully generate again the equation of balance of momentum. Instead, we will transform the more reliable version (4.24), based on the concept of material derivative, and interpret the result accordingly. The local rate of change of momentum of the constituent α in a fixed volume element in space is $\partial(\rho_\alpha v_\alpha)/\partial t$. We calculate

$$\frac{\partial(\rho_\alpha v_\alpha)}{\partial t} = \frac{D_\alpha(\rho_\alpha v_\alpha)}{Dt} - \frac{\partial(\rho_\alpha v_\alpha)}{\partial x}v_\alpha = \rho_\alpha\frac{D_\alpha v_\alpha}{Dt} + \frac{D_\alpha\rho_\alpha}{Dt}v_\alpha - \frac{\partial(\rho_\alpha v_\alpha)}{\partial x}v_\alpha, \qquad (4.25)$$

where we have just used the definition of material derivative, equation (4.5). Invoking the balance of mass (equation (4.17)) to transform the penultimate term, we obtain

$$\frac{\partial(\rho_\alpha v_\alpha)}{\partial t} = \rho_\alpha\frac{D_\alpha v_\alpha}{Dt} + \left(\pi_\alpha - \frac{\partial v_\alpha}{\partial x}\rho_\alpha\right)v_\alpha - \frac{\partial(\rho_\alpha v_\alpha)}{\partial x}v_\alpha \qquad (4.26)$$

or, finally,

$$\frac{\partial(\rho_\alpha v_\alpha)}{\partial t} = \rho_\alpha\frac{D_\alpha v_\alpha}{Dt} + \pi_\alpha v_\alpha - \frac{(\partial\rho_\alpha v_\alpha^2)}{\partial x}. \qquad (4.27)$$

Notice that a similar reasoning can be used for the mixture as a whole by invoking the balance of mass (4.20). The result is

$$\frac{\partial(\rho v)}{\partial t} = \rho\frac{Dv}{Dt} - \frac{(\partial\rho v^2)}{\partial x}. \qquad (4.28)$$

Returning to equation (4.27), we can now use the balance of momentum in the form of equation (4.24) to write

$$\frac{\partial(\rho_\alpha v_\alpha)}{\partial t} = b_\alpha + \frac{\partial(n_\alpha - \rho_\alpha v_\alpha^2)}{\partial x} + p_\alpha + \pi_\alpha v_\alpha. \qquad (4.29)$$

We may say that the influence of the flow of mass into the fixed spatial element is completely encompassed within an additive modification of the partial internal force n_α in the amount $-\rho_\alpha v_\alpha^2$. This term is not difficult to explain on intuitive grounds, since the amount of mass (per unit length and per unit time) entering the element through the left end is given by $\rho_\alpha v_\alpha$ and the corresponding momentum is obtained by multiplying this amount by v_α.

To obtain the equation of balance of momentum for the mixture, we start by adding up all the contributions (4.24) of the constituents, namely,

$$\sum_{\alpha=1}^{M}\frac{\partial(\rho_\alpha v_\alpha)}{\partial t} = \sum_{\alpha=1}^{M}b_\alpha + \sum_{\alpha=1}^{M}\frac{\partial(n_\alpha - \rho_\alpha v_\alpha^2)}{\partial x} + \sum_{\alpha=1}^{M}\left(p_\alpha + \pi_\alpha v_\alpha\right). \qquad (4.30)$$

We now define in a natural way the *total volume force* as the sum

$$b = \sum_{\alpha=1}^{M}b_\alpha. \qquad (4.31)$$

Recalling the definition of the mean velocity of the mixture (equation (4.10)), we can write equation (4.30) as

$$\frac{\partial(\rho v)}{\partial t} = b + \sum_{\alpha=1}^{M} \frac{\partial(n_\alpha - \rho_\alpha v_\alpha^2)}{\partial x} + \sum_{\alpha=1}^{M} \left(p_\alpha + \pi_\alpha v_\alpha\right). \tag{4.32}$$

By equation (4.28), we can write

$$\rho\frac{Dv}{Dt} - \frac{(\partial \rho v^2)}{\partial x} = b + \sum_{\alpha=1}^{M} \frac{\partial(n_\alpha - \rho_\alpha v_\alpha^2)}{\partial x} + \sum_{\alpha=1}^{M} \left(p_\alpha + \pi_\alpha v_\alpha\right). \tag{4.33}$$

Introducing the diffusion velocities, given by equation (4.11), we observe that

$$\sum_{\alpha=1}^{M} \rho_\alpha v_\alpha^2 = \sum_{\alpha=1}^{M} \rho_\alpha (v + u_\alpha)^2 = \rho v^2 + \sum_{\alpha=1}^{M} \rho_\alpha u_\alpha^2, \tag{4.34}$$

where equation (4.12) has been used to cancel a term. Substituting this expression into equation (4.33) yields

$$\rho\frac{Dv}{Dt} = b + \sum_{\alpha=1}^{M} \frac{\partial(n_\alpha - \rho_\alpha u_\alpha^2)}{\partial x} + \sum_{\alpha=1}^{M} \left(p_\alpha + \pi_\alpha v_\alpha\right). \tag{4.35}$$

Defining the *total internal force* of the mixture by

$$n = \sum_{\alpha=1}^{M} (n_\alpha - \rho_\alpha u_\alpha^2), \tag{4.36}$$

we can rewrite equation (4.35) as

$$\rho\frac{Dv}{Dt} = b + \frac{\partial n}{\partial x} + \sum_{\alpha=1}^{M} \left(p_\alpha + \pi_\alpha v_\alpha\right). \tag{4.37}$$

Comparing this result with the balance of momentum for an ordinary medium, equation (2.52), namely

$$\rho\frac{Dv}{Dt} = b + \frac{\partial n}{\partial x}, \tag{4.38}$$

we conclude that the mixture abides by the usual balance of momentum equation if, and only if, the following constraint is satisfied identically:

$$\sum_{\alpha=1}^{M} \left(p_\alpha + \pi_\alpha v_\alpha\right) = 0. \tag{4.39}$$

By virtue of equation (4.19), this condition can also be written in terms of the diffusion velocities as

$$\sum_{\alpha=1}^{M} \left(p_\alpha + \pi_\alpha u_\alpha\right) = 0. \tag{4.40}$$

Physically, this constraint expresses the fact that the internal affairs between the components of a mixture should not be discernible upon observing the motion of the mixture as a whole.

Remark 4.4.1   Diffusion and stress. The total stress of the mixture (or the total internal force n) is equal to the sum of the partial stresses (or the partial forces n_α) plus a correction term arising from the fact that the velocities of the components differ from the mean velocity and, therefore, contribute to the rate of momentum of the mixture. In the case where no diffusion is present, the correction term vanishes. Only in this case is the total stress equal to the sum of the partial stresses. Contrast this with the common notion of partial pressures and Dalton's law for a mixture of ideal gases.

4.5   Case Study: Confined Compression of Articular Cartilage

4.5.1   Introduction

Many solid materials of biological interest contain *voids* or *pores*, that is, gaps that complicate the connectivity without destroying the continuity of the solid. Roughly speaking, a pore is an entity (of the same dimensionality as the body) that may contain (and usually does contain) fluid as separate from the solid *matrix*. The definition of 'pore' is partly a matter of convention. It is, therefore, convenient to imagine that there is an underlying original, connected, continuous matrix $\mathcal{M}$ (that we will assume, by definition, without pores) and that pores have been later carved into it. A void is said to be *isolated* if it does not intersect the boundary $\partial\mathcal{M}$ of $\mathcal{M}$. Isolated pores are, therefore, ineffective for the passage of fluid through $\mathcal{M}$. All other pores are said to be *accessible*. Accessible pores that intersect the boundary $\partial\mathcal{M}$ only at a single (non-empty) connected set are also not effective in passing fluid. They are called *non-passing pores*. We may think of them as the finger holes in a bowling ball. All remaining accessible pores are said to be *passing*. A good everyday example of a *porous material* is a sponge.

Remark 4.5.1   Pores and dimension. Strictly speaking, the above definitions apply only in the three-dimensional context. In two dimensions, the notions of isolated and non-passing pores still make sense, but a passing pore would split the matrix into at least two separate pieces. In one dimension, not even an isolated pore makes physical sense. Nevertheless, when we speak of a two-dimensional model what we have in mind is a three-dimensional flat body whose thickness is relatively small when compared with its in-plane dimensions. In the case of a porous medium, we can think of it as a surface with thickness, or a (horizontal) sandwich, whose upper and lower surfaces are impermeable and which contains a network of interconnected horizontal pipes reaching the boundary curve of the surface. Similarly, a one-dimensional model corresponds to a cylinder with an impermeable lateral surface and containing pipes reaching from one end to the other of the cylinder.

To quantify the presence of pores, the notion of *mean porosity* is introduced as the ratio between the volume ΔV_p of all the pore space within a representative element of the porous medium and the volume ΔV of this element. Within the usual provisos regarding the meaning of the limiting process, the *porosity* at a point $X \in \mathcal{M}$ is given by

$$\phi = \lim_{\Delta V \to 0} \frac{\Delta V_p}{\Delta V}, \tag{4.41}$$

where X remains in the shrinking representative element throughout the limit process. From the definition it follows that $0 \leq \phi \leq 1$. It is wise to distinguish between the porosity and the *effective porosity*, which counts only the volume of the passing pores. In many applications, however, the amount of isolated and non-passing pores is very small and can be neglected, as we shall do from now on.

The treatment of porous media by means of mixture theory recognizes the presence of pores only through the porosity field, without paying attention to the detailed geometry of the pores themselves. Even when more geometric information is required (such as the degree of *tortuosity* of the conduits), this information is translated into *fields* defined at every point of the continuum $\mathcal{M}$. If the pores are filled with a fluid, therefore, the smearing process is consistent with the first basic tenet of mixture theory, according to which both the solid matrix and the fluid are present (in proportions determined by the porosity field) at every point of the body $\mathcal{M}$. As already remarked above, this picture makes sense even in lower-dimensional models (two- or one-dimensional idealizations of the kind we have been discussing in this book).

4.5.2 Empirical Facts

Skeletal joints are biological structures that provide a connection between adjacent bones in the skeleton. *Synovial joints* are skeletal joints that allow for a maximum degree of relative movement, particularly rotation, between the bones involved (Figure 4.1). To achieve this important functional objective, the bones are not directly connected to each other but rather converge into a common *synovial cavity* filled with *synovial fluid* and enveloped within a *synovial membrane*. A system of *ligaments*, running outside the synovial cavity, provides a loose attachment between the converging bones. At the contact area between each bone surface and the synovial membrane, the bone is equipped with a

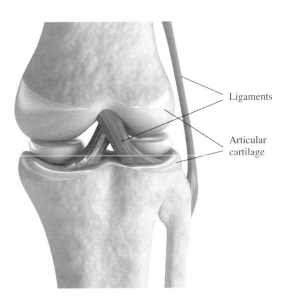

Ligaments

Articular
cartilage

Figure 4.1 Synovial joint

thin layer of *articular cartilage*, whose function is to distribute the load, to absorb shocks and to achieve a very low coefficient of friction.

The cellular components of articular cartilage are called *chondrocytes*. They lie sparsely distributed, like islands, within the *extracellular matrix*, which is itself produced through the metabolic activity of the chondrocytes. The extracellular matrix is a highly porous material comprising up to 95% of the total volume of the cartilage. The porosity of the matrix is of the order of 75% . The pores are filled with fluid that circulates freely within the cartilage and may also be exuded. The permeability of cartilage, however, is very low due to the relatively high drag forces between the moving fluid and the pore walls. An empirical equation, that can also be justified on theoretical grounds, establishes that the drag forces per unit volume are proportional to the diffusive velocity and to the square of the porosity. The coefficient of proportionality is the reciprocal of the so-called *permeability coefficient k*.

An important, and not surprising, observation of the mechanical response of articular cartilage is that, although the extracellular material is nearly elastic, the presence of fluid within its pores renders the behaviour of the mixture viscoelastic. For this reason, this type of response is known as *flow-induced viscoelasticty*, and should be distinguished from any possible intrinsic viscoelasticity of the matrix material. Any model of cartilage as a combination of a solid and a liquid should, therefore, account both qualitatively and quantitatively for the observed viscoelastic response. Historically, the first models of this type emerged from the field of soil mechanics, where the pioneering works of Terzaghi and Biot gave rise to a subdiscipline of solid mechanics that came to be known by the name of *poroelasticity*. In the case of articular cartilage, the use of the conceptual framework of mixture theory placed the model within the confines of modern continuum mechanics and gave rise to the *biphasic theory* of cartilage pioneered by Mow and his coworkers.

The main experimental facts concerning the viscoelastic response of cartilage can already be clearly discerned within the largely one-dimensional context of the *confined compression test*. A cylindrical plug of cartilage is placed snugly at the bottom of a rigid and impermeable vertical cylindrical chamber equipped with a rigid and completely permeable piston fitted above the cartilage sample, as shown schematically in Figure 4.2. In the *creep* experiment, a dead weight is applied to the piston and left in place for a long time period, while the displacement of the piston and the volume of liquid exuded

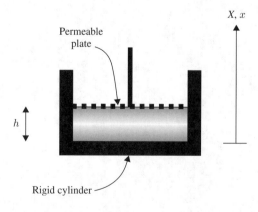

Figure 4.2   Confined compression test

are recorded as functions of time. Eventually, both the displacement and the exudation stop as an equilibrium state is reached. In the *stress relaxation* experiment, a downward displacement is applied rapidly and then stopped at a fixed value for a long time, while the force and the exudation are measured. The force is observed to increase rapidly and then relax slowly until an equilibrium state is attained. Finally, a *hysteresis experiment* consists of applying one cycle of harmonic motion, first downward and then upward, and recording the force. During the upward motion some of the fluid re-enters the cartilage. The loading and unloading force–displacement paths are different, the area contained between them representing the energy loss.

4.5.3   Field Equations

We start by rewriting the equations of a binary mixture explicitly, denoting the phases by the subscripts s and f, for the solid matrix and the fluid, respectively. The result is

$$\frac{\partial \rho_s}{\partial t} + \frac{\partial (\rho_s v_s)}{\partial x} = \pi_s, \tag{4.42}$$

$$\frac{\partial \rho_f}{\partial t} + \frac{\partial (\rho_f v_f)}{\partial x} = \pi_f, \tag{4.43}$$

$$\rho_s \frac{D_s v_s}{Dt} = b_s + \frac{\partial n_s}{\partial x} + p_s, \tag{4.44}$$

$$\rho_f \frac{D_f v_f}{Dt} = b_f + \frac{\partial n_f}{\partial x} + p_f. \tag{4.45}$$

We now make a number of simplifying assumptions leading to the *biphasic theory of articular cartilage*.[4] These assumptions are as follows:

1. The mixture is *chemically inert*, so that the mass supplies π_s and π_f vanish identically. The moment supplies p_s and p_f, on the other hand, are essential in capturing the viscous drag between the matrix and the interstitial fluid. In the simplest possible model, they are assumed to be of the form

$$p_s = K(v_f - v_s), \tag{4.46}$$

where K is a material coefficient that depends on the porosity ϕ, and

$$p_f = -p_s. \tag{4.47}$$

Note that, in the absence of mass supplies, this last equation is necessitated by equation (4.39).

2. The matrix is made of an *incompressible solid* material. In other words, the intrinsic density $\bar{\rho}_s$ of the matrix is constant in time. For simplicity, we also assume that the matrix is uniform (so that its intrinsic density is also constant throughout the body). The intrinsic incompressibility of the matrix does not imply the conservation of volume of the mixture, since the pores may change in volume. The relation between the intrinsic density and the solid density ρ_s is, accordingly, given by

$$\rho_s = (1 - \phi)\bar{\rho}_s. \tag{4.48}$$

---

[4] Mow et al. (1980).

3. The fluid filling the pores is an *incompressible liquid*. In the process of deformation, as the porosity changes, this liquid can be transferred from one part of the body to another. More importantly, liquid may ooze out of the matrix at the boundary $\partial \mathcal{M}(t)$, so that not even the total volume of the mixture is preserved. Denoting by $\bar{\rho}_f$ the intrinsic density of the liquid, we have

$$\rho_f = \phi \bar{\rho}_f. \tag{4.49}$$

4. Eliminating the porosity between equations (4.48) and (4.49), we obtain the *volume additivity constraint*,

$$\frac{\rho_s}{\bar{\rho}_s} + \frac{\rho_f}{\bar{\rho}_f} = 1. \tag{4.50}$$

This is a geometrical constraint. Just as the incompressibility of an ordinary body implies that the stress is determined by the deformation only up to an arbitrary additive pressure, the volume additivity constraint implies an indeterminacy of this kind. We will assume that the total stress of the mixture is determined up to an arbitrary pressure p and that this pressure[5] is apportioned to the solid and the fluid constituents according to $(1 - \phi)p$ and ϕp, respectively.

5. The liquid is *inviscid*, the only manifestation of a viscous effect being incorporated in the drag force p_s. As a consequence of this assumption, the constitutive law for the liquid is

$$n_f = -\phi p, \tag{4.51}$$

where the minus sign has been placed for consistency with the assumption that the pressure p is a positive number.

6. The solid is *elastic*. Its constitutive law is, therefore, given by

$$n_s = -(1 - \phi)p + f(e), \tag{4.52}$$

where $f(\cdot)$ is a constitutive function[6] and e is the engineering strain defined in equation (1.18). Notice in equation (4.52) how the deformation does not determine the stress in the solid completely, since the pressure p is in principle arbitrary (it can be determined only after solving each particular problem).

7. Although not strictly part of the theory, we will further assume that the body forces and, more drastically, the inertia forces are negligible.

On the basis of assumptions 1–7, the mass and momentum balance equations become

$$-\frac{\partial \phi}{\partial t} + \frac{\partial((1 - \phi)v_s)}{\partial x} = 0, \tag{4.53}$$

$$\frac{\partial \phi}{\partial t} + \frac{\partial(\phi v_f)}{\partial x} = 0, \tag{4.54}$$

$$-\frac{\partial((1 - \phi)p)}{\partial x} + \frac{df}{de}\frac{\partial e}{\partial x} + K(v_f - v_s) = 0, \tag{4.55}$$

$$-\frac{\partial(\phi p)}{\partial x} - K(v_f - v_s) = 0. \tag{4.56}$$

---

[5] The term 'pressure' should be understood in the one-dimensional context as a force, unless the cross-sectional area is brought into play.

[6] Recall that n_s is the partial internal force due to the solid phase within the mixture and not the intrinsic response of the matrix material. Thus, the constitutive function $f(e)$ depends in general on the initial porosity.

The four unknown fields are: the pressure p, the porosity ϕ, the solid displacement $\hat{u}_s$ and the fluid velocity v_f.[7] The solid velocity and the strain can be obtained as derivatives of the solid displacement field. In the Lagrangian formulation they are given by

$$v_s = \frac{\partial \hat{u}_s(X,t)}{\partial t} \qquad (4.57)$$

and

$$e = \frac{\partial \hat{u}_s(X,t)}{\partial X}. \qquad (4.58)$$

We proceed now to integrate exactly, as far as possible, the system of equations (4.53)–(4.56). It is remarkable that equation (4.53) can actually be integrated exactly to express the porosity as a function of the displacement $\hat{u}_s$. Consider the following physical reasoning. Since the matrix material is inherently incompressible, the change in volume $dv - dV$ from the reference to the current configuration must be entirely reflected in the change in porosity. If ϕ_0 is the porosity in the reference configuration, the current porosity is, therefore, given by the quotient

$$\phi = \frac{\phi_0 dV + (dv - dV)}{dv}. \qquad (4.59)$$

In the case of confined compression, the cross-sectional area remains constant, so that, according to equation (1.19),

$$\frac{dv}{dV} = \frac{dx}{dX} = 1 + e. \qquad (4.60)$$

Introducing this result in equation (4.57) we obtain

$$\phi = 1 - \frac{1 - \phi_0}{1 + e}. \qquad (4.61)$$

Exercise 4.5.2 Explicit expression for the porosity. Verify that this expression indeed satisfies the PDE (4.53). In doing so, extreme care must be exercised in switching between the Lagrangian and the Eulerian formulations.

Adding together the balance of mass equations (4.53) and (4.54) yields

$$\frac{\partial((1 - \phi)v_s + \phi v_f)}{\partial x} = 0, \qquad (4.62)$$

or, integrating with respect to x,

$$(1 - \phi)v_s + \phi v_f = g(t), \qquad (4.63)$$

where $g(t)$ is a function of integration to be determined from the boundary conditions. In the case of confined compression (Figure 4.2), since at all times the solid and fluid

---

[7] The symbol $\hat{u}_s$ for the solid displacement is introduced to avoid notational confusion with the diffusive velocity and with the internal energy density.

velocities at the bottom of the cylinder vanish, we conclude that $g(t) = 0$ identically. It follows that at each point we must have

$$v_f = -\frac{1 - \phi}{\phi} v_s. \tag{4.64}$$

In particular, if the permeable lid of the cylinder is moving downward with a prescribed velocity, liquid must be oozing upward at the lid in accordance with equation (4.64).

So far, we have obtained explicit expressions, equations (4.61) and (4.64), for two of the unknown fields, namely the porosity ϕ and the fluid velocity v_f, in terms of the solid displacement $\hat{u}_s$. Finally, adding equations (4.55) and (4.56), we obtain the balance of momentum for the mixture as

$$-\frac{\partial p}{\partial x} + \frac{df}{de}\frac{\partial e}{\partial x} = 0. \tag{4.65}$$

Integrating with respect to x, we obtain

$$-p(x, t) + f(e(x, t)) = g_1(t), \tag{4.66}$$

where $g_1(t)$ is a function to be determined. Equation (4.66) is a statement of the physical fact that the total internal force of the mixture is independent of x, as can be clearly understood from the equilibrium of a simple free-body diagram.

Exercise 4.5.3   A free-body diagram. Derive equation (4.66) directly from a free-body diagram obtained by cutting the cylinder with a horizontal plane.

Introducing equations (4.64) and (4.66) into (4.56), we write

$$-\frac{\partial f(e)}{\partial x}\phi - (f(e) - g_1(t))\frac{\partial \phi}{\partial x} + \frac{Kv_s}{\phi} = 0. \tag{4.67}$$

Since ϕ is given explicitly by (4.61), equation (4.67) can be regarded as a single second-order non-linear PDE for the displacement field $\hat{u}_s$. More explicitly, using the Lagrangian representation $\hat{u}_s = \hat{u}_s(X, t)$, this equation reads:

$$-\frac{df(e)}{de}\frac{\partial^2 \hat{u}_s}{\partial X^2}\frac{1}{1+e}\phi - (f(e) - g_1(t))\frac{\partial \phi}{\partial X}\frac{1}{1+e} + \frac{K}{\phi}\frac{\partial \hat{u}_s}{\partial t} = 0, \tag{4.68}$$

where e and ϕ are given, respectively, by (4.58) and (4.61). Assuming that the porosity ϕ_0 in the reference configuration is constant (i.e., independent of X), equation (4.68) can be written as

$$\left(\frac{df(e)}{de}\frac{\phi}{1+e} + \frac{f(e) - g_1(t)}{1+e}\frac{d\phi}{de}\right)\frac{\partial^2 \hat{u}_s}{\partial x^2} = \frac{K}{\phi}\frac{\partial \hat{u}_s}{\partial t}. \tag{4.69}$$

Exercise 4.5.4   The cartilage equation. Obtain equations (4.68) and (4.69) from equation (4.66).

4.5.4   Non-linear Creep

Equation (4.69) can be regarded as a *non-linear diffusion equation*, which we proceed to solve numerically to illustrate the creep experiment in a confined compression test, as described briefly in Section 4.5.2. In the creep experiment, a downward force $W(t)$ is applied starting at a zero value and increasing rapidly to a final value $W_{\max}$, at which it stays for an extended period of time. Clearly, by construction, in this case the applied force can be identified with the function $g_1(t)$:

$$g_1(t) = W(t). \tag{4.70}$$

With this condition in place, the field equation (4.69) becomes

$$\left(\frac{df(e)}{de} \frac{\phi}{1+e} + \frac{f(e) - W(t)}{1+e} \frac{d\phi}{de} \right) \frac{\partial^2 \hat{u}_s}{\partial x^2} = \frac{K}{\phi} \frac{\partial \hat{u}_s}{\partial t}. \tag{4.71}$$

The displacement $\hat{u}_s$ is measured from a reference configuration consisting of an unloaded matrix filled with fluid at atmospheric pressure. The variable p represents the additional pressure. The boundary condition for $\hat{u}_s$ at the cylinder base is

$$\hat{u}_s(0, t) = 0. \tag{4.72}$$

At the piston, since it has been assumed perfectly permeable, we must have $p = 0$ and, therefore, the entire load applied thereat must be carried by the matrix. It follows, therefore, from equation (4.66) that

$$f(e(h, t)) = g_1(t), \tag{4.73}$$

where we are using the Lagrangian coordinates X, t. Taking account of equation (4.58), equation (4.73) can be regarded as the (non-linear) boundary condition[8]

$$f\left(\left. \frac{\partial \hat{u}_s(X, t)}{\partial X} \right|_{X=h} \right) = g_1(t). \tag{4.74}$$

The initial condition is

$$\hat{u}_s(X, 0) = 0, \quad 0 \le X \le h. \tag{4.75}$$

To completely define a particular problem, in addition to the load history $W(t)$, we need to specify the reference porosity ϕ_0, the constitutive function $f(e)$ and the drag force $K(\phi)$. In their work on bovine articular cartilage, Ateshian et al. (1997) suggest the use of a constitutive law of the form

$$f(e) = \frac{1}{2} \lambda \left(\frac{(1+e)^2 - 1}{(1+e)^{2\beta+1}} \right) \exp[\beta((1+e)^2 - 1)], \tag{4.76}$$

where λ and β are material constants. Moreover, the drag force is given by

$$K = \frac{\phi^2}{k}, \tag{4.77}$$

---

[8] This boundary condition is of the *Neumann type*, as opposed to the boundary condition (4.72), which is of the *Dirichlet type*.

where k is a measure of the *permeability*, given by the expression

$$k = k_0 \left(\frac{(1 - \phi_0)\phi}{\phi_0(1 - \phi)} \right)^2 \exp[Me/2]. \tag{4.78}$$

In this expression, k_0 is the *permeability coefficient* in the reference configuration and M is another constitutive constant. In writing these constitutive formulas, we have converted the notation of Ateshian et al. (1997) into ours. On the other hand, it is to be noted that the field equation proposed by them differs from ours in the absence of the second term on the left-hand side of equation (4.68). This difference may be of some physical significance. Indeed, since the term in question involves the gradient of the porosity, its importance may be felt when the material has a non-uniform porosity in the reference configuration and/or under severe transient conditions.

On the basis of several experiments and the corresponding curve fittings to the suggested constitutive laws, Ateshian et al. (1997) propose a range of values for the four parameters λ, β, k_0, M. In the following example, we adopt the mean values

$$k_0 = 2.2 \times 10^{-15}\, \mathrm{m}^4\, \mathrm{N}^{-1}\, \mathrm{s}^{-1} \times A,$$

$$\lambda = 0.4\, \mathrm{MPa} \times A,$$

$$M = 0.4,$$

$$\beta = 0.35, \tag{4.79}$$

where, for the purposes of our one-dimensional formulation, A stands for the cross-sectional area of the sample. The reference porosity is $\phi_0 = 0.80$ and the thickness of the sample is $h = 1.4\,\mathrm{mm}$. The magnitude of the load is $W_{\max} = 0.2\,\mathrm{MPa} \times A$. The load is applied linearly within the first $10\,\mathrm{s}$ and left thereafter for $2000\,\mathrm{s}$. Figures 4.3 and 4.4 show, respectively, the elastic constitutive law and the permeability as functions of the

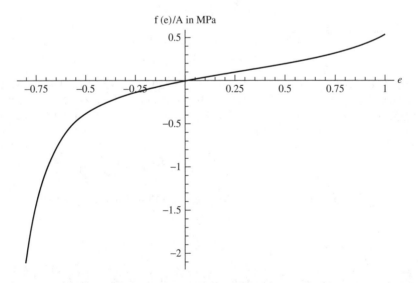

Figure 4.3   Elastic constitutive law

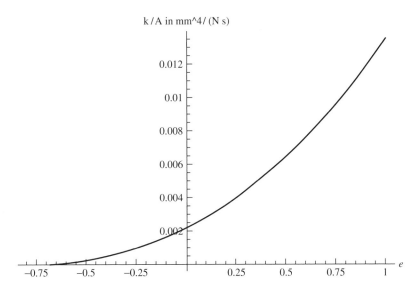

Figure 4.4 Permeability

```
phi0=0.8
h=1.4
k0=0.0022
lamda=0.4
m=0.4
bet=0.35
vamax=0.2
e=D[u[x,t],x]
eh=Derivative[1,0][u][h,t]
phi=1-(1-phi0)/(1+e)
k=k0*((1-phi0)*phi/(phi0*(1-phi)))^2*Exp[m*e/2]
fe=(1/2*lamda*(((1+e)^2-1)/(((1+eh)^(2*bet+1)))*Exp[bet*((1+e)^2-1)]
feh=(1/2*lamda*(((1+eh)^2-1)/((1+eh)^(2*bet+1)))*Exp[bet*((1+eh)^2-1)]
wt=If[t<10,-vamax*t/10,-vamax]
creep=NDSolve[{(D[phi*(fe-wt), x]/(1+e))==(phi/k)*D(u[x,t],t],
    u[0,t]==0,feh==wt,u[x,0]==0},u[x,t]{x,0,h},{t,0,2000}]
Plot3D[Evaluate[u[x, t]/.First[creep]],{x,0,h},{t,0,2000},PlotRange→All,
  AxesLabel→{"x (mm)"," t (s)", "us (mm) "}]
```

Figure 4.5 Mathematica® code for non-linear creep

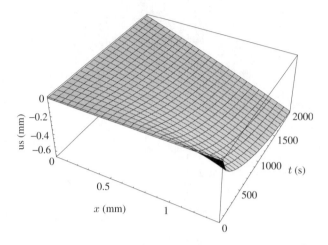

Figure 4.6   Creep displacement result

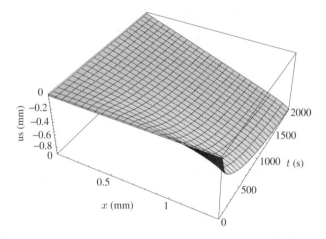

Figure 4.7   Creep displacement result with tripled load

engineering strain e for the numerical values adopted. Notice, in particular, that the compressive stress grows without bound as e approaches -1, as it should.

Figure 4.5 shows Mathematica® code that implements a numerical solution of our non-linear field equation. The corresponding results are illustrated in Figure 4.6. Notice that the solution reaches a steady state after a long time and that the distribution of the displacement attains a linear variation (constant strain). At this time, the system is at rest and liquid stops oozing from the piston. Notice the severe deformation, with strains of the order 50% . As an illustration of the importance of having a completely non-linear formulation, we triple the load. The results are shown in Figure 4.7, where the large stiffness exhibited in the highly compressive domain serves to check the deformation at a physically meaningful value. In the linearized version of the theory, as will be shown later, the results obtained for large loads would be meaningless.

4.5.5 *Hysteresis*

To exhibit the phenomenon of hysteresis, we apply a load given by

$$W(t) = 0.1(\cos(0.2\pi t) - 1), \tag{4.80}$$

for one cycle (10 s) and maintain it at a zero value thereafter for another 10 s. The results are shown in Figure 4.8, where it can be observed that the response, in terms of displacement, lags behind the acting force and decays in magnitude. Figure 4.9 shows the load–displacement diagram, with a typical hysteresis loop, whose area is a measure of the energy loss.

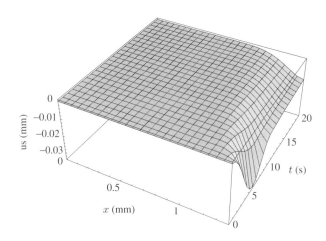

Figure 4.8   Displacement under one load cycle and beyond

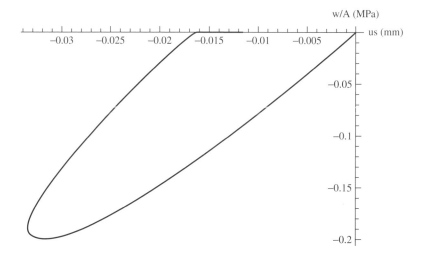

Figure 4.9   Hysteresis loop corresponding to Figure 4.8

4.5.6   The Linearized Theory

The linearized theory is obtained by expanding all the formulas in terms of a Taylor series in e keeping the lowest term of the expansion. Thus, for small strains ($e \ll 1$), equation (4.61) reduces to

$$\phi = \phi_0, \tag{4.81}$$

and equation (4.69) becomes

$$\frac{\partial^2 \hat{u}_s}{\partial x^2} = \frac{K}{\lambda \phi^2} \frac{\partial \hat{u}_s}{\partial t}, \tag{4.82}$$

where

$$\lambda = \left. \frac{df}{de} \right|_{e=0} \tag{4.83}$$

is an elastic constant.

Exercise 4.5.5   Linearized elasticity. Linearize the constitutive law (4.76) and show that the elastic constant λ of equation (4.83) is indeed the same as that of (4.76).

We mentioned earlier that the drag coefficient K may be a function of the porosity ϕ. This dependence is not universal as it arises from a variety of geometrical and physical factors. We have already indicated in Section 4.5.2 that a commonly accepted relation establishes that the drag coefficient is proportional to the square of the porosity, namely

$$K = \frac{\phi_0^2}{k_0}, \tag{4.84}$$

where k_0 is a constant known as the *permeability* of the matrix. We have already used this relation in the non-linear case, but now both the porosity and the permeability are considered as constants. The permeability is a measure of the ability of a porous medium to convey a fluid; it depends on the viscosity of the fluid. Articular cartilage has high porosity but low permeability, a feature that makes it ideally suited to act as a low-friction intermediate layer between articulating bones. In terms of the permeability, we can write equation (4.82) as

$$\frac{\partial^2 \hat{u}_s}{\partial x^2} = \frac{1}{\lambda k_0} \frac{\partial \hat{u}_s}{\partial t}, \tag{4.85}$$

We recognize in equation (4.85) once again the *diffusion equation* (or *heat equation*) that we encountered in earlier case studies. This equation is responsible for the viscous effects in cartilage. Note that these are purely *flow-induced* effects, as the matrix itself has been considered perfectly elastic. All viscous effects arise from the drag force between the fluid and the solid.

Using the values for the permeability and elasticity of bovine articular cartilage of our previous examples, the constant $1/(\lambda k_0)$ appearing on the right-hand side of equation (4.81) is $1136\,\mathrm{s\,mm^{-2}}$. Solutions of the diffusion equation can be constructed[9] from exponentials that decay with an exponent proportional to $-\lambda k t / h^2$, where h

---

[9] By the method of *separation of variables*.

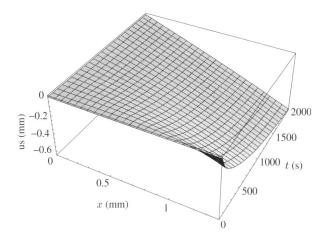

Figure 4.10 Creep displacement for the linearized theory

is a characteristic length (the cartilage thickness). For $h = 1.4\,\text{mm}$, we obtain the characteristic decay time $h^2/\lambda k$ of about 2200 s.

Neither the assumption of small displacements (and strains) nor the linear constitutive law are completely justified under realistic conditions. As an example, let us consider the linearized version of the creep example solved above (in Section 4.5.4). The result is shown in Figure 4.10. The agreement with the non-linear solution (Figure 4.6) is remarkably good, considering that the strains are as large as 50% . Nevertheless, since the equation has been linearized, tripling the load would triple the solution, leading to values of the displacement outside of the physically meaningful range (they would correspond to strains beyond -1, a physical impossibility). The non-linear solution, on the other hand, remains meaningful for any value of the load, as clearly shown in Figure 4.7.

4.6 Energy Balance

4.6.1 Constituent Balances

The energy balance for the constituent α can be stated as

$$
\frac{d}{dt}\int_{x(X_0,t)}^{x(X_1,t)} \rho_\alpha \left(\bar{u}_\alpha + \frac{1}{2}v_\alpha^2\right) dx = \int_{x(X_0,t)}^{x(X_1,t)} \left(b_\alpha v_\alpha + \rho_\alpha r_\alpha + p_\alpha v_\alpha + h_\alpha + \pi_\alpha \left(\bar{u}_\alpha + \frac{1}{2}v_\alpha^2\right)\right) dx
$$

$$
+ n_\alpha(x(X_1,t))v_\alpha(x(X_1,t)) - n_\alpha(x(X_0,t))v_\alpha(x(X_0,t))
$$

$$
- q_\alpha(x(X_1,t)) + q_\alpha(x(X_0,t)). \tag{4.86}
$$

The only novelty in this equation is the appearance of the *internal energy supply* h_α associated with the constituent α. The term $p_\alpha v_\alpha$ represents the power of the momentum supply, while the term containing π_α is the contribution to the internal and kinetic energies brought into play by the incoming mass. Except for the subscript α, the remaining terms have the same general meaning as in the energy balance of a simple substance (Section 2.9.2).

Exploiting the equations of mass and momentum balance for the constituent α, the local form of equation (4.86) can be written as

$$\rho_\alpha \frac{D_\alpha \bar{u}_\alpha}{DT} = \rho_\alpha r_\alpha + h_\alpha - \frac{\partial q_\alpha}{\partial x} + n_\alpha L_\alpha, \tag{4.87}$$

where L_α is the velocity gradient of the constituent α.

Exercise 4.6.1   Local Eulerian form of the energy balance for a constituent. Obtain equation (4.87) from (4.86) and the balances of mass and momentum for the constituent.

4.6.2   Mixture Balance

Just as we did in the case of mass and momentum balance, we want to recover the balance of energy for the mixture as a whole in the usual form of the energy balance for a single substance. This will require a combination of the equations of balance of energy for the constituents and an interpretation of the combined quantities (such as internal energy densities) in terms of the mixture as a whole. We have already seen that the total internal force (or stress) is not necessarily equal to the sum of the partial internal forces, but involves an extra term contributed by the diffusive velocities. Following a procedure similar to that used in the case of the balance of momentum, we express the individual constituent balances in terms of rates of change of the internal energy in a fixed spatial region. In this way, the individual contributions can be meaningfully added. The final result, obtained after a considerable amount of juggling, is indeed the following usual form (2.118) of the energy balance:

$$\rho \frac{D\bar{u}}{Dt} = \rho r + nL - \frac{\partial q}{\partial x}, \tag{4.88}$$

provided the following identifications are made:

$$\rho \bar{u} = \sum_{\alpha=1}^{M} \left(\rho_\alpha \bar{u}_\alpha + \frac{1}{2} \rho_\alpha u_\alpha^2 \right), \tag{4.89}$$

$$q = \sum_{\alpha=1}^{M} \left(q_\alpha - n_\alpha u_\alpha + \rho_\alpha \bar{u}_\alpha u_\alpha + \frac{1}{2} \rho u_\alpha^3 \right), \tag{4.90}$$

where u_α is the diffusion velocity of constituent α as defined in equation (4.11). Moreover, the following extra consistency conditions must be satisfied:

$$\sum_{\alpha=1}^{M} \left(h_\alpha + p_\alpha u_\alpha + \pi_\alpha (\bar{u}_\alpha + \frac{1}{2} u_\alpha^2) \right) = 0 \tag{4.91}$$

and

$$bv = -\sum_{\alpha=1}^{M} b_\alpha u_\alpha, \tag{4.92}$$

where b is given by equation (4.31). We start by noticing that, in the absence of diffusion (i.e., with $u_\alpha = 0$ for all α), equation (4.92) is automatically satisfied and equations (4.89)–(4.91) attain, respectively, the simpler forms

$$\rho \bar{u} = \sum_{\alpha=1}^{M} \rho_\alpha \bar{u}_\alpha, \tag{4.93}$$

$$q = \sum_{\alpha=1}^{M} q_\alpha, \tag{4.94}$$

and

$$\sum_{\alpha=1}^{M} h_\alpha = 0. \tag{4.95}$$

When diffusion is present, equation (4.92) deserves a special comment since, clearly, it cannot be satisfied identically. In other words, this condition is bound to be violated, unless a special relation is established between the partial body forces b_α and the total mixture body force b. It is not difficult to see that this special relation is

$$b_\alpha = \frac{\rho_\alpha}{\rho} b. \tag{4.96}$$

In physical terms, this relation establishes that the partial body forces *per unit mass* must be equal to each other, a condition that may or may not be fulfilled in practical applications.

Exercise 4.6.2   Energy balance for the mixture. By reversing the derivation process, verify that equation (4.88), together with conditions (4.89)–(4.92), is consistent with the sum of the energy balances of the individual constituents.

4.7   The Entropy Inequality

In addition to the many controversial issues raised by the formulation of the second law of thermodynamics in continuous media in general, there are specific questions pertaining to the theories of mixtures in particular. Two of these questions are the following: (i) Should an entropy inequality, in some form or another, be required for each constituent individually or just for the mixture as a whole? (ii) Should each component be assigned, at any given location and time, its own temperature, or should a single temperature be used for all the components? In our presentation, we will opt for the second choice in the answers to each of these questions, that is to say, we will formulate a single entropy inequality for the whole mixture and adopt a single temperature field for all the constituents. We will nevertheless assign to each constituent its own entropy density s_α per unit mass of the constituent, and we will define the entropy density per unit mass of the mixture as

$$s = \frac{1}{\rho} \sum_{\alpha=1}^{M} \rho_\alpha s_\alpha. \tag{4.97}$$

The entropy inequality for the mixture will be postulated as

$$\rho \frac{Ds}{Dt} \geq \frac{\rho r}{\theta} - \frac{\partial}{\partial x}\left(\frac{\hat{q}}{\theta}\right). \tag{4.98}$$

which differs from the usual form given by equation (2.125) only in that the entropy
flux term uses the following modified heat flux:

$$\hat{q} = \sum_{\alpha=1}^{M} \left(q_\alpha + \rho_\alpha \theta s_\alpha u_\alpha \right).$$ (4.99)

4.8   Chemical Aspects

Mixture theory has so far not achieved the wide use that it deserves in applications
involving biological systems. Yet, most living tissues can be regarded as chemically
reacting mixtures whose inner workings result in the very preservation or destruction of
life. One of the reasons for the fact that the use of mixture theory is still in its infancy
may be found, perhaps, in the intricacy of the theory. A second reason can be traced to the
difficulty in measuring the increased number of parameters required to describe a mixture
as compared with a single-substance formulation. Finally, the incorporation of the details
of complex chemical reactions must be mentioned as a stumbling block in formulating
viable models that can be used in realistic applications. Fortunately, the last few years
have witnessed a significant increase in activity aimed at overcoming all these and other
obstacles. The biphasic theory of cartilage is one of the success stories in the application
of mixture theory, as we have learned in Section 4.5. But, impressive as it is, the biphasic
theory deals with an *inert* mixture. It might as well be applied to the modelling of the
behaviour of porous media in general, as has been done in the fields of soil mechanics
and reservoir engineering. In this section, therefore, with a view to formulating models of
chemically reacting mixtures, we will briefly review some basic notions of chemistry that
may be relevant to the understanding of biological processes such as bone remodeling
and growth, or force production in muscle, or metabolic processes in general.

4.8.1   Stoichiometry

For the purpose of this review, we will take the notion of *chemical element* as a primitive.
There exist about 119 chemical elements, as listed in the periodic table. The *atom* constitutes the smallest identifiable unit of an element. Elements and their atoms are denoted by
a capital letter (e.g., H for hydrogen, C for carbon, I for iodine) or by a capital letter followed by a lower-case letter (e.g., Cl for chloride, Pb for lead, Na for sodium). In generic
formulas, we will use just a single capital letter A, B, C, . . . to designate elements or atoms.

A *substance* is a definite combination of one or more elements. The *molecule* is the
smallest identifiable unit of a substance. A molecule of a substance is made up of a definite, invariable number of atoms of the participating elements. The *formula* of a molecule
consists of a string of the symbols of the participating elements, each symbol followed by
a subscript indicating the number of atoms. The number 1 is usually omitted. Thus, A indicates a substance consisting of molecules made up of a single atom of element A, while
A_2BC_3 denotes a substance each of whose molecules consists of two atoms of A, one
atom of B and three atoms of C. The atoms in a molecule are held together by *chemical
bonds* and the number and type of atoms participating in a molecule are dictated by the
nature of these bonds.

In a *chemical reaction* a number of substances, called the *reactants*, are combined and
transformed into one or more substances, called the *products* of the reaction. A chemical

reaction takes place by breaking the chemical bonds of the reactants and establishing new chemical bonds to form the products. A chemical reaction is expressed by means of a *chemical equation*, one side of which represents the reactants while the other represents the products. Each reactant and product is preceded by a numerical coefficient, which is a natural number representing the corresponding number of molecules participating in the reaction. Instead of an equals sign to relate both sides of the equation, it is customary to place an arrow, the tip of which points toward the products of the reaction. A chemical equation cannot be any arbitrary combination of symbols but must satisfy certain physical principles. An equation satisfying these principles is said to be *balanced*. The algebraic discipline governing the correct balancing of chemical equations is called *stoichiometry*.

The basic principles of stoichiometry are the following:

1. *Indestructibility of elements*. According to this principle, the elements constituting the reactants cannot be transmuted into other elements and, therefore, the products must be made up of exactly the same elements as the reactants, and vice versa.
2. *Indestructibility of matter*. The total number of atoms of each element on one side of a chemical equation must be exactly equal to the number of atoms of that element on the other side of the equation. Strictly speaking, the validity of this principle implies the validity of the previous one.

Since any amount of substance participating in a chemical reaction consists ultimately of a large collection of molecules, the principle of indestructibility of matter can also be interpreted as a principle of conservation of mass for each element.

A familiar example of a chemical reaction is the formation of water from hydrogen (H) and oxygen (O). The hydrogen molecule is *diatomic* and its symbol is, therefore, H_2. The oxygen molecule, O_2, is also diatomic. The chemical formula of the water molecule (as dictated by the appropriate chemical bonds) is H_2O. The reaction whereby water is obtained from hydrogen and oxygen can be written as

$$mH_2 + nO_2 \rightarrow pH_2O, \tag{4.100}$$

where m, n and p are numerical coefficients to be determined by stoichiometry. We first ascertain that the elements appearing on the left-hand side (H and O) are the same as those appearing on the right-hand side. Moreover, it is easy to check in this case that the smallest natural numbers that would satisfy the principle of indestructibility of matter are $m = 2$, $n = 1$ and $p = 2$. The balanced equation reads:

$$2H_2 + O_2 \rightarrow 2H_2O. \tag{4.101}$$

From the mathematical point of view, stoichiometry is an application of the theory of *linear Diophantine equations*, namely, linear equations with integer coefficients whose solutions are also integer. In the elementary example of equation (4.100), we have the system of equations:

$$\begin{cases} 2m = 2p \\ 2n = p. \end{cases} \tag{4.102}$$

The first and second equations establish, respectively, the preservation of the number of atoms of hydrogen and oxygen. Any integer multiple of the above solution ($m = 2$, $n = 1$, $p = 2$) is also a solution. In any realistic chemical reaction the number of molecules involved is very large, but the essential fact is that the proportions between the numbers of molecules of each substance involved in a reaction are determined, once and for all, by the coefficients of a stoichiometrically balanced equation. A convenient unit to measure the amount of a substance is the *mole*. A mole of a substance consists of approximately $6.022\,142 \times 10^{23}$ molecules of that substance. This number is known as the *Avogadro number*.

Remark 4.8.1 Molar masses. To obtain the mass of 1 mole of a given substance, it is only necessary to know the masses of the atoms involved. These, in turn, can be obtained from the approximate picture of the atom as made from three kinds of *subatomic particles*: *protons* (positively charged), *neutrons* (not electrically charged) and *electrons* (negatively charged). Since atoms (as opposed to *ions*) are electrically neutral and since the charge of an electron is equal to the charge of a proton, the number of electrons must be equal to the number of protons. The actual mass of an atom, however, is (except in the case of hydrogen) always less than the direct sum of the masses of its subatomic particles because of the *binding energy* required to keep the nucleus together. This energy is a relativistic effect at the expense of a small fraction of the mass of the nucleus. The masses of the atoms are determined, therefore, experimentally.

The atomic masses are usually expressed in daltons, abbreviated as Da, with $1\,\text{Da} = 1.660\,538\,8 \times 10^{-27}$ kg. Multiplying this value by the Avogadro number results in exactly 1 g, thus revealing the convenient origin of the dalton as a mass unit. The atomic masses of the elements are usually listed (in daltons) in the periodic table. The mass of a hydrogen atom is 1.007 94 Da and that of oxygen is listed as 15.9994 Da. Since both these molecules are diatomic, the respective masses of the molecules are double these values. Thus, a mole of hydrogen molecules has a mass of 2.015 88 g and a mole of oxygen molecules has a mass of 31.9988 g. Equation (4.101) can be interpreted as stating that combining 4.031 76g of hydrogen with 31.9988g of oxygen results in 36.030 56g of water.

Remark 4.8.2 If certain amounts of reactants are put together, they may or may not react. If they do, a balanced chemical equation such as (4.101) only expresses the fact that the quantities of reactants and products that participate in the reaction do so in the proportions established by the coefficients of the equation. Moreover, the reaction may come to a stop either because one of the reactants has been fully utilized or because a situation has developed whereby the reaction takes place in both directions and a state of thermodynamic equilibrium has been attained.

The example of equation (4.100), in which two elements are involved in a single reaction comprising a total of three substances, is particularly simple. Specifically, its solution, given by equation (4.102), uniquely determines the ratio $m : n : p$. In a more general case, a reaction may involve any number of elements and substances. Moreover, it is possible (at least in theory) that several competing reactions take place simultaneously. The algebraic analysis of the most general conditions that can be imposed on the coefficients of these equations is beyond the scope of this review.[10] Nevertheless, even if working with just one equation, we would like to establish at least some link between

---

[10] For a clear and complete treatment, see Bowen (1976). An even more elegant geometric treatment can be found in Bowen (1968a).

the chemical conceptual framework and that of mixture theory. Let a generic balanced chemical equation be expressed as

$$\sum_{\alpha=1}^{M} \nu_{\alpha} S^{\alpha} = 0. \tag{4.103}$$

In this equation S^{α} ($\alpha = 1, \ldots, M$) are M substances which we identify with the constituents of our mixture. The distinction between reactants and products of the given reaction is, by convention, subsumed under the sign of the integer coefficients ν_{α}, namely: a negative sign corresponds to a reactant and a positive sign to a product. In a fixed spatial element dx and a time interval dt, the mass of each constituent α will change proportionally to the product ν_{α} times the molar mass of the substance S^{α}. Denoting the molar mass of the substance α by M^{α} we can write

$$\pi_{\alpha} = k \nu_{\alpha} M^{\alpha}, \tag{4.104}$$

where π_{α} stands for the mass supply and k is some constant measured in moles per unit volume and per unit time. This equation imposes a restriction on any theory of chemically reacting mixtures, since it establishes that the mass supplies are controlled by the stoichiometric coefficients. The coefficient k is known as the *reaction rate*.

As far as the participating elements are concerned, say N in number, let them be denoted by E^i ($i = 1, \ldots, N$). Each substance S^{α} is given by some chemical formula that assigns a coefficient T_i^{α} to each element E^i. This coefficient will be zero (if the element i does not enter the formula) or positive and equal to the corresponding numerical subscript. The numbers T_i^{α} form an $M \times N$ matrix that we call the *composition matrix*. In the example of equation (4.101), if we number the participating substances in the order H_2, O_2, H_2O and the elements in the order H, O, the composition matrix is

$$[T] = \begin{bmatrix} 2 & 0 \\ 0 & 2 \\ 2 & 1 \end{bmatrix}. \tag{4.105}$$

Notice that the composition matrix relates the molar masses M^{α} of the substances to the molar masses M^i of the elements according to

$$M^{\alpha} = \sum_{i=1}^{N} T_i^{\alpha} M^i. \tag{4.106}$$

The stoichiometric equations express the conservation of the number of atoms of each element, namely

$$\sum_{\alpha=1}^{M} \nu_{\alpha} T_i^{\alpha} = 0, \quad i = 1, \ldots, N. \tag{4.107}$$

Exercise 4.8.3 Verify that equation (4.107) reproduces the system of equations (4.102) in the case of the reaction (4.101).

In view of equation (4.104), equation (4.107) can be written as

$$\sum_{\alpha=1}^{M} \frac{\pi_\alpha}{M^\alpha} T_i^\alpha = 0, \quad i = 1,\ldots,N. \tag{4.108}$$

Multiplying this equation by M^i and summing over i, we obtain

$$0 = \sum_{i=1}^{N} M^i \left(\sum_{\alpha=1}^{M} \frac{\pi_\alpha}{M^\alpha} T_i^\alpha \right) = \sum_{\alpha=1}^{M} \frac{\pi_\alpha}{M^\alpha} \left(\sum_{i=1}^{N} T_i^\alpha M^i \right) = \sum_{\alpha=1}^{M} \pi_\alpha, \tag{4.109}$$

where we have used equation (4.106). Thus, we recover equation (4.19) of mixture theory.

If there is no diffusion, it is possible to define the concept of *extent of a reaction* ξ at a point of the mixture and at time t. Denoting by t_0 an initial reference time, the extent of the reaction is defined by

$$\xi = \int_{t_0}^{t} \frac{k(\tau)}{\rho} d\tau, \tag{4.110}$$

where k is from equation (4.104) and where the integration is carried out following a material particle. Recalling equation (4.21), and setting $u_\alpha = 0$, we can write equation (4.104) as

$$\rho \frac{Dc_\alpha}{Dt} = k v_\alpha M^\alpha. \tag{4.111}$$

Integrating with respect to time (following a particle), we obtain

$$c_\alpha(t) - c_\alpha(t_0) = v_\alpha M^\alpha \xi. \tag{4.112}$$

The extent of the reaction, therefore, is a single number (expressed in moles per unit mass of the mixture) that controls the changes in concentration of each and every constituent as time goes on. An alternative convenient way to write equation (4.112) results from introducing *the number of moles* of constituent α per unit mass of the mixture as

$$\gamma_\alpha = \frac{c_\alpha}{M^\alpha}. \tag{4.113}$$

In terms of these modified quantities, we can write equation (4.112) as

$$\gamma_\alpha(t) - \gamma_\alpha(t_0) = v_\alpha \xi. \tag{4.114}$$

Remark 4.8.4   In a more general situation, there exist as many extents of reaction as the number of independent chemical reactions that can occur between the given substances. In many applications, however, even when in theory a number of such reactions can coexist, only one is selected as actually taking place.

4.8.2   Thermodynamics of Homogeneous Systems

Many of the thermodynamic considerations pertaining to chemical reactions are made under the explicit or implicit assumption that the processes involved are spatially homogeneous. Heat conduction as such, for example, is ruled out, since in a homogeneous process there can be no temperature gradient. Under such simplified conditions, the whole apparatus of mixture theory and, indeed, of continuum mechanics boils down to statements about a few time-dependent quantities. Instead of the internal force n it is customary to refer to the *pressure* p.[11] Similarly, the deformation $x(X,t)$ is replaced by the *volume* $V(t)$. The *state* of a system is assumed to be determined entirely by the volume and the temperature (and, for more complex systems, some other parameters such as concentrations). The notion of the constitutive equation is replaced with that of the *function of state*. The art of the continuum mechanics adept is a balancing act whose result should be the incorporation of the wealth of information and the considerable thermodynamical apparatus developed by chemists and biochemists in the language of their disciplines[12] into the rather demanding theoretical constructs of continuum mechanics and, in particular, of the theory of chemically reacting mixtures.

Consider a *closed system*[13] made of a single substance. Its functions of state are given by

$$p = p(V, \theta), \tag{4.115}$$

$$U = U(V, \theta), \tag{4.116}$$

$$S = S(V, \theta), \tag{4.117}$$

where U and S denote the internal energy and the entropy, respectively.

Remark 4.8.5   In terms of our one-dimensional continuum mechanics formulation, the following identifications can be made assuming that the internal force n, the current cross sectional area $\hat{s}$, the internal energy density $\bar{u}$ and the entropy density s are independent of x:

$$p(t) = -n/\hat{s}, \tag{4.118}$$

$$V(t) = \hat{s}(t) \left(x(X_1, t) - x(X_0, t) \right), \tag{4.119}$$

$$U(t) = \hat{\rho}\bar{u}V, \tag{4.120}$$

$$S(t) = \hat{\rho} s V. \tag{4.121}$$

For homogeneous processes the laws of mechanical and thermodynamical balance simplify considerably due to the fact that the integrations over the body become trivial. If we

---

[11] Even in the three-dimensional treatment and even when the substance involved is a solid, many thermodynamic considerations are made on the basis of 'the pressure', rather than the fully-fledged stress tensor.

[12] Among the many excellent texts available in this field, we single out Prigogine and Defay (1950), Eisenberg and Crothers (1979) Guggenheim (1977).

[13] A system is *closed* if it exchanges no mass with the exterior.

consider cases in which the forces of inertia are negligible, we do not need to worry about the conservation of momentum. The conservation of mass is just the statement

$$\hat{\rho}V = \text{constant.} \qquad (4.122)$$

The power of the internal forces is given by

$$W_{\text{int}} = p\frac{dV}{dt}. \qquad (4.123)$$

Since we are neglecting the kinetic energy, the power of the external forces W_{ext} is equal to $-W_{\text{int}}$, in accordance with equation (2.109), while the non-mechanical (i.e., thermal) power is

$$W_{\text{th}} = \hat{\rho}rV. \qquad (4.124)$$

Note that, for consistency with the assumption that the process is spatially homogeneous, the thermal power must be assigned entirely to the volumetric (radiation) term r. According to equation (2.113), the first law of thermodynamics for a homogeneous process is

$$\frac{dU}{dt} = W_{\text{th}} - p\frac{dV}{dt}. \qquad (4.125)$$

The second law of thermodynamics for a homogeneous process reduces to the inequality

$$\frac{dS}{dt} \geq \frac{W_{\text{th}}}{\theta}. \qquad (4.126)$$

Beyond the three functions (4.115)–(4.117), it is convenient to define other functions of state. We have already encountered the *Helmholtz free energy* per unit mass, ψ, defined in the general case by equation (3.23). In the homogeneous case, we use the total Helmholtz free energy Ψ obtained as

$$\Psi = \hat{\rho}\psi V = U - \theta S. \qquad (4.127)$$

Another useful function of state is the *enthalpy* H defined as

$$H = U + pV. \qquad (4.128)$$

Finally, the *Gibbs free energy* is given by

$$G = H - \theta S = \Psi + pV. \qquad (4.129)$$

Unlike the Helmholtz free-energy function, neither the enthalpy nor the Gibbs free energy (also called the *free enthalpy*) lend themselves to a straightforward generalization for a deformable continuum, a fact that in no way detracts from their important role in chemical considerations.

4.8.3 Enthalpy and Heats of Reaction

The meaning of the enthalpy can be gathered from the first law (4.125), written via (4.128) as

$$\frac{d(H - pV)}{dt} = W_{\text{th}} - p\frac{dV}{dt}, \qquad (4.130)$$

whence

$$\frac{dH}{dt} = W_{\text{th}} + V\frac{dp}{dt}. \tag{4.131}$$

In an *isobaric* (i.e., constant-pressure) process the increase in enthalpy between two times $t_1 < t_2$ is equal to the total heat-energy input, namely

$$H_2 - H_1 = \int_{t_1}^{t_2} W_{\text{th}}\, dt. \tag{4.132}$$

Remark 4.8.6 In practice, thermodynamicists use the volume–temperature state variables even for processes that are not strictly homogeneous, such as a fluid agitated by a stirrer. In such cases, the external working input is divided into a part that can be accounted for by $p\, dV$ and a part that cannot, so that equation (4.132) is not directly applicable.

Many chemical reactions of interest take place under atmospheric conditions and can, therefore, be considered as approximately isobaric processes. We usually also assume that the reaction takes place at constant temperature. As the reaction proceeds, the enthalpy contents of the reactants and the products vary continuously, as the mass of the reactants diminishes and the mass of the products increases with time. Over a given time interval $\Delta t > 0$, if the enthalpy changes of the reactants and the products are denoted, respectively, by $\Delta H_{\text{reactants}}$ and $\Delta H_{\text{products}}$, the total enthalpy change is given by

$$\Delta H = \Delta H_{\text{products}} - \Delta H_{\text{reactants}}. \tag{4.133}$$

If, more logically, a chemical reaction is written in the form of equation (4.103), the change in enthalpy associated with the reaction can be written as the algebraic sum:

$$\Delta H = \sum_{\alpha=1}^{M} \nu_\alpha \Delta H^\alpha, \tag{4.134}$$

where ΔH^α is the enthalpy content of 1 mole of the substance α. The magnitude and sign of the stoichiometric coefficients ν_α take care of the rest.

A reaction is called *exothermic* if heat is released and *endothermic* if heat is absorbed by the reacting system. From equation (4.132), it follows that in an exothermic reaction $\Delta H < 0$ and in an endothermic reaction $\Delta H > 0$. In other words, according to equation (4.133), a reaction is exothermic if the increase in enthalpy of the products is less than the decrease in enthalpy of the reactants.

Remark 4.8.7 The notion of enthalpy of reaction. To make the definitions somewhat more precise, we must consider the enthalpy (as well as other functions of state of a chemically reacting mixture) as functions of the pressure, the temperature and the extent of the reaction. In this way, we may introduce the *enthalpy of reaction* or *heat of reaction* as the derivative of the total enthalpy of the system with respect to the extent of the reaction at constant pressure and temperature. A reaction is exothermic if the enthalpy of reaction thus defined is negative.

Choosing a reference pressure and temperature (usually 1 atm and 25°C), the *heat of formation* is defined as the enthalpy change of the reaction that produces (1 mole of) a substance out of its component elements. These values, which can be tabulated once and for all, are used to calculate and predict (approximately) the heat emitted or absorbed by a given reaction at constant pressure (and temperature). The heat of formation depends on whether the substance produced is in the solid, liquid or gaseous state. When giving a chemical reaction it is, therefore, important to indicate these states by a bracketed argument following the formula of the substance: (s), (l) or (g), respectively. Other possibilities exist and can be so indicated (e.g., different crystal arrangements of the same solid substance). To pass from one form of a substance to another (such as from liquid to vapour) it is necessary to know the corresponding *latent heat* (of, say, vaporization). The elements themselves are, by convention, assigned a zero value of heat of formation (in their most stable natural state).

Example 4.8.8   Heat of reaction.   Calculate the heat of reaction of

$$2H_2S \ (g) + 3O_2 \ (g) \rightarrow 2H_2O \ (l) + 2SO_2 \ (g), \qquad (4.135)$$

at standard conditions of pressure and temperature. The following values have been obtained from a table of heats of formation:

Substance	ΔH (kJ mol^{-1})
H_2S (g)	-20.6
O_2 (g)	0.0
H_2O (g)	-241.8
SO_2 (g)	-296.8

Since the heat of formation of water has been given for the gaseous state, whereas in the given chemical equation water appears in the liquid state, we need to know the latent heat of vaporization of water. At the given conditions of pressure and temperature this value is 44 kJ mol^{-1}. Since we need to convert vapour into liquid, heat has to be extracted and, therefore, this value should be taken with a negative sign. We obtain, accordingly, the heat of formation of liquid water at the given conditions as -241.8 kJ mol$^{-1} - 44$ kJ mol$^{-1} = -285.8$ kJ mol^{-1}. We calculate:

$$\Delta H_{reactants} = 2 \, mol \times (-20.6 \, kJ \, mol^{-1}) + 3 \, mol \times 0.0 = -41.2 \, kJ \qquad (4.136)$$

and

$$\Delta H_{products} = 2 \, mol \times (-285.8 \, kJ \, mol^{-1}) + 2 \, mol \times (-296.8 \, kJ \, mol^{-1}) = -1165.2 \, kJ. \qquad (4.137)$$

According to equation (4.133), we obtain

$$\Delta H = -1165.2 \, kJ - (-41.2 \, kJ) = -1124 \, kJ. \qquad (4.138)$$

The reaction is exothermic. The numerical result obtained corresponds to the complete consumption of 2 moles of hydrogen sulphide (H_2S). Since the molar masses of atoms of sulphur (S) and hydrogen (H) are, respectively, 32.064 and 1.00797, this corresponds to the consumption of $2 \times (32.064 + 2 \times 1.00797) = 68.16$ grams of H_2S. The result for other values of the mass can be calculated proportionately. For other temperatures, heat capacities[14] at constant pressure can be used *a posteriori*. The effect of pressure on enthalpy is often negligible within the usual ranges.

Exercise 4.8.9   The unexplained energy in muscle contraction. The question of *unexplained energy* in skeletal muscle contraction arises in the attempt to verify the satisfaction of the first law of thermodynamics by independent measurements of chemical energy, mechanical work and heat. Although a number of simplifying assumptions must be made along the way, it is generally agreed[15] that, even taking into consideration the margins of error introduced, there appears to be a deficit in the net energy supplied by the identifiable chemical reactions involved. The main reaction consists of the hydrolysis of adenosine triphosphate (ATP). Like most organic compounds, the complex structure of ATP (whose organic part is the adenosine) cannot be faithfully represented in a formula (involving atoms of phosphorus, carbon, hydrogen and oxygen). Hydrolysis is a reaction of a substance with water molecules which, as a result, are split into positive and negative ions. In the process, adenosine diphosphate (ADP) and an inorganic phosphate are formed:

$$ATP + H_2O \rightarrow ADP + phosphate. \tag{4.139}$$

This is an exothermic reaction. The exact value of the heat of reaction has been somewhat controversial in the past, but the accepted value is of the order of $-30\,kJ\,mol^{-1}$. In Woledge et al. (1985) this enthalpy is equated to the measured heat and work of the muscle. Whether or not this is a fair implementation of the first law (which equates the rate of *internal energy* to the mechanical and thermal power input), it is pleasing to observe that *under steady-state conditions* or, more specifically, under conditions of steady isometric tetanus once the heat released per unit time has attained a constant value (see Figure 3.5) and no mechanical power is present, this measured value of the heat rate is matched quite well by the rate of enthalpy of the reaction. In other words, under these conditions there is no appreciable unexplained energy. During the initial transient phase, on the other hand, when the unexplained energy question arises, the precise calculation and measurement of the various factors contributing to the implementation of the first law become somewhat fuzzier. At any rate, assuming that the unexplained energy has been legitimately identified and measured, its most probable cause can be attributed to the presence of other exothermic chemical reactions. It is an illuminating exercise (hereby suggested to the reader) to peruse the treatment of this problem in Woledge et al. (1985), which contains an abundance of data and of interesting critical comments.

4.8.4   *The Meaning of the Helmholtz Free Energy*

Combining the statement of the second law (equation (4.126)) with the definition of the Helmholtz free energy (equation (4.127)) and the first law (equation (4.125)), we obtain the inequality

$$\frac{d\Psi}{dt} + S\frac{d\theta}{dt} + p\frac{dV}{dt} \leq 0. \tag{4.140}$$

---

[14] See equation (2.135).
[15] See Woledge et al. (1985).

This equation expresses the reduced Clausius–Duhem inequality (3.24) for homogeneous processes. For isothermal processes (constant θ), integrating equation (4.140) between two times $t_1 < t_2$, we can write

$$\int_{t_1}^{t_2} p\,dV \leq \Psi_1 - \Psi_2. \tag{4.141}$$

We conclude that the mechanical work that can be extracted from a closed system in an isothermal process is limited by the decrease in the Helmholtz free energy. The maximum value corresponds to a reversible process.

Exercise 4.8.10   Reversibility. For a substance undergoing homogeneous processes and strictly abiding by the constitutive laws (4.115)–(4.117), show that all processes are reversible and that the free energy of Helmholtz acts as a potential function for the pressure and the entropy, that is,

$$p = -\frac{\partial \Psi}{\partial V} \tag{4.142}$$

and

$$S = -\frac{\partial \Psi}{\partial \theta}. \tag{4.143}$$

[Hint: use the arguments adduced for similar situations in the general continuum case, as shown in Section 3.5.]

Exercise 4.8.11   Maxwell's law. Under the same assumptions as in the previous exercise, prove *Maxwell's law*:

$$\frac{\partial p}{\partial \theta} = \frac{\partial S}{\partial V}. \tag{4.144}$$

[Hint: use the symmetry of mixed partial derivatives and bear in mind that in equation (4.144) the independent variables are V and θ.]

Exercise 4.8.12   Damping. Discuss the implications of the dissipation inequality for a fluid whose constitutive equations depend on the pressure, the volume and the time derivative of the volume. Analyse in more depth the particular case in which the dependence on the time derivative of the volume is linear.

4.8.5   Homogeneous Mixtures

If the medium under consideration consists of a mixture of several substances undergoing chemical reactions, the Helmholtz free energy of the mixture will depend also on the molar concentrations of the constituents. These concentrations we assume in the present treatment as spatially homogeneous, but they are certainly functions of time. We define the *chemical potential* μ_α of the constituent α as

$$\mu_\alpha = \frac{\partial \hat{\Psi}(V, \theta, \gamma_1, \ldots, \gamma_M)}{\partial \gamma_\alpha}. \tag{4.145}$$

Remark 4.8.13 The quantity γ_α introduced in equation (4.113) represents the number of moles of constituent α per unit mass of the mixture. In a homogeneous process, it is expedient to replace this quantity by the total number of moles of constituent α or, more elegantly, to consider that all the extensive quantities involved (including the Helmholtz free energy) pertain to a unit mass of the mixture. A more important point to highlight is that the function $\hat{\Psi}$ is considered here as a function of all the concentrations γ_α and, therefore, the partial derivative in equation (4.145) is obtained by *holding all the other concentrations at a fixed value*. In other words, the stoichiometry of the reaction is not enforced while calculating this derivative.

Instead of considering each and every concentration as an independent variable, we proceed now to extend the space of arguments of the constitutive laws to include only the extent of the reaction, ξ.[16] In other words, the stoichiometric relations between the concentrations are enforced. Moreover, an extra constitutive law, called an *evolution equation*, needs to be specified to couple the progress of the reaction to the remaining field variables.[17] It is customary to assume that this evolution law takes the form of an ordinary first-order differential equation. Summarizing, the constitutive equations of the theory in the presence of chemical reactions are:

$$p = p(V, \theta, \xi), \tag{4.146}$$

$$U = U(V, \theta, \xi), \tag{4.147}$$

$$S = S(V, \theta, \xi), \tag{4.148}$$

$$\frac{d\xi}{dt} = \omega(V, \theta, \xi), \tag{4.149}$$

where ω represents the anticipated evolution law. This form of the equations has been specialized for homogeneous processes. For a general mixture, the same general ideas apply, namely: the space of constitutive arguments is augmented to include the extent of reaction and an extra evolution equation is supplied in the form of an ordinary first-order differential equation.

We may substitute the Helmholtz free energy Ψ for the internal energy U. The corresponding constitutive equation reads:

$$\Psi = \Psi(V, \theta, \xi). \tag{4.150}$$

Remark 4.8.14 In equation (4.150) we have indulged in a flagrant, albeit habitual, abuse of notation, which we have mitigated by use of the 'hat' notation for the Helmholtz free-energy function in equation (4.145). The precise relation between the functions appearing in equations (4.145) and (4.150) is

$$\hat{\Psi}\left((V, \theta, \gamma_1, \ldots, \gamma_M)\right) = \hat{\Psi}\left(V, \theta, \left((\gamma_1(t_0) + \nu_1\xi), \ldots, (\gamma_M(t_0) + \nu_M\xi)\right)\right) = \Psi(V, \theta, \xi). \tag{4.151}$$

---

[16] In the present treatment, the extent of the reaction ξ is a scalar quantity. On the other hand, as we have already pointed out, when several chemical reactions are at play, the extent of reaction can be considered as a vector in a space whose dimension equals the number of *independent* chemical reactions.

[17] This is a particular case of a *theory with internal state variables*.

The derivative of the Helmholtz free energy with respect to the extent of the reaction is an important quantity known as the *chemical affinity*,

$$\sigma = -\frac{\partial \Psi}{\partial \xi}, \tag{4.152}$$

where the minus sign is introduced for convenience of interpretation. We can express the affinity in terms of the chemical potentials of the various species as

$$\sigma = -\sum_{\alpha=1}^{M} \mu_\alpha \nu_\alpha, \tag{4.153}$$

where equations (4.145) and (4.151) have been used.

4.8.6   Equilibrium and Stability

A state $\{V_0, \theta_0, \xi_0\}$ is an *equilibrium state* if

$$\omega(V_0, \theta_0, \xi_0) = 0. \tag{4.154}$$

The meaning of this concept of equilibrium is, clearly, a state in which the reaction rate vanishes. Suppose now that, when crossing an equilibrium state (by keeping the volume and temperature constant while varying ξ continuously in some open interval), the constitutive function ω changes in sign. We call this state an *ordinary equilibrium state*. If ω decreases with increasing ξ the ordinary equilibrium is stable.

Exercise 4.8.15   Explain the (admittedly restricted) notion of stability just introduced. [Hint: consider small perturbations $\Delta\xi$ of the equilibrium state. Compare with the equilibria of a pendulum.]

Theorem 4.8.16   *If $\{V_0, \theta_0, \xi_0\}$ is an ordinary equilibrium state, then the affinity necessarily vanishes in that state, namely*

$$\sigma(V_0, \theta_0, \xi_0) = 0. \tag{4.155}$$

Proof. Introducing our constitutive equations (4.146)–(4.149) into the dissipation inequality (4.140) and reasoning as in Exercise 4.8.10, we obtain the residual inequality

$$\sigma\omega \geq 0. \tag{4.156}$$

Assuming that the function σ depends continuously on ξ, it must vanish at $\{V_0, \theta_0, \xi_0\}$. Indeed, if it did not, it would have a constant sign on an open neighbourhood of this point. Given that ω changes in sign in this neighbourhood, the residual inequality would be contradicted.   ∎

If we assume that the constitutive functions are differentiable, an ordinary equilibrium corresponds to the non-vanishing of the partial derivative of the evolution function ω with respect to the extent of reaction in the equilibrium state. Notice that the theorem can also be reversed in the sense that if the affinity vanishes at a point while its ξ-partial derivative does not, then the point is an equilibrium state.[18]

Corollary 4.8.17   *At a state of ordinary equilibrium, the combined chemical potential of the reactants is equal to that of the products.*

Proof. The proof follows directly from Theorem 4.8.16 and equation (4.153).   ■

If the chemical affinity in some state is positive, it follows from equation (4.156) that the reaction rate is non-negative. The physical meaning of this corollary is that the chemical affinity acts as the driving force for the spontaneous advance of chemical reactions, just as the potential energy drives the motion of a falling rock. The negative sign in the definition of chemical affinity ensures that, if the volume and temperature are held fixed, the total Helmholtz free energy decreases with the advance of the reaction.

Exercise 4.8.18   Prove that at a stable equilibrium the Helmholtz free energy has a minimum with respect to the variable ξ.

Although we have adopted the usual point of view in most chemical textbooks, whereby the concepts are presented within the context of homogeneous processes, similar ideas can be implemented in the more general context of mixture theory within the terminology of continuum mechanics.[19]

4.8.7   *The Gibbs Free Energy as a Legendre Transformation*

The various state functions introduced in Section 4.8.2 as alternatives to the internal energy U, namely the enthalpy H, the Helmholtz free energy Ψ and the Gibbs free energy G, can, on the one hand, be regarded as mere combinations of the internal energy with various products of the pressure, volume, entropy and absolute temperature. We have seen, moreover, how the enthalpy and the Helmholtz free energy are convenient for expressing certain physical facts (such as the heat of reaction and chemical potential) in a perhaps more direct manner than by means of the internal energy. On the other hand, it turns out that there is a deeper and more meaningful relationship between these functions of state than would appear at first sight. In particular, we have so far used the volume, temperature and species concentrations as the independent variables of these functions of state and we may ask whether other choices (such as replacing the volume by the pressure as an independent variable) might be fruitful. The answer to this question turns out to be that each of these functions is best expressed in terms of a particular choice of independent variables and that, in fact, the passage from one function to another, including the choice of the appropriate independent variables, is a particular case of a *Legendre transformation* (see Window 4.1).

---

[18] In the general case, a distinction can be drawn between equilibria that do and those that do not satisfy the vanishing of the affinity. This interesting topic, as well as the notion of stability, both beyond the scope of our review, are fully discussed in the excellent article by Bowen (1968b).

[19] See Bowen (1968b).

Window 4.1   The Legendre transformation
Let $F(x)$ be a *convex* real function defined on an open interval of a real variable
x. The notion of convexity can be introduced in a rather general way (roughly,
a function on an open interval is convex if its graph is never above its secants).
Convex functions are necessarily continuous. Since we are interested in applications
where the function $F(x)$ is at least twice differentiable, we can use the following well-
known criterion to characterize (or, indeed, to define) convexity: a function is convex
in an open interval if, and only if, its second derivative is positive semi-definite at
every point of the interval,

$$\frac{d^2 F(x)}{dx^2} \geq 0.$$

We will require a somewhat stronger condition (*strict convexity*), namely

$$\frac{d^2 F(x)}{dx^2} > 0.$$

The first derivative (the *slope* $s = s(x)$) of a strictly convex function $F(x)$ is strictly
increasing and, therefore, injective (see Figure 1.9). The essential point here is that
the slope function is *invertible* over its range. We denote this well-defined inverse
by $x(s)$. The idea behind the Legendre transform is to create a new function $L(s)$
which conveys essentially *the same information* as $F(x)$ while satisfying an additional *involutivity property*, namely: the iterated application of the Legendre transform
reproduces the initial function! The Legendre transform is given by the formula

$$L(s) = sx(s) - F(x(s)). \tag{4.157}$$

Exercise 4.8.19   Involutivity. Prove the involutivity property of the Legendre transform. In particular, the slope function of the transform reproduces the corresponding
value of the original independent variable x.

With functions of many variables, it is meaningful to define the Legendre transform
with respect to just one of the variables.

Let us produce the Legendre transform of the Helmholtz free-energy function with
respect to the variable V (while implicitly assuming a convexity condition). The new
independent variable is the slope

$$\frac{\partial \Psi(V, \theta, \xi)}{\partial V} = -p, \tag{4.158}$$

where we have used equation (4.142), as in Exercise 4.8.10 and Theorem 4.8.16. The
Legendre transform of $\hat{\Psi}$ is, according to the general prescription of equation (4.157):

$$-pV(p, \theta, \xi) - \hat{\Psi}(V(p, \theta, \xi), \theta, \xi) = -G(p, \theta, \xi). \tag{4.159}$$

We conclude that, except for an immaterial change of sign, the Gibbs free energy is indeed the Legendre transform of the Helmholtz free energy with respect to the volume. The new independent variables are the pressure, the temperature and the molar concentrations. An unavoidable abuse of notation in equation (4.159) should not obscure the fact that, after finding out that the transformed independent variable is indeed the pressure, every appearance of the 'old' volume variable implies the inversion of equation (4.158) to express V in terms of p, θ and $\gamma_1, \ldots, \gamma_M$.

Exercise 4.8.20   Legendre transforms. Obtain the enthalpy and the internal energy as Legendre transforms of the Helmholtz free energy. Interpret the corresponding independent variables.

The following exercise is an important result:

Exercise 4.8.21   Chemical potentials. Prove that the chemical potentials are given in terms of the Gibbs free energy by

$$\mu_\alpha = \frac{\partial \hat{G}(p, \theta, \gamma_1, \ldots, \gamma_M)}{\partial \gamma_\alpha}, \qquad (4.160)$$

a formula closely resembling (4.145). Conclude that a similar result applies to the definition of the affinity. Notice that $\hat{G}$ and G are related by a formula of the type (4.151). [Hint: carefully use the chain rule of differentiation in equation (4.159).]

Exercise 4.8.22   Express the chemical potentials in terms of the internal energy U.

All the conditions at equilibrium remain valid when expressed in terms of the Gibbs free energy. Moreover, the criterion of spontaneity under conditions of constant *pressure* and temperature is that the Gibbs free energy must instantaneously decrease. One of the reasons for the usefulness of the Gibbs free energy (as opposed to its Helmholtz counterpart) for assessing spontaneity is that many chemical reactions (e.g., those involving a sudden expansion) take place at approximately constant temperature and pressure (rather than constant temperature and volume).

4.9   Ideal Mixtures

4.9.1   The Ideal Gas Paradigm

The *ideal gas* has played in the past and continues to play an important role in the development and understanding of thermodynamic theories. Moreover, it provides a link between phenomenological and statistical formulations, both in the classical and in the quantum mechanics contexts. In this section, we will review some of the main features of the classical theory of mixtures of ideal gases, a theory that serves as a paradigm for the more general theory of *ideal mixtures*.

Exercise 4.9.1   The Helmholtz free energy of an ideal gas. The constitutive law for the pressure of an ideal gas is the well-known formula

$$p = \frac{\gamma R \theta}{V}, \qquad (4.161)$$

where γ is the number of moles and R is the *universal gas constant*,

$$R = 8.3144 \, \text{J} \, \text{K}^{-1} \, \text{mol}^{-1}. \tag{4.162}$$

Obtain the Helmholtz free energy of an ideal gas as

$$\Psi = \gamma \left(f(\theta) - R\theta \ln \frac{V}{V_0} \right), \tag{4.163}$$

where $f(\theta)$ is a function that characterizes the gas and V_0 is a reference volume.[20] [Hint: use the result of Exercise 4.8.10.]

Exercise 4.9.2 The reference volume V_0 plays the role of a constant of integration. It is chosen so that the function $f(\theta)$ will represent the Helmholtz free energy of 1 mole of gas kept at the constant volume V_0 as the temperature varies. Strictly speaking, therefore, the function $f(\theta)$ appearing in equation (4.163) depends on the adopted reference volume V_0. Find the general rule for the transformation of $f(\theta)$ under a change of V_0.

It is customary to adopt standard reference values p_0 and θ_0 for the pressure and temperature, respectively, and to obtain the corresponding reference volume from equation (4.161). For $p_0 = 1 \, \text{atm} = 101.325 \, \text{kPa}$ and $\theta_0 = 25°\text{C} = 298 \, \text{K}$, one obtains $V_0 = \gamma \times 24.4531$ litres per mole. For the same pressure and $\theta_0 = 0°\text{C} = 273 \, \text{K}$ the result is $V_0 = \gamma \times 22.4017$ litres per mole.

Exercise 4.9.3 Internal energy. Prove that, for an ideal gas, the internal energy per mole depends on the temperature only. Find its explicit expression in terms of the function $f(\theta)$. [Hint: first obtain the entropy using equation (4.143).]

Exercise 4.9.4 Heat capacity at constant volume. The *heat capacity at constant volume* is defined as

$$c_v = \frac{1}{\gamma} \frac{\partial U}{\partial \theta}. \tag{4.164}$$

Show that for an ideal gas,

$$c_v = -\theta \frac{d^2 f(\theta)}{d\theta^2}. \tag{4.165}$$

On the basis of statistical mechanics considerations it can be shown that (within certain wide temperature bounds) c_v is constant. From this observation, obtain the most general form of the function $f(\theta)$ under the assumption of constant c_v.

Exercise 4.9.5 The Gibbs free-energy function. Show that the Gibbs free-energy function of an ideal gas is given by

$$G(p, \theta) = \gamma \left(g(\theta) + R\theta \ln \frac{p}{p_0} \right), \tag{4.166}$$

---

[20] The function f can be further expressed in terms of the heat capacity at constant volume.

with

$$g(\theta) = f(\theta) + R\theta \left(1 - \ln \frac{\theta}{\theta_0}\right). \tag{4.167}$$

The function $g(\theta)$ represents the Gibbs free energy of 1 mole of gas kept at the constant pressure p_0 while the temperature varies.

4.9.2 Mixtures of Ideal Gases

The fundamental postulate of the theory of (ideal) mixtures of ideal gases is that any one, and therefore each one, of the functions of state of the mixture is the sum of the corresponding functions of state of the constituents as if each of them were occupying the total volume of the mixture at the given temperature. In other words, there is a complete decoupling of the functions of state, as if the fact that the mixture had occurred made no difference to the behaviour of its components.

Consider, for example, the Helmholtz free energy of the mixture. According to the axiom just stated, its value is

$$\Psi = \sum_{\alpha=1}^{M} \Psi_\alpha = \sum_{\alpha=1}^{M} \gamma_\alpha \left(f_\alpha(\theta) - R\theta \ln \frac{V}{V_0}\right). \tag{4.168}$$

According to equation (4.142), the derivative of this function with respect to the volume should render (minus) the pressure, namely

$$p = \frac{R\theta}{V} \sum_{\alpha=1}^{M} \gamma_\alpha = \frac{\gamma R\theta}{V}, \tag{4.169}$$

where

$$\gamma = \sum_{\alpha=1}^{M} \gamma_\alpha \tag{4.170}$$

is the total number of moles, regardless of substance. Equation (4.169) can be interpreted physically as stating that the mixture behaves as an ideal gas. Alternatively, it states that the total pressure of the mixture is equal to the sum of the *partial pressures*, namely the pressure that each constituent would experience if it alone were occupying the total volume at the given temperature. This statement is known as *Dalton's law*. Defining the *molar fractions* as

$$x_\alpha = \frac{\gamma_\alpha}{\gamma}, \tag{4.171}$$

the partial pressures can be expressed as

$$p_\alpha = x_\alpha p. \tag{4.172}$$

Remark 4.9.6 From equation (4.171) it follows that

$$\sum_{\alpha=1}^{M} x_\alpha = 1 \tag{4.173}$$

identically, regardless of whether chemical reactions are taking place or not. On the other hand, even for a closed system, the total number of moles γ is not conserved in general. For example, 2 moles of hydrogen molecules and 1 mole of oxygen molecules react to produce 2 moles of water molecules.

The Gibbs free-energy function of the mixture can be calculated as

$$G = \sum_{\alpha=1}^{M} G_\alpha = \sum_{\alpha=1}^{M} \gamma_\alpha \left(g_\alpha(\theta) + R\theta \ln \frac{p_\alpha}{p_0} \right) = \sum_{\alpha=1}^{M} \gamma_\alpha \left(g_\alpha(\theta) + R\theta \ln \frac{x_\alpha p}{p_0} \right). \quad (4.174)$$

Proposition 4.9.7   Chemical potentials. *The chemical potential of the constituent α of an ideal gas mixture is given by*

$$\mu_\alpha = g_\alpha(\theta) + R\theta \ln \frac{p_\alpha}{p_0} = g_\alpha(\theta) + R\theta \ln \frac{x_\alpha p}{p_0}. \quad (4.175)$$

Proof. What is remarkable about this result is that the chemical potential of a constituent is oblivious to the presence of the other constituents, as long as one calculates it at the corresponding partial pressure p_α. From equation (4.160), we obtain

$$\mu_\alpha = \frac{\partial G(p,\theta,\gamma_1,\ldots,\gamma_M)}{\partial \gamma_\alpha} = \left(g_\alpha(\theta) + R\theta \ln \frac{x_\alpha p}{p_0} \right) + R\theta \sum_{\beta=1}^{M} \gamma_\beta \frac{\partial \ln x_\beta}{\partial \gamma_\alpha}. \quad (4.176)$$

Comparing with the desired result (equation (4.175)), it is clear that we need to prove that

$$\sum_{\beta=1}^{M} \gamma_\beta \frac{\partial \ln x_\beta}{\partial \gamma_\alpha} = 0, \quad (4.177)$$

a result that, at first sight, seems unlikely. But, recalling (4.170),

$$\frac{\partial \ln x_\beta}{\partial \gamma_\alpha} = \frac{\partial (\ln \gamma_\beta - \ln \gamma)}{\partial \gamma_\alpha} = \frac{\delta_{\beta\alpha}}{\gamma_\beta} - \frac{1}{\gamma}, \quad (4.178)$$

where $\delta_{\beta\alpha}$ is the Kronecker symbol (see equation (5.31)). The result (4.177) follows suit.   ∎

Consider a mixture of M ideal gases, S^α ($\alpha = 1,\ldots,M$), undergoing the chemical reaction

$$\sum_{\alpha=1}^{M} \nu_\alpha S^\alpha, \quad (4.179)$$

where the notation and sign convention introduced in equation (4.103) have been used. The chemical affinity is obtained from equation (4.153) as

$$\sigma = -\sum_{\alpha=1}^{M} \mu_\alpha \nu_\alpha = -\sum_{\alpha=1}^{M} \nu_\alpha \left(g_\alpha(\theta) + R\theta \ln \frac{x_\alpha p}{p_0} \right). \quad (4.180)$$

For fixed temperature and pressure, the affinity depends on the molar fractions x_α, which are ultimately controlled by the extent of the reaction ξ. The second law implies, according to equation (4.156), that the reaction progresses in the direction of decreasing free energy. In accordance with Theorem 4.8.16, equilibrium is achieved when the affinity σ vanishes. Equating the affinity to zero, we obtain from equation (4.180) the following important condition of equilibrium:

$$\prod_{\alpha=1}^{M} \left(\frac{x_\alpha p}{p_0}\right)^{\nu_\alpha} = K(\theta). \tag{4.181}$$

The function K, defined as

$$K(\theta) = \exp\left(-\frac{\sum \nu_\alpha g_\alpha}{R\theta}\right), \tag{4.182}$$

depends on temperature alone. For a given temperature, it is known as the *equilibrium constant* of the reaction. It is a dimensionless quantity. Equation (4.181) can be solved numerically for the extent of the reaction at equilibrium.

Remark 4.9.8 Writing the equilibrium condition (4.181) as

$$\left(\frac{p}{p_0}\right)^{\nu} \prod_{\alpha=1}^{M} (x_\alpha)^{\nu_\alpha} = K(\theta), \tag{4.183}$$

where

$$\nu = \sum_{\alpha=1}^{M} \nu_\alpha, \tag{4.184}$$

we observe that, if the total number of moles happens to be preserved (i.e., if $\nu = 0$), the equilibrium condition reduces to the condition

$$\prod_{\alpha=1}^{M} (\gamma_\alpha)^{\nu_\alpha} = K(\theta). \tag{4.185}$$

In the general case ($\nu \neq 0$), the equilibrium constant, although non-dimensional, is scaled by the choice of p_0. In practice, p_0 is always chosen as 1 atm.

Exercise 4.9.9 It is customary to write the equilibrium condition (4.181) in the form

$$\frac{\prod_{\text{products}} \left(\dfrac{x_\alpha p}{p_0}\right)^{\nu_{\text{products}}}}{\prod_{\text{reactants}} \left(\dfrac{x_\alpha p}{p_0}\right)^{|\nu|_{\text{reactants}}}} = K(\theta). \tag{4.186}$$

After making sure that this is a legitimate substitution, interpret the physical relevance of the equilibrium constant in terms of its magnitude. What does a very large value of the equilibrium constant imply?

Just as for the case of enthalpy, the values of the Gibbs free energy of formation of many substances at standard conditions of pressure ($p_0 = 1$ atm) and temperature ($\theta_0 = 298$ K) have been tabulated. For a given chemical reaction (4.179), we can calculate the associated change in Gibbs free energy,

$$\Delta G_0 = \sum_{\alpha=1}^{M} \nu_\alpha \Delta G_0^\alpha, \tag{4.187}$$

where G_0^α is the Gibbs free energy of formation of the substance S_α at the standard conditions p_0 and θ_0. It is important to remark, however, that this ΔG_0 is not the change of Gibbs free energy in the actual mixture, since the partial pressures are necessarily smaller than the mixture pressure (p_0). On the other hand, it follows from equation (4.166) that the tabulated values must be identified with

$$\Delta G_0^\alpha = g_\alpha(\theta). \tag{4.188}$$

Combining this result with (4.188) and (4.182), we obtain the useful result

$$K(\theta) = \exp\left(-\frac{\Delta G_0}{R\theta}\right). \tag{4.189}$$

The implication of this result is that the equilibrium constant at the standard temperature can be directly calculated from tabulated values. For other temperatures, the equilibrium constant can be estimated[21] or measured experimentally.

Exercise 4.9.10   Calculate the equilibrium constant of the following reaction at standard conditions:

$$2CO(g) + 2NO(g) \rightarrow 2CO_2(g) + N_2(g). \tag{4.190}$$

The following values have been obtained from a table of Gibbs free energies of formation at standard conditions:

Substance	ΔG_0 (kJ mol^{-1})
CO (g)	-137.3
NO (g)	86.7
CO$_2$ (g)	-394.4
N$_2$ (g)	0.0

Comment on the result.

4.9.3   Other Ideal Mixtures

Mixtures of ideal gases are amenable, as we have shown in Section 4.9.2, to an elegant treatment due to the availability of explicit constitutive equations and to the additivity property. There are many other systems, such as dilute solutions, for which some or all

---

[21] By the use of the *Van't Hoff equation*.

of the formulas developed for the ideal gas case can be reasonably imported, at least as a first approximation. We call these systems *ideal mixtures*.[22]

4.10 Case Study: Bone as a Chemically Reacting Mixture

Generally speaking, bone consists of *bone cells* and extracellular *matrix*, which constitutes the overwhelming part of the mass of the bone. The main three types of bone cells are *osteoblasts*, *osteocytes*, and *osteoclasts* (see Figure 7.10). The function of the osteoblasts is to manufacture *osteoid*, a protein that later mineralizes and becomes part of the matrix. Osteoblasts reside on the surface of pores. As they mature, they migrate into the bone matrix, are entrapped by it and become osteocytes. Finally, osteoclasts perform the opposite function, namely, they partially destroy the bone matrix in a process known as *bone resorption*. Thus, in normal healthy bone, a state of dynamic equilibrium, known as *homeostasis*, is reached whereby the rate of bone production by the osteoblasts equals the rate of bone resorption by the osteoclasts.

The bone matrix is made of a mixture of inorganic and organic components. The inorganic component, or *bone mineral*, forming about 65% of the bone mass, is mainly a crystalline calcium compound, the repository of almost all of the calcium in the animal body, known as *hydroxyapatite* or *hydroxylapatite*. Its chemical formula is $Ca_{10}(PO_4)_6(OH)_2$.[23] The remainder of the bone matrix is organic material, mainly Type 1 collagen, a protein.

Following Rouhi (2006),[24] we will present a simplified model of the phenomenon of resorption only, a phenomenon that is of importance in the understanding of osteoporosis, a metabolic disease responsible for loss of bone mass and increased susceptibility to fracture (see Figure 7.5). The main reaction involved in the dissolution of hydroxyapatite has been studied in great detail (mainly because of its relevance to the phenomenon of tooth decay). When placed in an aqueous acid environment (an aqueous solution with pH less than 7) the hydroxyapatite reacts with the free positive ions (protons) of hydrogen. A simplified version of the reactions that take place is embodied in the following reaction:

$$Ca_{10}(PO_4)_6(OH)_2 + 2H^+ \rightarrow 10Ca^{2+} + 6(PO_4)^{3-} + 2H_2O. \qquad (4.191)$$

Note that this equation is stoichiometrically balanced from both the mass and the electrical charge points of view. The phosphate ions $(PO_4)^{3-}$ and the calcium ions Ca^{2+} released are dissolved in the water together with the hydrogen ions to form a dilute solution. Experimental measurements[25] on the concentrations of these ions as functions of time are instrumental in obtaining information on the extent of the reaction (4.191) and in formulating constitutive equations for the mixture.

The model considers bone as a biphasic mixture made of a solid phase (the bone matrix) and a fluid phase (the bone fluid). The transfer of mass between these two phases is due exclusively to the chemical reaction (4.191). Because of the time scale involved in the

---

[22] Many biological systems can be treated as combinations of ideal solutions in combination with solid constituents. In chemistry textbooks it is often argued that solid constituents can be disregarded in the calculation of the equilibrium constant since they are not affected appreciably under a change of pressure. It is clear that a sweeping statement of this nature is, more than anything else, a recognition of the limitations of the standard treatment when it comes to solids.

[23] The subscripts in the formula are all even numbers, so that, theoretically, the formula can also be written as $Ca_5(PO_4)_3(OH)$, but the typical crystal unit cell is made of two such units.

[24] See also Rouhi et al. (2007) and Silva and Ulm (2002).

[25] Margolis and Moreno (1992).

resorption phenomenon (hours, days) it will be assumed that the process is isothermal (no heat flux) and non-diffusive. The constitutive equations are of the form

$$\psi = \psi(\theta, F, \xi), \tag{4.192}$$

$$s = s(\theta, F, \xi), \tag{4.193}$$

$$n = n(\theta, F, \xi), \tag{4.194}$$

$$\frac{d\xi}{dt} = \omega(\theta, F, \xi). \tag{4.195}$$

Notice that the absence of diffusion greatly reduces the number of variables. By the standard arguments introduced in Section 3.5 to exploit the constitutive consequences of the Clausius–Duhem inequality, we obtain the relations

$$n = \rho_0 \frac{\partial \psi}{\partial F}, \tag{4.196}$$

$$s = -\frac{\partial \psi}{\partial \theta}, \tag{4.197}$$

and the residual inequality

$$\omega \, \frac{\partial \psi}{\partial \xi} \leq 0. \tag{4.198}$$

By virtue of equations (4.196) and (4.197), we only need to specify the constitutive functions ψ and ω. It is precisely in these two constitutive functions that the chemo-mechanical coupling will find its expression.

We assume the solid phase (namely, the composite of hydroxyapatite and collagen) to be thermoelastic with a known (intrinsic) Helmholtz free-energy function $\bar{\psi}_s(F, \theta)$, where the deformation gradient is measured with respect to some fixed (possibly stress-free) reference configuration for the solid substance. This function is given per unit mass of the solid. The Helmholtz free energy of the solid phase per unit mass of the mixture is, therefore, given by

$$\psi_s = c_s \, \bar{\psi}_s, \tag{4.199}$$

where c_s is the current concentration of the solid.

In attempting to establish a reference configuration for the mixture, however, it is necessary to point out that, as the reaction progresses, the mass of the solid steadily decreases. We will assume, however, that this time-varying reference configuration[26] for the mixture preserves the remaining solid phase in its unaltered state. Concomitantly, the fluid phase will have a time-varying (or, more precisely, a ξ-dependent) reference density. Moreover, we assume that the addition of the dissolved ions to the fluid phase does not affect the mechanical response. Denoting by $\bar{\rho}^0_{H_2O}$ the intrinsic reference density of the fluid (excluding the mass of the ions) at time $t = 0$ (when $\xi = 0$), its value $\bar{\rho}^R_{H_2O}$ corresponding to any subsequent value of the extent of reaction ξ is calculated in the following exercise.

---

[26] A more detailed treatment of problems of this kind will be offered in Chapter 7.

The change in mass of solid per unit mass of the mixture at a value ξ of the extent of the reaction is

$$\Delta m_s = \nu_s \xi M^s = -1004.643\,14\xi. \tag{4.200}$$

The change in the mass of liquid water per unit mass of the mixture is

$$\Delta m_{H_2O} = \nu_{H_2O}\xi M^{H_2O} = 36.030\,68\xi. \tag{4.201}$$

The volume vacated per unit mass of mixture is

$$\Delta V = \frac{|\Delta m_s|}{\bar{\rho}_s} = \frac{1004.643\,14\xi}{\bar{\rho}_s}, \tag{4.202}$$

where $\bar{\rho}_s$ is the (assumed already time-independent) intrinsic reference density of the solid. The new intrinsic density of the water phase in the reference configuration is, therefore,

$$\bar{\rho}^R_{H_2O} = \frac{c^0_{H_2O} + \Delta m_{H_2O}}{\frac{c^0_{H_2O}}{\bar{\rho}^0_{H_2O}} + \Delta V} = \frac{c^0_{H_2O} + 36\xi}{\frac{c^0_{H_2O}}{\bar{\rho}^0_{H_2O}} + \frac{1005\xi}{\bar{\rho}_s}}, \tag{4.203}$$

where $c^0_{H_2O}$ is the initial concentration of water. Assuming ξ to be small, we can approximate this expression by

$$\bar{\rho}^R_{H_2O} = \bar{\rho}^0_{H_2O} \left(1 + \left(36 - 1005\frac{\bar{\rho}^0_{H_2O}}{\bar{\rho}_s}\right)\frac{\xi}{c^0_{H_2O}}\right). \tag{4.204}$$

The relative density of hydroxyapatite being approximately 3.1, we can further approximate this formula to obtain

$$\bar{\rho}^R_{H_2O} = \bar{\rho}^0_{H_2O} \left(1 - 290\frac{\xi}{c^0_{H_2O}}\right). \tag{4.205}$$

A further application of the deformation gradient F (the same for both components of the mixture) renders the spatial fluid density as

$$\bar{\rho}_{H_2O} = \bar{\rho}^R_{H_2O} F^{-1} = \bar{\rho}^0_{H_2O} \left(1 - 290\frac{\xi}{c^0_{H_2O}}\right) F^{-1}. \tag{4.206}$$

It is this value of the fluid density which will naturally enter the constitutive equation of the fluid. The interstitial fluid (essentially water with ions) is assumed to be an inviscid, compressible, thermoelastic liquid with (intrinsic) Helmholtz free-energy function $\bar{\psi}_{H_2O}(\bar{\rho}_{H_2O}, \theta)$, per unit mass of water. Since the liquid phase acts also as a solvent for the various ions, we need to add their contribution to the free energy. Denoting by $f(\theta)_\alpha$ the Helmholtz free energy associated with these solutes at some reference concentrations (e.g., those at time $t = 0$) and assuming an ideal solution behaviour, the Helmholtz free energy of the liquid phase per unit mass of the mixture can, therefore, be written as

$$\psi_f = c_{H_2O}\,\bar{\psi}_{H_2O} + \sum_\alpha (\gamma^0_\alpha + \nu_\alpha \xi)\left(f_\alpha(\theta) + R\theta\,\ln\left(\frac{x_\alpha}{x^0_\alpha}\right)\right), \tag{4.207}$$

where a superscript 0 refers to the reference concentrations and where $\alpha = Ca^{2+}$, $(PO_4)^{3-}, H^+$.

We construct the Helmholtz free-energy function ψ out of the direct addition of the solid and the fluid phases, namely

$$\psi = \psi_s + \psi_f \tag{4.208}$$

$$= c_s \, \bar{\psi}_s + c_{H_2O} \, \bar{\psi}_{H_2O} + \sum_\alpha (\gamma_\alpha^0 + \nu_\alpha \xi) \left(f_\alpha(\theta) + R\theta \, \ln \left(\frac{x_\alpha}{x_\alpha^0} \right) \right).$$

Assuming a dilute solution, the concentration ratios can be written as

$$\frac{x_\alpha}{x_\alpha^0} = \frac{\gamma_\alpha^0 + \nu_\alpha \xi}{\gamma_\alpha^0}, \quad \alpha = Ca^{2+}, (PO_4)^{3-}, H^+, \tag{4.209}$$

where equation (4.114) has been used. Thus, equation (4.208) furnishes the total Helmholtz free energy as a function of F, θ and ξ, as expected.

As far as the evolution function ω is concerned, we observe that it must satisfy the residual inequality (4.198) identically. This inequality can be rewritten as

$$\omega \, \sigma \geq 0, \tag{4.210}$$

where σ is the chemical affinity (per unit mass of the mixture) following equation (4.152). A possible way to satisfy this inequality is to stipulate the evolution function as

$$\omega(F, \theta, \xi) = C\sigma, \tag{4.211}$$

where C is a non-negative constant (or function). Notice that, when calculating the affinity, care should be taken in using the chain rule of differentiation, particularly taking into consideration that the argument of $\bar{\psi}_{H_2O}$ is $\bar{\rho}_{H_2O}$, as given by equation (4.206). It is this detail that will bring into play the fluid pressure together with the energy density in the expression for the chemical affinity. From this vantage point, we are in the presence of an example of a *configurational force* in continuum mechanics.

References

Ateshian, G.A., Warden, W.H., Kim, J.J., Grelsamer, R.P. and Mow, V.C. (1997) Finite deformation bipha-sic material properties of bovine articular cartilage from confined compression experiments. *Journal of Biomechanics* **30**(11–12), 1157–1164.

Bowen, R.M. (1968a) On the stoichiometry of chemically reacting mixtures. *Archive for Rational Mechanics and Analysis* **29**, 114–124.

Bowen, R.M. (1968b) Thermochemistry of reacting materials. *Journal of Chemical Physics* **49**(4), 1625–1637.

Bowen, R.M. (1976) Theory of mixtures. In A.C. Eringen (ed.), *Continuum Physics*, Vol. III, pp. 1–127. New York: Academic Press.

Eisenberg, D. and Crothers, D. (1979) *Physical Chemistry with Applications to the Life Sciences*. Menlo Park, CA: Benjamin/Cummings.

Guggenheim, E.A. (1977) *Thermodynamics: An Advanced Treatment for Chemists and Physicists*. Amsterdam: North-Holland.

Margolis, H.C. and Moreno, E.C. (1992) Kinetics of hydroxyapatite dissolution in acetic, lactic and phosphoric acid solutions. *Calcified Tissue International* **50**(2), 137–143.

Mow, V.C., Kuei, S.C., Lai, W.M. and Armstrong, C.G. (1980) Biphasic creep and stress relaxation of articular cartilage in compression: Theory and experiments. *Journal of Biomechanical Engineering* **102**, 73150-84.

Prigogine, I. and Defay, R. (1950) *Traité de Thermodynamique: Thermodynamique Chimique*. Liège: Desoer. Translated into English (1954) by D. Everett as *Treatise on Themodynamics*. London: Longmans Green.

Rajagopal, K.R. and Tao, L. (1995) *Mechanics of Mixtures*. Singapore: World Scientific.

Rouhi, G.R. (2006) Theoretical aspects of bone remodeling and resorption processes. PhD thesis, University of Calgary, Canada.

Rouhi, G.R., Epstein, M., Sudak, L. and Herzog, W. (2007) Modeling bone resorption using mixture theory with chemical reaction. *Journal of Mechanics of Materials and Structures* **2**(6), 1141–1155.

Silva, E.C.C.M. and Ulm, F.J. (2002) A bio-chemo-mechanics approach to bone resorption and fracture. In B.L. Karihaloo (ed.), *IUTAM Symposium on Analytical and Computational Fracture Mechanics of Non-homogeneous Materials*, pp. 355–366. Dordrecht: Kluwer.

Truesdell, C. and Toupin, R. (1960) *The Classical Field Theories*. *Handbuch der Physik* (S. Flügge, ed.), Vol. III/1, pp. 226–858. Berlin: Springer-Verlag.

Woledge, R.C., Curtin, N.A. and Homsher, E. (1985) *Energetic Aspects of Muscle Contraction*. London: Academic Press.

Part II

Towards Three Spatial Dimensions

5

Geometry and Kinematics

5.1 Introduction

Having by now become familiar with the scope and aims of continuum mechanics, intentionally restricted so far to the one-dimensional context, the passage to the conceptual framework of the more realistic three-dimensional formulation should be natural and almost self-evident. Some of the technicalities involved, however, are more than mere generalizations. In one spatial dimension, the notion of *rotation*, for example, is completely absent and, therefore, the three-dimensional formulation must tackle the problem of how to factor out rigid-body rotations from the process of deformation. Another essential difference is brought about by the fact that the boundary of a one-dimensional domain consists of a disjoint collection (of two boundary points), whereas in two or more spatial dimensions the boundary is itself a continuous domain, a fact that has repercussions on the formulation of the equations of balance and on the methods of solution of the governing partial differential equations. Finally, among other important differences, it is worthwhile mentioning the notion of *material symmetries*, such as *isotropy*, which in a one-dimensional formulation can be almost completely ignored.

5.2 Vectors and Tensors

5.2.1 Why Linear Algebra?

It is not our intention to undertake a thorough review of vectors and tensors, the subject of *linear algebra*, but only to offer a list of a few basic notions. Linear algebra deals with vector spaces and linear transformations between them. At the outset, one may wonder why one needs to worry about *linear* transformations when the very essence of modern continuum mechanics is *non-linearity*. Indeed, lack of proportionality between causes and effects is evident in many aspects of continuum mechanics in general, and in many applications pertaining to biological materials in particular (see Window 5.1).

The Elements of Continuum Biomechanics, First Edition. Marcelo Epstein.
© 2012 John Wiley & Sons, Ltd. Published 2012 by John Wiley & Sons, Ltd.

Window 5.1   Sources of non-linearity

Typically, there are two types of non-linearity in the mechanical response of structural elements. The first kind, known as *material non-linearity*, stems from the lack of proportionality that many materials exhibit between forces and elongations at the local level. If we take a small length of tendon and subject it to a tensile force, doubling or tripling the force will in general result in less than double or triple the elongation. The second kind, known as *geometrical non-linearity*, is more subtle. Assume that you pluck a string of a guitar or a lute by applying a transverse force with the finger. Regardless of whether or not the material of the string exhibits material non-linearity, the relation between the transverse force applied and the resulting transverse deflection will be non-linear (i.e., doubling the force will not result in doubling the deflection). This somewhat unexpected result becomes clearer when one considers that the equilibrium of forces must be established not on the original, straight, configuration of the string, but rather on the unknown deformed configuration, itself depending on the force applied! In many cases, the geometrical difference between these two configurations can be neglected, but not if large deformations are at play. In our example, a moment's reflection reveals that it would be impossible to balance the applied transverse force by means of two collinear forces at right angles with it and, as a consequence, relatively large deflections need to take place so that adequately large angles develop. Another everyday example of a geometrical non-linearity is the bending of a very flexible bar or plate, such as a ruler or a sheet of paper.[a]

Exercise 5.2.1   Plucking a string. A string of length L is fixed at both ends under an initial tension T_0. If a transverse force P is applied at the midpoint, calculate and plot the deflection under the force as a function of the force applied. Assume the material to be linearly elastic, the coefficient of proportionality between tension and elongation being given. Consider, moreover, two extreme cases: (a) The initial tension vanishes. Note that in this case a linear approximation does not even exist. (b) The initial tension is very large. In this case one may neglect small changes in tension and assume that it remains constant. Plot the force–deflection curve for both cases and comment on the results.

_____

[a]The non-linear effect in the pitch of a lute string was probably first observed by Vincenzo Galilei (1520–1591), father of Galileo. The post-buckling shapes of a slender bar were first quantified by Leonhard Euler (1707–1783).

So, why study linear algebra when doing continuum mechanics? One of the answers to this question is rooted in the concept of *vector*. Galileo Galilei observed that the motion of a particle under constant gravity could be obtained by adding, according to the parallelogram rule, the horizontal and vertical velocities.[1] Later, vectors were used

_____

[1] This result can, of course, be attributed to the linearity of Newton's second law of motion. The usefulness of vectors in physics is not limited to mechanics. It was, in fact, Maxwell's definitive formulation of electromagnetism that launched the new era of vector algebra and analysis. Maxwell made use of the *quaternion* concept, invented by Hamilton. The pioneering works of Tait, Gibbs and Heaviside established definitively the vectorial approach.

systematically as convenient representations of physical quantities. Their decomposition and addition, moreover, turned out to have meaning in physical contexts.

A second reason for the importance of linear algebra stems from the basic notion of differential calculus, namely, the *differential* of a function, which is nothing but a linear approximation to a function around a given point. The differential is, therefore, a *linear operator* that maps vectors in the space of independent variables into vectors in the space of dependent variables. Most physical laws are formulated *locally* in terms of these differentials. The curved trajectory of a planet, for example, is the result of putting together (i.e., *integrating*) an infinite number of infinitesimally small straight pieces. The concept of the deformation gradient, which we have already encountered in previous chapters and which we will develop further in this, is part of a similar conceptual structure.

Finally, an important part of the machinery of linear algebra is the concept of *duality*. Given a space of vectors, another space emerges automatically from the shadows. This so-called *dual space*, consisting of linear functions over the original space, is invariably the carrier of significance. Thus, forces are precisely the dwellers of the dual space of the space of displacements (or velocities), over which they perform work (or power) in a linear fashion.

Another interesting feature of linear algebra is its power to replicate linear structures and to create new ones from old by means of a process that can be metaphorically regarded as a mathematical mitosis. In addition to the notion of dual space, we can consider linear maps between vector spaces. These maps, or *tensors*, constitute themselves a vector space whose dual can be defined. With these new vector spaces, we can again consider linear maps and continue this iterative process *ad infinitum*.

5.2.2 Vector Spaces

We will deal with *real vector spaces* only, namely, vector spaces that presuppose the existence of the set $\mathbb{R}$ of real numbers (or *scalars*), which will play a role in the definition. A real vector space consists of a set V (whose elements are called *vectors*) endowed with two operations: *vector addition* and *multiplication of a vector by a scalar*. The results of these operations are vectors. They enjoy the following properties.

1. *Associativity of vector addition*:

$$(\mathbf{u} + \mathbf{v}) + \mathbf{w} = \mathbf{u} + (\mathbf{v} + \mathbf{w}), \quad \text{for all } \mathbf{u}, \mathbf{v}, \mathbf{w} \in V. \tag{5.1}$$

2. *Existence of a zero vector*:

$$\text{There exists } \mathbf{0} \in V \text{ such that } \mathbf{0} + \mathbf{v} = \mathbf{v} + \mathbf{0} = \mathbf{v}, \quad \text{for all } \mathbf{v} \in V. \tag{5.2}$$

3. *Existence of an inverse*:

$$\text{For each } \mathbf{v} \in V, \text{ there exists } (-\mathbf{v}) \in V \text{ such that } \mathbf{v} + (-\mathbf{v}) = (-\mathbf{v}) + \mathbf{v} = \mathbf{0}. \tag{5.3}$$

4. *Commutativity of the addition*:

$$\mathbf{u} + \mathbf{v} = \mathbf{v} + \mathbf{u}, \quad \text{for all } \mathbf{u}, \mathbf{v} \in V. \tag{5.4}$$

5. *Distributivity to vector addition*:

$$\alpha(\mathbf{u} + \mathbf{v}) = \alpha\mathbf{u} + \alpha\mathbf{v}, \quad \text{for all } \alpha \in \mathbb{R}, \ \mathbf{u}, \mathbf{v} \in V. \tag{5.5}$$

6. *Distributivity to scalar addition*:

$$(\alpha + \beta)\mathbf{u} = \alpha\mathbf{u} + \beta\mathbf{u}, \quad \text{for all } \alpha, \beta \in \mathbb{R}, \ \mathbf{u} \in V. \tag{5.6}$$

7. *Associativity to multiplication of scalars*:

$$(\alpha\beta)\mathbf{u} = \alpha(\beta\mathbf{u}), \quad \text{for all } \alpha, \beta \in \mathbb{R}, \ \mathbf{u} \in V. \tag{5.7}$$

8. *Multiplicative unit consistency*:

$$1\,\mathbf{u} = \mathbf{u}, \quad \text{for all } \mathbf{u} \in V. \tag{5.8}$$

Note that vector addition is denoted with the usual symbol '+' and multiplication by a scalar with simple apposition. This abuse of notation is justified by the properties themselves, which resemble those of the real numbers, so that one can confidently operate on vectors (open brackets, etc.) as if they were numbers.

Exercise 5.2.2 Some derived properties. Show that: (a) the zero vector is unique; (b) the inverse of any given vector is unique; (c) $0\mathbf{u} = \mathbf{0}$, for all $\mathbf{u} \in V$; (d) $(-1)\mathbf{u} = (-\mathbf{u})$, for all $\mathbf{u} \in V$. Make sure that you use just the eight properties above and that you are not carried away by the notation. In fact, this exercise amply justifies this being carried away from now on. In particular, it justifies the 'subtraction' of vectors: $\mathbf{u} - \mathbf{v} := \mathbf{u} + (-\mathbf{v})$.

Window 5.2 Examples of vector spaces

The standard example, and the most common in continuum mechanics, is the space of oriented segments issuing from a point in space. Vector addition is carried out by the parallelogram rule. Multiplication by a scalar magnifies the segment accordingly, turning it around if the scalar is negative. Other examples are as follows:

1. *Real polynomials of degree n*. Consider the collection of polynomials of the form $P(x) = a_0 + a_1 x + a_2 x^2 + \ldots + a_n x^n$, where $a_0, \ldots, a_n$ are real constants and x is a real variable. Defining the addition of two such polynomials as the polynomial resulting from adding the corresponding coefficients (of equal powers of x), and multiplication of a polynomial by a scalar as the polynomial resulting from multiplying each coefficient by the given scalar, all eight properties of our definition are satisfied. The importance of this example resides in the fact that, unlike the standard one, no concept of length or angle is used. An equivalent example is the collection of all ordered $(n + 1)$-tuples of real numbers with the appropriate operations. Thus, $\mathbb{R}^n$ has a natural vector-space structure.

2. *Continuous functions*. Given an (open or closed) interval of the real line, we look at the collection of all continuous real functions defined over this interval. By pointwise addition and pointwise multiplication by a scalar, it is a straightforward matter to verify that this collection forms a vector space. Again, as in the case of the polynomials, the vector-space structure is independent of any metric notion. We will soon see that this example differs from the previous one in a fundamental respect.

5.2.3 *Linear Independence and Dimension*

Linear Combinations

A *linear combination* of a finite subset $\{\mathbf{v}_1, \mathbf{v}_2, \ldots, \mathbf{v}_k\} \subset V$ is an expression of the form:

$$\sum_{i=1}^{k} \alpha_i \mathbf{v}_k = \alpha_1 \mathbf{v}_1 + \ldots + \alpha_k \mathbf{v}_k, \tag{5.9}$$

where the real numbers α_i are called the *coefficients* of the linear combination. When evaluated with given coefficients, a linear combination produces a vector. By the properties of vector operations, a linear combination will always vanish if each of its coefficients is zero. Such a linear combination (namely, one all of whose coefficients are zero), is called *trivial*. On the other hand, a linear combination may vanish even if some of its coefficients are not zero. For instance, if the zero vector is a member of our subset, then we can simply choose *its* coefficient as, say, 1, and the remaining coefficients as 0. By the properties of the vector operations, we will thus obtain a vanishing non-trivial linear combination.

Linear Independence, Bases

A finite subset $\{\mathbf{v}_1, \mathbf{v}_2, \ldots, \mathbf{v}_k\} \subset V$ is said to be *linearly independent* (equivalently, the members of this subset are said to be mutually linearly independent) if there exists no vanishing non-trivial linear combination of the given vectors. Put in other words, the subset is linearly independent if, and only if, the only zero linear combination is the trivial one (namely, the one with all coefficients equal to zero). A subset that is not linearly independent is said to be *linearly dependent*. A linearly independent subset is said to be *maximal* if the addition of *any* (arbitrary) vector to the subset renders it linearly dependent. There is no *a priori* reason for the existence of maximal linearly independent subsets.

A maximal linearly independent subset (if it exists) is called a *basis* of the vector space. The reason for this terminology is contained in the following theorem:

Theorem 5.2.3 *Every vector can be expressed uniquely as a linear combination of the elements of a basis.*

Proof. Let $\{\mathbf{e}_1, \ldots, \mathbf{e}_k\}$ be a basis and let $\mathbf{v} \in V$. Since the basis is, by definition, maximal, the augmented subset $\{\mathbf{v}, \mathbf{e}_1, \ldots, \mathbf{e}_k\}$ must be linearly dependent. This means that there exists a non-trivial linear combination that vanishes. Moreover, in this non-trivial linear combination the coefficient of $\mathbf{v}$ must be different from zero. Otherwise, the linear combination would be trivial, since the elements in the basis are already linearly independent. Let, therefore, the vanishing non-trivial linear combination be given by

$$\beta \mathbf{v} + \alpha_1 \mathbf{e}_1 + \ldots + \alpha_k \mathbf{e}_k = \mathbf{0}, \tag{5.10}$$

with $\beta \neq 0$. Dividing throughout by β and using the algebraic properties of the operations, we can therefore write

$$\mathbf{v} = -(\alpha_1/\beta)\mathbf{e}_1 - \ldots - (\alpha_k/\beta)\mathbf{e}_k, \tag{5.11}$$

which proves that $\mathbf{v}$ is a linear combination of the basis. The proof of uniqueness is left as an exercise. ∎

Exercise 5.2.4 Uniqueness. Complete the proof of the theorem by showing that the linear combination just obtained is unique. [Hint: assume that there exists a different one, subtract and invoke linear independence].

The unique coefficients of the linear combination just obtained are called the *components* of $\mathbf{v}$ in the given basis. They are denoted by v^i, $i = 1, \ldots, k$, and we write

$$\mathbf{v} = v^1 \mathbf{e}_1 + \ldots + v^n \mathbf{e}_n = \sum_{i=1}^{n} v^i \mathbf{e}_i. \tag{5.12}$$

An important corollary of this theorem is that all bases of a vector space (if they exist) have the same number of elements. This common number is called the *dimension* of the vector space. If there are no (finite) maximal linearly independent subsets, the vector space is said to be of *infinite* dimension.

Exercise 5.2.5 Examples. Determine the dimension of each of the vector spaces in Window 5.2.

Exercise 5.2.6 Show that if every vector is expressible as a linear combination of a fixed finite set of linearly independent vectors, then these vectors form a basis of the vector space.

Subspaces

A *subspace* S of a vector space V is a subset $S \subset V$ such that every linear combination of elements of S is also in S. In other words, a subspace is closed under the two operations defined in the vector space V. It is a trivial exercise to show that a subspace S is itself a vector space, hence its name.

Given any subset U of a vector space V, we define the *span* of U as the subset $sp(U)$ of all linear combinations of elements of U. Clearly, $sp(U) \subset V$ is a subspace of V.

Exercise 5.2.7 Show that if the subset of departure $U \subset V$ consists of k linearly independent vectors, then the dimension of $sp(U)$ is k.

5.2.4 Linear Operators, Tensors and Matrices

A *linear operator* $\mathbf{T}$ between two vector spaces U and V is a map $\mathbf{T} : U \longrightarrow V$ that respects the vector-space structure. More precisely,

$$\mathbf{T}(\alpha \mathbf{u}_1 + \beta \mathbf{u}_2) = \alpha \mathbf{T}(\mathbf{u}_1) + \beta \mathbf{T}(\mathbf{u}_2), \quad \text{for all } \alpha, \beta \in \mathbb{R}, \ \mathbf{u}_1, \mathbf{u}_2 \in U, \tag{5.13}$$

where the operations are understood in the corresponding vector spaces. When the source and target vector spaces coincide, the linear operator is called a *tensor*. Occasionally, the terminology of *two-point tensor* (or just tensor) is also used for the general case, particularly when the dimension of both spaces is the same. We will use these terms (linear operator, linear map, tensor, and so on) liberally. The action $\mathbf{T}(\mathbf{u})$ of a tensor $\mathbf{T}$ on a vector $\mathbf{u}$ is denoted also as $\mathbf{T}\mathbf{u}$.

Remark 5.2.8 **The principle of superposition.** If we regard the source and target vector spaces of a linear operator as representing, respectively, 'causes' and 'effects', while the operator represents the 'law' that connects them, then the linearity of the operator can be interpreted as a manifestation of the 'principle of superposition'. This states that the combined effect of two causes acting simultaneously is equal to the sum of the effects produced by the individual causes acting separately. This 'principle' is not a general law of nature, but is often invoked as a simplifying assumption on more or less legitimate physical grounds. Consider, for instance, the space of voltages that can be applied to a conducting wire at a given temperature to produce electric current. Ohm's law asserts that the current is proportional to the applied voltage. In a three-dimensional setting, Ohm's law relates the current density with the electric field, both vectorial quantities, by means of a resistivity tensor, operating on the former to produce the latter. This law, of course, is applicable only to certain materials and certain ranges of currents, much in the same way as Hooke's law of elasticity and Fourier's law of heat conduction.

We want to see how a linear operator looks when we use components. Let U and V be of finite dimensions m and n, respectively, and let $\{\mathbf{e}_\alpha\}$, $\alpha = 1, \ldots, m$, and $\{\mathbf{f}_i\}$, $i = 1, \ldots, n$, be the respective bases. Given $\mathbf{u} \in U$, we apply the linear operator $\mathbf{T}$ and obtain a vector $\mathbf{v} \in V$ as

$$\mathbf{v} = \mathbf{T}\,\mathbf{u} = \mathbf{T}\left(\sum_{\alpha=1}^{n} u^\alpha \mathbf{e}_\alpha\right) = \sum_{\alpha=1}^{n} u^\alpha\left(\mathbf{T}\,\mathbf{e}_\alpha\right), \tag{5.14}$$

where we have used equations (5.12) and (5.13). Now, each vector $\mathbf{T}\,\mathbf{e}_\alpha$ is a vector in V and, consequently, it can be expressed uniquely in terms of the chosen basis therein. Namely, there exist n real numbers, which we denote as $T_\alpha{}^i$, $i = 1, \ldots, n$, such that

$$\mathbf{T}\,\mathbf{e}_\alpha = \sum_{i=1}^{n} T_\alpha{}^i \mathbf{e}_i. \tag{5.15}$$

These $m \times n$ numbers are defined once and for all and depend only on the operator $\mathbf{T}$ and the bases chosen to represent it. We call these numbers the *components* of $\mathbf{T}$ in those bases. If we choose to express the vectors $\mathbf{u}$ and $\mathbf{v}$ as the columns of their respective ordered components, and if we place the components of the linear operator $\mathbf{T}$ in the form of a matrix, the lower index (in this case) indicating the rows and the upper index the columns, equation (5.15) can be written as

$$\begin{Bmatrix} v^1 \\ v^2 \\ \cdot \\ \cdot \\ v^n \end{Bmatrix} = \begin{bmatrix} T_1{}^1 & T_1{}^2 & \cdots & T_1{}^m \\ T_2{}^1 & T_2{}^2 & \cdots & T_2{}^m \\ \cdot & \cdot & \cdots & \cdot \\ \cdot & \cdot & \cdots & \cdot \\ T_n{}^1 & T_n{}^2 & \cdots & T_n{}^m \end{bmatrix} \begin{Bmatrix} u^1 \\ u^2 \\ \cdot \\ \cdot \\ u^m \end{Bmatrix}. \tag{5.16}$$

More compactly,

$$\{\mathbf{v}\} = [\mathbf{T}]\,\{\mathbf{u}\}. \tag{5.17}$$

The 'mysterious' origin of the rule for matrix multiplication is thus unveiled.

Remark 5.2.9   Tensors and matrices. The notion of *tensor* should not be confused with that of *matrix*. The latter is the representation of the former in a particular choice of bases. Notice that not every matrix is the representation of a tensor. Consider, for example, a train schedule or a table of distances between cities.

Exercise 5.2.10   Map composition and matrix multiplication. The composition $\mathbf{Z} \circ \mathbf{T}$ of the linear maps $\mathbf{Z}$ and $\mathbf{T}$ is usually indicated simply as $\mathbf{ZT}$. If U, V and W are vector spaces of dimensions m, n and p, respectively, and if $\mathbf{T} : U \to V$ and $\mathbf{Z} : V \to W$ are linear maps, show that the composition $\mathbf{ZT} : U \to W$ is a linear map. Choosing bases in each vector space, show that the component expression for $\mathbf{ZT}$ is precisely the matrix product of the corresponding matrices.

Isomorphisms

A bijective linear map $\mathbf{T} : U \to V$ is called an *isomorphism* between the vector spaces U and V. The spaces U and V are said to be *isomorphic* if there exists an isomorphism between them.

Exercise 5.2.11   Show that all vector spaces of the same finite dimension are mutually isomorphic. [Hint: choose a basis in each space and consider the linear function that maps basis to basis.]

Tensors as Vectors

It should not come as a surprise that we can consider the collection $L(U, V)$ of *all* linear maps between two given vector spaces, and endow it with the natural structure of a vector space. To do so, we define the sum of two linear operators $\mathbf{S}$ and $\mathbf{T}$ as the linear operator $\mathbf{S} + \mathbf{T}$ whose action on an arbitrary vector $\mathbf{u} \in U$ is given by

$$(\mathbf{S} + \mathbf{T})(\mathbf{u}) := \mathbf{S}(\mathbf{u}) + \mathbf{T}(\mathbf{u}). \tag{5.18}$$

Similarly, we define the product of a scalar α by a linear operator $\mathbf{T}$ as the linear operator $\alpha\mathbf{T}$ given by

$$(\alpha\mathbf{T})(\mathbf{u}) := \alpha\mathbf{T}(\mathbf{u}). \tag{5.19}$$

It is a straightforward matter to verify that the set $L(U, V)$, with these two operations, is a vector space. Such is the power of linearity. If the dimensions of U and V are m and n, the dimension of $L(U, V)$ is $m \times n$.

The Dual Space

A particular case of the above construction occurs when the target vector space is chosen as the one-dimensional vector space $\mathbb{R}$. In that case, the collection of all such real-valued linear operators is called the *dual space* of the source space U, and is denoted by U^*. Its dimension is the same as the dimension of the source. The elements of U^* are sometimes called *covectors*. Since a covector is a map, it can be evaluated on the vectors of the original vector space U. If these vectors represent velocities, the covectors represent forces. The evaluation of a force on a velocity has the physical meaning of *power*.

Example 5.2.12   The dual basis. A trivial example of a covector is the zero operator $\mathbf{0} : V \rightarrow \mathbb{R}$. It assigns to each vector of V the number zero. This operator is clearly linear. A constructive way to obtain non-trivial covectors is the following. Let $\mathbf{e}_1, \ldots, \mathbf{e}_m$ be a basis of V. For each $\alpha = 1, \ldots, m$ we define a real-valued operator $\mathbf{e}^\alpha$ as

$$\mathbf{e}^\alpha : V \rightarrow \mathbb{R}$$

$$\mathbf{v} \mapsto v^\alpha. \tag{5.20}$$

In other words, the operator $\mathbf{e}^\alpha$ assigns to each vector $\mathbf{v} \in V$ its component number α in the given basis. It is easy to verify that this operator is actually linear. We conclude, therefore, that $\mathbf{e}^\alpha \in V^*$, according to the very definition of the dual space V^*. It can be shown, moreover, that the m covectors $\mathbf{e}^\alpha$ constitute a basis of the dual space V^*. This basis is known as the *dual basis* of the original basis $\mathbf{e}_\alpha$ of V.

5.2.5   Inner-product Spaces

We have come a long way without the need to speak about metric concepts, such as the length of a vector or the angle between two vectors. That even the concept of power of a force can be introduced without any metric background may seem somewhat surprising, particularly when we are accustomed to hear about 'the magnitude of the force multiplied by the magnitude of the velocity and by the cosine of the angle they form'. To provide this kind of interpretation, we must somehow make the elements of the dual space (forces, say) cohabit with those of the original space (velocities). This identification, however, necessitates the artificial introduction of further structure into the vector space.[2] One way to achieve this objective is the introduction of a new operation called an *inner product* (or a *dot product* or, less felicitously, a *scalar product*).

A vector space V is said to be an *inner-product space* if it is endowed with a real-valued operation,

$$\cdot : V \times V \rightarrow \mathbb{R}$$

$$(\mathbf{u}, \mathbf{v}) \mapsto \mathbf{u} \cdot \mathbf{v}, \tag{5.21}$$

called an *inner product*, satisfying the following properties.

1. *Commutativity*:

$$\mathbf{u} \cdot \mathbf{v} = \mathbf{v} \cdot \mathbf{u}, \quad \text{for all } \mathbf{u}, \mathbf{v} \in V. \tag{5.22}$$

2. *Bilinearity*:[3]

$$(\alpha \mathbf{u}_1 + \beta \mathbf{u}_2) \cdot \mathbf{v} = \alpha(\mathbf{u}_1 \cdot \mathbf{v}) + \beta(\mathbf{u}_2 \cdot \mathbf{v}), \quad \text{for all } \alpha, \beta \in \mathbb{R}, \ \mathbf{u}_1, \mathbf{u}_2, \mathbf{v} \in V. \tag{5.23}$$

---

[2] It is often the case in applications to particular fields (mechanics, theoretical physics, chemistry, engineering, and so on) that there is much more structure to go around than really needed to formulate the basic concepts.

[3] The term 'bilinearity' refers to the fact that the inner product is linear in each of its two arguments. Nevertheless, given that we have already assumed commutativity, we only need to show linearity with respect to one of the arguments.

3. *Positive definiteness*:

$$\mathbf{v} \neq \mathbf{0} \implies \mathbf{v} \cdot \mathbf{v} > 0. \tag{5.24}$$

Exercise 5.2.13 Show that $\mathbf{0} \cdot \mathbf{v} = 0$, for all $\mathbf{v}$.

The *magnitude* or *length* of a vector $\mathbf{v}$ is defined as the non-negative number $\sqrt{\mathbf{v} \cdot \mathbf{v}}$. Two vectors $\mathbf{u}, \mathbf{v} \in V$ are called *orthogonal* (or *perpendicular*) to each other if $\mathbf{u} \cdot \mathbf{v} = 0$.

Exercise 5.2.14 Show that the zero vector is perpendicular to all vectors, and that it is the only vector with this property.

An inner product in a finite-dimensional vector space V induces an isomorphism between V and its dual V^*. This is a *natural isomorphism* in the sense that, once the existence of an inner product has been declared, the isomorphism between V and V^* can be established independently of any basis. To see that this is the case, let $\mathbf{v} \in V$ be a fixed element of V. By the linearity of the inner product, the product $\mathbf{v} \cdot \mathbf{u}$ is linear in the second argument. This means that to each vector $\mathbf{v} \in V$ we can assign, via the dot product, a linear operator or, more concisely, a covector. Following this line of thought, it can be rigorously proved that in fact *every* linear function on V (namely, every element of V^*) can be represented uniquely in this way. We conclude that in an inner product space there is no need to distinguish between vectors and covectors.

5.2.6 The Reciprocal Basis

Let $\mathbf{e}_i$, $i = 1, \ldots, n$, be a basis of the inner-product space V. We form the inner products

$$g_{ij} = \mathbf{e}_i \cdot \mathbf{e}_j. \tag{5.25}$$

These n^2 numbers form a matrix called the *metric matrix* of V relative to the chosen basis. By construction, the metric matrix is symmetric.

Exercise 5.2.15 The inner product in components. Show that the inner product of two vectors is expressible as

$$\mathbf{u} \cdot \mathbf{v} = \sum_{i=1}^{n} \sum_{j=1}^{n} g_{ij} u^i v^j. \tag{5.26}$$

In matrix notation,

$$\mathbf{u} \cdot \mathbf{v} = \{\mathbf{u}\}^T [\mathbf{g}] \{\mathbf{v}\}, \tag{5.27}$$

where T denotes the transpose of a matrix.

Since the vectors of a basis are, by definition, linearly independent, the metric matrix can be shown to be always invertible. The elements of its inverse are denoted by g^{ij}. If we define the n vectors

$$\mathbf{e}^i = \sum_{j=1}^{n} g^{ij} \mathbf{e}_j, \tag{5.28}$$

bases. Show that: (a) the metric matrix of an
rix; (b) the reciprocal of an orthonormal basis coincides
orthonormal basis, covariant and contravariant compo-
an orthonormal basis, the inner product of two vectors is

$$\mathbf{u} \cdot \mathbf{v} = \{\mathbf{u}\}^T \{\mathbf{v}\}, \tag{5.36}$$

$$\mathbf{u} \cdot \mathbf{v} = \sum_{i=1}^{n} u^i v^i. \tag{5.37}$$

ator $\mathbf{T} : U \longrightarrow V$ between two inner-product spaces is
$/$ defined by property

$$\mathbf{u} \cdot \mathbf{T}^T(\mathbf{v}), \quad \text{for all } \mathbf{u} \in U, \ \mathbf{v} \in V. \tag{5.38}$$

ition to define a symmetric tensor. Let $\mathbf{T} : U \longrightarrow U$ be a
ict space into itself. If $\mathbf{T}^T = \mathbf{T}$ the operator is called *sym-*
alled *skew-symmetric* or *antisymmetric*. Not surprisingly,
rices representing symmetric (skew-symmetric) operators
).

an arbitrary linear operator between inner-product spaces
$\mathbf{T}$ and $\mathbf{TT}^T$ are symmetric. On which space does each of

ery tensor $\mathbf{T} : U \to U$ can be *uniquely* decomposed as the
w-symmetric tensor. [Hint: consider the tensors $\mathbf{T} \pm \mathbf{T}^T$.]

ical Space-time

t, all one needs to know about classical space-time is that
the real line and that space admits systems of Cartesian
f an *observer* as a materialization of this concept, namely, a
i rigid orthogonal frame made up of three identical measuring
ers will measure different coordinates for the same physical
 servers are in general in motion with respect to each other.
le column vectors of coordinates recorded by these observers
the relation between them is given by the formula

$$\{\mathbf{x}'\} = \{\mathbf{c}(t)\} + [\mathbf{Q}(t)]\{\mathbf{x}\}, \tag{5.39}$$

are a time-dependent vector and a time-dependent *orthogonal*
vector $\{\mathbf{c}(t)\}$ represents the relative translation of the origins of
while the orthogonal matrix $[\mathbf{Q}(t)]$ measures their relative rotation
a reflection about a plane). Recall that a matrix is *orthogonal* if its

it can be shown that they are linearly independent. Theref(
of V called the *reciprocal basis* of the original.

Exercise 5.2.16 Prove that the vectors e^i are indeed
moreover, that

$$e^i \cdot e^j = g^{ij}$$

and

$$e^i \cdot e_j = \delta^i_j.$$

where δ^i_j is the *Kronecker symbol*,

$$\delta^i_j = \begin{cases} 1 & \text{if } i = j \\ 0 & \text{if } i \neq j. \end{cases}$$

Given a vector $v \in V$, its components v^i in a basi
contravariant components, while its components in the re
are called *covariant components* of v, and are indicated w

$$v = \sum_{i=1}^n v^i e_i = \sum_{i=1}^n v_i e^i.$$

Exercise 5.2.17 Covariant and contravariant compon

$$v_i = \sum_{j=1}^n g_{ij} v^j,$$

$$v^i = v \cdot e^i$$

and

$$v_i = v \cdot e_i.$$

Exercise 5.2.18 A numerical example. Consider the spa
$e_1 = (1,0)$, $e_2 = (0,1)$. Show that the vectors $g_1 = 2e_1$
new (covariant) basis of $\mathbb{R}^2$. Draw these vectors in the
inner product (dot product), construct the reciprocal (contrav
this basis, clearly indicating any orthogonality with the c
vector $v = 3e_1 + 3e_2$, find its contravariant and covariant co
respect to the new basis. Do this by using the parallelogram
equations (5.34) and (5.35) hold true.

A basis of an inner-product space is called *orthonormal* if a
length and mutually orthogonal.

---

[4] This terminology arises from the way the components behave with a change of
discussing at this point. The original base vectors are called covariant and the reci
contravariant for the same reason. Covariant (contravariant) quantities are indicated w

columns are mutually orthogonal unit vectors. A reader who is satisfied by this treatment may skip the remainder of Section 5.3. The following subsections, on the other hand, contain a more detailed investigation of the geometric underpinnings of classical space-time.

5.3.2 $\mathbb{R}^3$ as a Vector Space

The set $\mathbb{R}^3$ consists of all the *ordered triples* of real numbers. A typical element $\{\mathbf{v}\}$ of $\mathbb{R}^3$ can be represented as a column

$$\{\mathbf{v}\} = \left\{ \begin{array}{c} v^1 \\ v^2 \\ v^3 \end{array} \right\}, \tag{5.40}$$

where we have used superscripts to denote the ordered entries. Any element $\{\mathbf{v}\}$ of $\mathbb{R}^3$ can be multiplied by a real number α to produce another element $\alpha\{\mathbf{v}\}$ of $\mathbb{R}^3$ defined as

$$\alpha\{\mathbf{v}\} = \left\{ \begin{array}{c} \alpha v^1 \\ \alpha v^2 \\ \alpha v^3 \end{array} \right\}. \tag{5.41}$$

Moreover, given two elements, $\{\mathbf{v}\}$ and $\{\mathbf{w}\}$, of $\mathbb{R}^3$, their *sum*, $\{\mathbf{v}\} + \{\mathbf{w}\}$, is the element of $\mathbb{R}^3$ defined as

$$\{\mathbf{v}\} + \{\mathbf{w}\} = \left\{ \begin{array}{c} v^1 + w^1 \\ v^2 + w^2 \\ v^3 + w^3 \end{array} \right\}. \tag{5.42}$$

With these two operations (*multiplication by a scalar* and *vector addition*), the set $\mathbb{R}^3$ acquires the structure of a *vector space*. Its elements are called *vectors* in $\mathbb{R}^3$. The element of $\mathbb{R}^3$ all of whose entries are zero is the *zero vector* of $\mathbb{R}^3$.

The three special vectors

$$\{\mathbf{e}_1\} = \left\{ \begin{array}{c} 1 \\ 0 \\ 0 \end{array} \right\}, \quad \{\mathbf{e}_2\} = \left\{ \begin{array}{c} 0 \\ 1 \\ 0 \end{array} \right\}, \quad \{\mathbf{e}_3\} = \left\{ \begin{array}{c} 0 \\ 0 \\ 1 \end{array} \right\} \tag{5.43}$$

constitute the *natural basis* of $\mathbb{R}^3$. Clearly, every vector $\{\mathbf{v}\}$ is uniquely expressible as the linear combination

$$\{\mathbf{v}\} = \sum_{i=1}^{3} v^i \{\mathbf{e}_i\}. \tag{5.44}$$

Thus, the entries of $\{\mathbf{v}\}$, as the ordered triple introduced in equation (5.40), can be also regarded as the *components* of the vector $\{\mathbf{v}\}$ in the natural basis.

In the vector space $\mathbb{R}^3$, there exists a *natural inner product* given, in matrix notation, by

$$\{\mathbf{v}\} \cdot \{\mathbf{w}\} = \{\mathbf{v}\}^T \{\mathbf{w}\} = \{\mathbf{w}\}^T \{\mathbf{v}\}, \tag{5.45}$$

where T indicates the transpose operator, which turns columns into rows. The inner product generates a real number out of a pair of vectors. The *length* $|\{\mathbf{v}\}|$ of the vector {**v**} is defined as

$$|\{\mathbf{v}\}| = \sqrt{\{\mathbf{v}\} \cdot \{\mathbf{v}\}} = \sqrt{\{\mathbf{v}\}^T \{\mathbf{v}\}}. \tag{5.46}$$

Two vectors are said to be *orthogonal* if their inner product vanishes. The natural base vectors ($\{\mathbf{e}_1\}, \{\mathbf{e}_2\}, \{\mathbf{e}_3\}$) form an *orthonormal triad*, in the sense that they are mutually orthogonal and of unit length, as can be verified directly.

5.3.3 $\mathbb{E}^3$ as an Affine Space

The physical space of classical (non-relativistic) continuum mechanics, the stage on which all instantaneous events are supposed to take place, is the three-dimensional *Euclidean space* $\mathbb{E}^3$. Unlike the case of $\mathbb{R}^3$, in $\mathbb{E}^3$ there are neither preferred points nor preferred directions, although (just as in $\mathbb{R}^3$) there is a specific way to measure distances between points and angles between directions. The intimate connection between $\mathbb{E}^3$ and $\mathbb{R}^3$ will now be elucidated.

An *affine space* consists of a set $\mathcal{A}$ of *points*, a vector space V and a map that assigns to each ordered pair of points $p, q \in \mathcal{A}$ a vector in V, denoted variously as $\{\mathbf{v}\}_{pq}$, $\overrightarrow{pq}$ or $q - p$. This assignment, however, is not arbitrary. It must enjoy the following properties:

1. *Anticommutativity*:

$$\overrightarrow{pq} = -\overrightarrow{qp}, \quad \text{for all } p, q \in \mathcal{A}. \tag{5.47}$$

2. *Triangle rule*:

$$\overrightarrow{pr} = \overrightarrow{pq} + \overrightarrow{qr}, \quad \text{for all } p, q, r \in \mathcal{A}. \tag{5.48}$$

3. *Arbitrary origin*: For each $p_0 \in \mathcal{A}$ and for each vector $\{\mathbf{v}\}$, there exists a *unique* point $q \in \mathcal{A}$ such that $\overrightarrow{p_0 q} = \{\mathbf{v}\}$.

The vector space V is the *underlying vector space* of the affine space $\mathcal{A}$. The *dimension* of the affine space $\mathcal{A}$ is, by definition, the dimension of the vector space V. By virtue of Property 3, the choice of an arbitrary point p_0 as an *origin* renders the affine space a vector space, namely, to each point there corresponds a vector and, vice versa, to each vector a point. For this reason, it is customary to say that an affine space is a vector space devoid of origin.

In an affine space one can define the notion of straight line. Moreover, given a straight line one can uniquely define the parallel line through any point of the affine space. Affine spaces are, therefore, very special sets.[5] A *Euclidean space* is even more special. It is an affine space whose underlying vector space is an inner-product space. Thus, a Euclidean space is also a *metric space*, namely, a space with a notion of *distance*. Specifically, the *distance* between two points, p and q, of $\mathbb{E}^3$ is, by definition, the length of the associated vector $\overrightarrow{pq}$.

We denote by $\mathbb{E}^3$ a three-dimensional Euclidean space whose underlying vector space is $\mathbb{R}^3$. This is the space of Euclidean geometry and the backbone of Newtonian physics.

---

[5] The surface of a sphere, for example, is not an affine space.

5.3.4 Frames

We have already pointed out that in the physical space $\mathbb{E}^3$ there should be no preferred points or directions. On the other hand, the vector space $\mathbb{R}^3$ has a specific origin and a natural basis. The choice of a specific origin p_0 in $\mathbb{E}^3$ and the assignation of three other points, p_1, p_2, p_3, to the natural base vectors of $\mathbb{R}^3$ constitutes a *frame* of $\mathbb{E}^3$. In principle, this assignation can be quite arbitrary, but we will assume that there is a physically meaningful ruler so that these three points are at a unit distance from the chosen origin according to this ruler. Moreover, we can also assume that there is a physical way to establish (by means of a carpenter's square, say) that the three directions $p_0 p_1$, $p_0 p_2$ and $p_0 p_3$ are at right angles with each other. Once such an assignation has been agreed upon, the application of the properties of affine spaces completely determines a one-to-one correspondence between the points of $\mathbb{E}^3$ and the vectors of $\mathbb{R}^3$. In other words, the defining relation between the affine space $\mathbb{E}^3$ and the underlying vector space $\mathbb{R}^3$ can be established in different, but ultimately equivalent, ways. It is for this reason that the choice of a frame in $\mathbb{E}^3$ is tantamount to a *Cartesian coordinate system* therein.

What is the most general relation between two frames of $\mathbb{E}^3$? Figure 5.1 shows two arbitrary origins, p_0 and p_0', and corresponding choices of frame points. An event represented by a point q would be assigned the vectors $\{\mathbf{x}\}$ and $\{\mathbf{x}'\}$ in $\mathbb{R}^3$ corresponding, respectively, to $\overrightarrow{p_0 q}$ and $\overrightarrow{p_0' q}$, in their respective frames. Our question can be rephrased as follows: how are $\{\mathbf{x}\}$ and $\{\mathbf{x}'\}$ related? We start by noting that, from the point of view of the first (unprimed) frame, the three vectors $\overrightarrow{p_0 p_i}$ are nothing but the natural vectors $\{\mathbf{e}_i\}$. On the other hand, always from the point of view of the first frame, the three vectors $\{\mathbf{f}_i\} = \overrightarrow{p_0' p_i'}$ can be expressed uniquely in terms of some components Q_i^j as

$$\{\mathbf{f}_i\} = \sum_{j=1}^{3} Q_i^j \{\mathbf{e}_j\}. \tag{5.49}$$

But, since these vectors $\{\mathbf{f}_i\}$ are of unit length and mutually orthogonal, we must have, in accordance with equation (5.37),

$$\sum_{k=1}^{3} Q_i^k Q_j^k = \delta_{ij}. \tag{5.50}$$

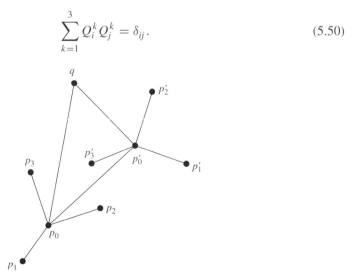

Figure 5.1 Frame transformation

If we arrange the numbers Q_i^k in a 3×3 matrix $[\mathbf{Q}]$, the subscript denoting the row and the superscript the column, equation (5.50) can be expressed as

$$[\mathbf{Q}]\,[\mathbf{Q}]^T = [\mathbf{I}], \tag{5.51}$$

where $[\mathbf{I}]$ is the 3×3 unit matrix. A matrix enjoying this property is called an *orthogonal matrix*. Its inverse (which always exists) is equal to its transpose:

$$[\mathbf{Q}]^{-1} = [\mathbf{Q}]^T. \tag{5.52}$$

Moreover, by the triangle property, we have

$$\overrightarrow{p_0'q} = \overrightarrow{p_0'p_0} + \overrightarrow{p_0q}. \tag{5.53}$$

The vector $\overrightarrow{p_0'p_0}$, joining the two origins, can be expressed in terms of some components c^i as

$$\overrightarrow{p_0'p_0} = \sum_{i=1}^{3} c^i \{\mathbf{f}_i\}. \tag{5.54}$$

Accordingly, we obtain

$$\sum_{i=1}^{3} x'^{\,i}\{\mathbf{f}_i\} = \sum_{i=1}^{3} c^i \{\mathbf{f}_i\} + \sum_{i=1}^{3} x^i \{\mathbf{e}_i\}, \tag{5.55}$$

or, using equation (5.49) and equating components,

$$\sum_{i=1}^{3} x'^{\,i} Q_i^j = \sum_{i=1}^{3} c^i Q_i^j + x^j. \tag{5.56}$$

In matrix notation, we obtain

$$\{\mathbf{x}'\}^T [\mathbf{Q}] = \{\mathbf{c}\}^T [\mathbf{Q}] + \{\mathbf{x}\}^T, \tag{5.57}$$

which, by virtue of (5.52), can also be written as

$$\{\mathbf{x}'\} = \{\mathbf{c}\} + [\mathbf{Q}]\{\mathbf{x}\}. \tag{5.58}$$

This is the desired relation between $\{\mathbf{x}\}$ and $\{\mathbf{x}'\}$. Thus, the most general *frame transformation* of Euclidean space is governed by an arbitrary vector $\{\mathbf{c}\}$ and an arbitrary orthogonal matrix $[\mathbf{Q}]$, according to equation (5.58). The physical meaning of a frame is that of an *observer* at one particular instant of time.

Exercise 5.3.1 Show that equation (5.52) is equivalent to the statement that a matrix is orthogonal if, and only if, its rows (and columns) are mutually orthogonal unit vectors.

Exercise 5.3.2 Length preservation. Let p and q be two points of $\mathbb{E}^3$. Show that the length of the vector $\overrightarrow{pq}$ is preserved under a change of frame.

Exercise 5.3.3   Kronecker symbol. Verify the *filtering property* of the Kronecker symbol, namely

$$a_{ijk\ldots} = \sum_{m=1}^{3} \delta_i^m \, a_{mjk\ldots},\tag{5.59}$$

for any indexed quantity $a_{ijk\ldots}$.

5.3.5   *Space-time and Observers*

The incorporation of the time variable into the physical picture is not as straightforward as one might imagine. A rather naïve representation, the *Aristotelian space-time* that dominated the thought of antiquity, would conceive of space and time as two completely separate entities. Two different observers would agree not only on whether or not two events happened simultaneously but also on whether or not they happened at the same place at different times.

The Galileo–Newtonian revolution, while preserving the notion of simultaneity, can be said to have done away with the notion of absolute space. Thus, two observers will in general disagree as to the location of two events that did not happen simultaneously. An observer carrying a flashlight emitting a light pulse at regular intervals would describe those light flashes as happening consistently 'here'. Another observer, if in motion with respect to the first, would see the flashes happening at different locations as time goes on. In this picture of the world, to be sure, the observers cannot be completely arbitrary – to paraphrase George Orwell's *Animal Farm*, some observers are more equal than others. But among the privileged class of observers, called *inertial observers*, there is no physical way to distinguish who is at rest and who is moving.

Let us start with the notion of time and let us identify it with the one-dimensional Euclidean space $\mathbb{E}$, that is, something that looks exactly like the real line $\mathbb{R}$, but without a specific origin. Let us assume, moreover, that all observers agree on the rhythm of the passage of time, which is to say that they all agree on some standard clock mechanism provided by a mutually agreed periodic natural phenomenon. If this is the case, the only difference between observers, as far as time is concerned, is in the choice of a time origin. Denoting by t the time recorded by one observer and by t' the time recorded by another, the relation between these quantities is given by

$$t' = t + a,\tag{5.60}$$

where a is a real constant. Clearly, observers agree on time *intervals* between events, namely

$$\Delta t' = \Delta t,\tag{5.61}$$

since the constant a cancels upon subtraction. All observers, therefore, agree on the collection of all events that happen simultaneously. And what can this collection be but a copy of $\mathbb{E}^3$? For each instant of time, accordingly, there is a copy of the typical three-dimensional space $\mathbb{E}^3$ attached, as it were, like a balloon by a string to the corresponding point of the time line. This curious picture, mathematically called a *fibre bundle*, is schematically represented in Figure 5.2.

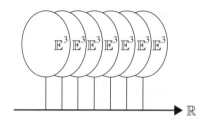

Figure 5.2   Space-time bundle

We have already established that, even if disagreeing on the location of simultaneous events, two observers agree on the distance between them. On the basis of this idea, we have found the mathematical relation between frames at a given time. This is precisely the content of equation (5.58). At a different time, the quantities $\{c\}$ and $[Q]$ will in general be different. We conclude that the most general relation between the measurements of location of events as reckoned by two different observers is

$$\{\mathbf{x}'\} = \{\mathbf{c}(t)\} + [\mathbf{Q}(t)]\{\mathbf{x}\}, \tag{5.62}$$

where $\{\mathbf{c}(t)\}$ and $[\mathbf{Q}(t)]$ are arbitrary functions of time, of which we will demand sufficient differentiability.

Two observers are said to be *inertially related* if

$$\frac{d^2\{\mathbf{c}(t)\}}{dt^2} = 0 \tag{5.63}$$

and

$$\frac{d[\mathbf{Q}(t)]}{dt} = 0. \tag{5.64}$$

In physical terms, the observers are moving at a constant relative translational velocity with a vanishing relative angular velocity. All observers can, therefore, be divided into equivalence classes according to whether or not they are inertially related. Of all these classes, Newton has singled out one class of *inertial observers* as the one for which the laws of mechanics acquire a particularly simple form (force equals mass times acceleration). The laws of motion of continuum mechanics are formulated in inertial frames. The laws of material response, however, hold true for arbitrary, not necessarily inertial, observers.

5.3.6   Fields and the Divergence Theorem

Introduction

A *field* over a continuum $\mathcal{C}$ is an assignment to each point of $\mathcal{C}$ of an object of a certain kind. In continuum mechanics, the most common kinds of objects of interest are scalars, vectors and tensors. Correspondingly, one speaks of *scalar*, *vector* and *tensor fields*. The underlying continuum $\mathcal{C}$ can be the whole of space-time, a part thereof, a material body, or some other suitably chosen domain.

Technically speaking, a continuum $\mathcal{C}$ is a *differentiable manifold*, the main object of study of differential geometry. At this level of generality, without further assumptions, a powerful result, known as *Stokes' theorem*, establishes a precise relation between an integral over the n-dimensional manifold C and an integral over its $(n-1)$-dimensional boundary ∂C. The objects being integrated are known as *differential forms* and the relation between the integrands is given by an operation known as *exterior differentiation*. Stokes' theorem can be regarded as a generalization of the fundamental theorem of calculus ('integration is the inverse of differentiation'). For the sake of general interest, and without explaining the notation or precisely defining the objects of the discourse, the following is the formulaic statement of Stokes' theorem:

$$\int_C d\omega = \int_{\partial C} \omega. \tag{5.65}$$

This compact and elegant formula encompasses and greatly generalizes all of the classical results collectively known as *vector calculus*, whose derivation is based on the extra structure provided by the Euclidean geometry of $\mathbb{R}^n$ and, more particularly, of $\mathbb{R}^3$. Since this classical approach is more than sufficient for our present purposes, we will abandon, with some regret, the all-encompassing power and elegance of equation (5.65).

Let us, therefore, consider the space $\mathbb{R}^n$ with its natural coordinate system $x^1,\ldots,x^n$ and natural basis $\mathbf{e}_1,\ldots,\mathbf{e}_n$. A scalar field ψ is then represented by a function

$$\psi = \psi(x^1,\ldots,x^n), \tag{5.66}$$

and a vector field $\mathbf{f}$ is represented by n functions

$$f^i = f^i(x^1,\ldots,x^n), \quad i = 1,\ldots,n. \tag{5.67}$$

A field is continuous (differentiable, smooth) if its representative functions are continuous (differentiable, smooth).

The Gradient of a Scalar Field

If a scalar field ψ is continuously differentiable, we can define its *gradient* as the vector field grad ψ with component functions

$$(\text{grad } \psi)_i = \frac{\partial \psi}{\partial x^i}, \quad i = 1,\ldots,n. \tag{5.68}$$

An alternative notation for grad ψ is $\nabla \psi$.

Remark 5.3.4 Note the subtlety that the gradient is really a covector, as indicated by the lower placement of the index. Because of the Euclidean structure prevalent in $\mathbb{R}^n$, this fact is irrelevant.

The equation

$$\psi(x^1,\ldots,x^n) = \text{constant} \tag{5.69}$$

represents an $(n-1)$-dimensional *hypersurface*, a term derived from the particular case $n = 3$, whereby equation (5.69) represents a surface in space. The name *gradient* is an allusion to the fact that, at each point of $\mathbb{R}^n$, the vector grad ψ is orthogonal to the hypersurface of constant ψ passing through that point. Indeed, consider a small vector

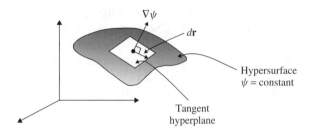

Figure 5.3   The gradient

$d\mathbf{r}$ tangent to that hypersurface, as shown schematically in Figure 5.3 for the case $n = 3$. Since, by construction, ψ is constant on the hypersurface, we must have that

$$d\psi = \frac{\partial \psi}{\partial x^1}dx^1 + \ldots + \frac{\partial \psi}{\partial x^n}dx^n = 0, \tag{5.70}$$

where $dx^1, \ldots, dx^n$ are the components of $d\mathbf{r}$. This equation can be rewritten as

$$(\operatorname{grad} \psi) \cdot (d\mathbf{r}) = 0, \tag{5.71}$$

which shows that grad ψ is perpendicular to *every* vector tangent to the hypersurface. Thus, the gradient points in the direction of shortest distance towards the neighbouring hypersurfaces of constant ψ. For any unit vector $\mathbf{m}$, the dot product

$$\frac{\partial \psi}{\partial m} := \nabla \psi \cdot \mathbf{m} \tag{5.72}$$

is the directional derivative of ψ in the direction of $\mathbf{m}$. Therefore, since

$$\mathbf{n} = \frac{\nabla \psi}{\sqrt{\nabla \psi \cdot \nabla \psi}} \tag{5.73}$$

is the unit vector pointing in the direction of $\nabla \psi$, we conclude that the gradient points in the direction at which ψ increases faster than in any other direction (direction of *steepest ascent*).

Flux of a Vector field through an Oriented Hypersurface

Given a (continuous) vector field $\mathbf{f}$ and a hypersurface element da with unit normal $\mathbf{n}$, the *elementary flux* of $\mathbf{f}$ through da is defined as the dot product

$$\mathcal{F}_{\mathbf{f},da} := \mathbf{f} \cdot \mathbf{n}, \tag{5.74}$$

where $\mathbf{f}$ is evaluated at da. Note that if the opposite unit normal is chosen, then the elementary flux reverses its sign. Thus, a flux distinguishes one side of the surface element from the other ('what goes into one side comes out from the other side'). The dot product indicates that, in calculating the flux, only the component of the vector field normal to the surface makes a contribution. This idea, as well as the terminology, is clearly inspired by the case of flow of water through an area dA. If the velocity vector $\mathbf{v}$ is not perpendicular to the area, as shown in Figure 5.4, the amount of water per unit time flowing into the

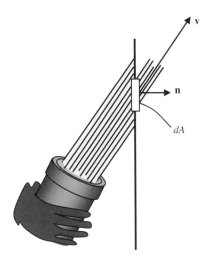

Figure 5.4   Flow of water through an opening

side of the area indicated by the arrow of **n** is, obviously, $\mathbf{v} \cdot \mathbf{n} \, dA$. In particular, if the velocity and the normal are at right angles to each other, no net flow takes place. Notice also that, if we were to consider in Figure 5.4 the other side of the surface (i.e., if we were to replace **n** by $-\mathbf{n}$) the result would be negative, namely a loss of water on that side. Thus, the two results are physically consistent with each other.

The flux of **f** through an oriented hypersurface a with unit normal field **n** is defined as

$$\mathcal{F}_{\mathbf{f},a} := \int_a \mathbf{f} \cdot \mathbf{n} \, da. \tag{5.75}$$

The unit vector **n**, which varies smoothly[6] from point to point of the surface, is uniquely defined by the assumption that the surface is oriented (unlike, for example, the Möbius band[7]) which implies that we have made an explicit choice of one of the two sides of the surface. Although we use an economic notation, the integral in equation (5.75) is evaluated as a multiple integral over an appropriate domain of $n-1$ curvilinear coordinates on the hypersurface.

The Divergence Theorem

The slogan 'integration is the inverse of differentiation' is an informal statement of the *fundamental theorem of calculus*. In particular, if $y = f(x)$ is a differentiable function of x, the theorem stipulates that

$$\int_a^b \frac{dy}{dx} \, dx = f(b) - f(a). \tag{5.76}$$

---

[6] Much less than smoothness is needed, but we will not detain ourselves to contemplate such subtleties at this point.
[7] The Möbius band is the non-orientable surface obtained by taking a plane strip, rotating one of its ends out of plane by half a turn and then gluing the two ends together in a ring-like fashion. One can then pass from one side of any surface element to the other side by moving continuously around the strip.

From the applications point of view, as we have already seen in Section 2.2, the fundamental theorem of calculus establishes the equivalence between the evaluation of the integral of a function over an interval $[a, b]$ and the evaluation of another function (a 'primitive') on the boundary of this interval, a boundary that consists of just two points. Our question now is whether this result can be generalized for multiple integrals. If so, we expect that, for example, a certain integral over a three-dimensional domain can be replaced by the evaluation of another integral over its two-dimensional boundary. The affirmative answer to this question is the content of the *divergence theorem*.

We define the *divergence* of the differentiable vector field $\mathbf{f}$ as the *scalar* function div $\mathbf{f}$ given by

$$\text{div } \mathbf{f} = \frac{\partial f^1}{\partial x^1} + \frac{\partial f^2}{\partial x^2} + \ldots + \frac{\partial f^n}{\partial x^n} = \sum_{i=1}^{n} \frac{\partial f^i}{\partial x^i}. \tag{5.77}$$

Exercise 5.3.5   The Laplacian. Calculate the divergence of the gradient of a scalar field ψ. The resulting scalar is called the *Laplacian* of ψ and is denoted by $\nabla^2 \psi$.

Exercise 5.3.6   Show that the divergence of a vector field is invariant under changes of Cartesian coordinates.

With the notions of flux and divergence in place, we are in a position to formulate the following important theorem:

Theorem 5.3.7   Divergence theorem. *Let C be a bounded domain in $\mathbb{R}^n$, whose boundary we denote by ∂C. Let ∂C be piecewise differentiable.[8] If $\mathbf{f}$ is a vector field defined over $C \cup \partial C$ with a continuous derivative, then the integral over C of the divergence of this vector field is equal to the flux of the vector field over the boundary ∂C oriented by the exterior unit normal:*

$$\int_C \text{div } \mathbf{f} \, dC = \int_{\partial C} \mathbf{f} \cdot \mathbf{n} \, da. \tag{5.78}$$

Proof. Without giving a rigorous proof, it is nevertheless a useful exercise to investigate intuitively the source of the final result. To this effect, consider a differential hypercube with edges parallel to the coordinate axes, as shown in Figure 5.5 for the case $n = 3$. If $\mathbf{f}$ is the value of the vector field at the centre of the cube, we can approximate its value at the centre of each of the faces. Let us denote by p_1^+ and p_2^- the centres of the two opposite faces parallel to the x^1-axis, the sign indicating the direction of the respective exterior unit normal (pointing forward like x^1 for one of the faces and backward for the other face). Then, we can write the first-order approximations

$$\mathbf{f}(p_1^+) = \mathbf{f} + \frac{\partial \mathbf{f}}{\partial x^1} \frac{dx^1}{2} \tag{5.79}$$

---

[8] By 'piecewise differentiable' we mean that the boundary may be made up of a finite number of pieces each of which has a well-defined tangent plane at each interior point. Thus, the boundary ∂C may have a few sharp edges and corners.

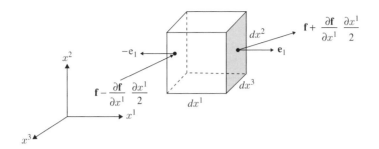

Figure 5.5 Sketch of proof of the divergence theorem

and

$$\mathbf{f}(p_1^-) = \mathbf{f} - \frac{\partial \mathbf{f}}{\partial x^1} \frac{dx^1}{2}. \tag{5.80}$$

The contribution of these two faces to the total flux through the boundary of this cube is, therefore, approximated by

$$\mathcal{F}_{\mathbf{f},1} = \mathbf{f}(p_1^+) \cdot \mathbf{e}_1 dx^2 \ldots dx^n + \mathbf{f}(p_1^-) \cdot (-\mathbf{e}_1) dx^2 \ldots dx^n = \frac{\partial f_1}{\partial x^1} dx^1 dx^2 \ldots dx^n. \tag{5.81}$$

The contribution of all the pairs of faces constitutes the approximate flux through the boundary of this cube, namely

$$\mathcal{F}_{\mathbf{f},1\ldots n} = \operatorname{div} \mathbf{f} \, dx^1 \ldots dx^n. \tag{5.82}$$

Considering that the (Riemann) integral can be approximated by a subdivision of the domain of integration into small hypercubes of the kind just considered, and taking into consideration that adjacent hypercubes share just one face (with opposite unit normals), a massive cancelling out of the contributions of all the interior cubes takes place, the only survivors being those faces (of some hypercubes) at the boundary of the region $\mathcal{C}$. The statement of the theorem follows suit. ∎

Just like the fundamental theorem of calculus, the divergence theorem establishes that the integral of some differential expression (in this case, the divergence of a vector field) over, say, a three-dimensional domain is equal to the integral over its two-dimensional boundary of another quantity (the projection of the vector field over the exterior unit normal to the boundary).

Exercise 5.3.8 Specialize the divergence theorem for a two-dimensional domain in $\mathbb{R}^2$ and for a one-dimensional domain in $\mathbb{R}$. Check that, with the correct interpretation of the 'exterior unit normal', this last case reduces indeed to the fundamental theorem of calculus.

Exercise 5.3.9 Harmonic functions. A function $\psi(x^1, \ldots, x^n)$ is *harmonic* in a given domain of $\mathbb{R}^n$ if it satisfies the *Laplace equation*

$$\nabla^2 \psi = 0, \tag{5.83}$$

where ∇^2 is the Laplacian operator introduced in Exercise 5.3.5. Show that the flux, over any closed surface, of the gradient of a harmonic function vanishes.

Exercise 5.3.10   Balls in n dimensions. The n-dimensional ball $B_{n,r}$ of radius r is the domain of $\mathbb{R}^n$ defined by

$$\sum_{i=1}^{n} \left(x^i\right)^2 \leq r^2. \tag{5.84}$$

The n-dimensional volume $V_{n,r}$ of $B_{n,r}$ is proportional to r^n, while the $(n-1)$-dimensional area $A_{n,r}$ of its boundary is proportional to r^{n-1}. Show that

$$n\, V_{n,r} = r\, A_{n,r}. \tag{5.85}$$

[Hint: consider a radially linear vector field.]

5.4   Eigenvalues and Eigenvectors

5.4.1   General Concepts

A matrix $[\mathbf{A}]$ can be regarded as a representation of a *linear operator* $\mathbf{A}$, a notion introduced in Section 5.2.4. We are focusing on the case of real square matrices of order 3 which, acting on vectors in $\mathbb{R}^3$, produce vectors in $\mathbb{R}^3$ according to the linear formula

$$\{\mathbf{w}\} = [\mathbf{A}]\{\mathbf{v}\}. \tag{5.86}$$

Since the input and the output vectors belong to the same vector space $\mathbb{R}^3$, it is legitimate to evaluate their inner product, that is,

$$\{\mathbf{v}\}^T\{\mathbf{w}\} = \{\mathbf{v}\}^T[\mathbf{A}]\{\mathbf{v}\}. \tag{5.87}$$

The number thus obtained may be positive, negative or zero. In geometrical terms, these three possibilities correspond, respectively, to the input and output vectors forming an acute, obtuse or right angle. The last possibility also includes the case in which at least one of the vectors vanishes. Clearly, if the input vector $\{\mathbf{v}\}$ vanishes, so does the output $\{\mathbf{w}\}$, regardless of the matrix $[\mathbf{A}]$ under consideration. On the other hand, it is possible for a matrix to produce a vanishing output out of a non-vanishing input. If that is the case, the matrix is said to be *singular*. A theorem in linear algebra establishes that a matrix is singular if, and only if, its determinant vanishes.

An interesting question that may be asked is the following: given a matrix $[\mathbf{A}]$, are there any particular non-zero input vectors $\{\mathbf{u}\}$ such that the output $[\mathbf{A}]\{\mathbf{u}\}$ is strictly *aligned* with the input? Such vectors, if they exist, are called *characteristic vectors* or *eigenvectors* of $[\mathbf{A}]$, and usually have an important physical meaning according to the context in which the operator $\mathbf{A}$ arises, from acoustics to quantum mechanics. Assume that an eigenvector $\{\mathbf{u}\}$ of $[\mathbf{A}]$ has been found, Then, by interpreting the expression 'strictly aligned', we conclude that there must exist a (real) number λ such that

$$[\mathbf{A}]\{\mathbf{u}\} = \lambda\{\mathbf{u}\}. \tag{5.88}$$

This number λ is called the *characteristic value* or *eigenvalue* associated with the eigenvector $\{\mathbf{u}\}$. It should be clear from equation (5.88) that if $\{\mathbf{u}\}$ is an eigenvector of $[\mathbf{A}]$ so is any non-zero multiple of $\{\mathbf{u}\}$. In other words, it is not the *vector* $\{\mathbf{u}\}$ that is a characteristic quantity, but rather its *line of action*.

To find the eigenvectors of a matrix it turns out that it is convenient to start by finding the eigenvalues. Equation (5.88) can be suggestively written as

$$([\mathbf{A}] - \lambda[\mathbf{I}])\{\mathbf{u}\} = \{\mathbf{0}\}, \tag{5.89}$$

where $[\mathbf{I}]$ is the unit matrix. We conclude that the matrix $[\mathbf{A}] - \lambda[\mathbf{I}]$ must be singular, since it produces a zero output from a putatively non-zero input. Therefore, according to the theorem of linear algebra mentioned above, its determinant must vanish. We thus obtain the condition

$$\det([\mathbf{A}] - \lambda[\mathbf{I}]) = 0, \tag{5.90}$$

or, more explicitly,

$$\begin{vmatrix} a_1^1 - \lambda & a_1^2 & a_1^3 \\ a_2^1 & a_2^2 - \lambda & a_2^3 \\ a_3^1 & a_3^2 & a_3^3 - \lambda \end{vmatrix} = 0. \tag{5.91}$$

Expanding this determinant, we obtain the following expression:

$$-\lambda^3 + I_A\lambda^2 - II_A\lambda + III_A = 0, \tag{5.92}$$

with

$$I_A = \text{trace } [\mathbf{A}] = a_1^1 + a_2^2 + a_3^3, \tag{5.93}$$

$$II_A = \frac{1}{2}\left((\text{trace } [\mathbf{A}])^2 - \text{trace } [\mathbf{A}]^2\right) = a_1^1 a_2^2 + a_1^1 a_3^3 + a_2^2 a_3^3 - a_1^2 a_2^1 - a_1^3 a_3^1$$
$$- a_2^3 a_3^2, \tag{5.94}$$

and

$$III_A = \det[\mathbf{A}] = a_1^1 a_2^2 a_3^3 + a_2^1 a_3^2 a_1^3 + a_3^1 a_2^3 a_1^2 - a_1^1 a_3^2 a_2^3 - a_2^1 a_3^3 a_1^2 - a_3^1 a_2^2 a_1^2. \tag{5.95}$$

The scalars I_A, II_A and III_A are the *characteristic invariants* or *principal invariants* of $[\mathbf{A}]$ and the expression on the left-hand side of equation (5.92) is the *characteristic polynomial* of $[\mathbf{A}]$. According to the fundamental theorem of algebra, a cubic polynomial with real coefficients has either just one real root or three real roots.[9] In other words, a 3×3 matrix has at least one real eigenvalue. Once a real eigenvalue has been found, the corresponding eigenvectors can be found by solving the system of equations (5.89) for the entries of the vector $\{\mathbf{u}\}$. This is a linear homogeneous system with vanishing determinant. This means that at least one of the equations is a linear combination of the others and can be eliminated. The number of independent eigenvectors depends on the rank of the coefficient matrix.

Exercise 5.4.1 Expand the determinant on the left-hand side of equation (5.91) by means of the well-known Cramer rule, and obtain the result of equations (5.92)–(5.95).

---

[9] The roots may not be distinct. For example, the polynomial $(\lambda - 1)(\lambda - 3)^2$ has roots 1, 3 and 3. In this example, 3 is a *double root* or a root with *multiplicity 2*.

5.4.2 More on Principal Invariants

The terminology of principal or characteristic invariants alludes to the following fact: if a matrix $[\mathbf{A}]$ is multiplied on the left and on the right by an orthogonal matrix $[\mathbf{Q}]$ and its transpose, the resulting matrix $[\mathbf{Q}][\mathbf{A}][\mathbf{Q}]^T$ has the same principal invariants as the original matrix $[\mathbf{A}]$. Thus, these numbers are *invariant* with respect to that particular action of the orthogonal matrix group.[10]

The property just described can be made more explicit by bringing to the fore some properties of the trace and the determinant of a matrix. The trace of a matrix is defined as the sum of its diagonal entries, as already shown in equation (5.93), namely

$$\text{trace}[\mathbf{A}] = \sum_{i=1}^{3} a_i^i. \tag{5.96}$$

Consider now the trace of the product $[\mathbf{A}][\mathbf{B}]$ of two matrices. The result is

$$\text{trace}\,([\mathbf{A}][\mathbf{B}]) = \sum_{i=1}^{3}\sum_{k=1}^{3} a_k^i b_i^k = \sum_{k=1}^{3}\sum_{i=1}^{3} b_i^k a_k^i = \sum_{i=1}^{3}\sum_{k=1}^{3} b_k^i a_i^k = \text{trace}\,([\mathbf{B}][\mathbf{A}])\,. \tag{5.97}$$

We conclude that the trace of the product of two matrices is indifferent to the order of the multiplication. To generalize this property for the product of a string of matrices, it should be enough to see what happens with three. Indeed, using (5.97), we obtain

$$\text{trace}\,([\mathbf{A}][\mathbf{B}][\mathbf{C}]) = \text{trace}\,(([\mathbf{A}][\mathbf{B}])[\mathbf{C}]) = \text{trace}\,([\mathbf{C}]([\mathbf{A}][\mathbf{B}])) = \text{trace}\,([\mathbf{C}][\mathbf{A}][\mathbf{B}])\,. \tag{5.98}$$

The trace of the product of three or more matrices is preserved under a *cyclic* permutation, whereby the last factor is moved to the beginning of the chain. This process can, naturally, be repeated.

Let us apply the property just discovered to calculate the first principal invariant of the matrix $[\mathbf{Q}][\mathbf{A}][\mathbf{Q}]^T$. We obtain

$$I_{[\mathbf{Q}][\mathbf{A}][\mathbf{Q}]^T} = \text{trace}\,\big([\mathbf{Q}][\mathbf{A}][\mathbf{Q}]^T\big) = \text{trace}\,\big([\mathbf{Q}]^T[\mathbf{Q}][\mathbf{A}]\big) = \text{trace}[\mathbf{A}] = I_A, \tag{5.99}$$

where we have exploited the fact that $[\mathbf{Q}]$ is an orthogonal matrix and, therefore, $[\mathbf{Q}]^T[\mathbf{Q}] = [\mathbf{I}]$. This shows that I_A is indeed an orthogonal invariant. Moreover, the square of $[\mathbf{Q}][\mathbf{A}][\mathbf{Q}]^T$ is $[\mathbf{Q}][\mathbf{A}]^2[\mathbf{Q}]^T$, so that a direct application of equation (5.94) yields the result that II_A is also an orthogonal invariant.

As far as III_A is concerned, we can use the well-known result that the determinant function commutes with the matrix product, namely

$$\det([\mathbf{A}][\mathbf{B}]) = (\det[\mathbf{A}])(\det[\mathbf{B}]), \tag{5.100}$$

whence

$$III_{[\mathbf{Q}][\mathbf{A}][\mathbf{Q}]^T} = \det([\mathbf{Q}][\mathbf{A}][\mathbf{Q}]^T) = \det[\mathbf{A}] = III_A. \tag{5.101}$$

---

[10] For a definition of the notion of group, see Section 6.4.1.

For later use, we calculate the derivatives of the principal invariants of a matrix with respect to the matrix itself. From equation (5.96),

$$\frac{\partial I_A}{\partial a_n^m} = \sum_{i=1}^{3} \frac{\partial a_i^i}{\partial a_n^m} = \sum_{i=1}^{3} \delta_m^i \delta_i^n = \delta_m^n, \tag{5.102}$$

or equivalently,

$$\frac{\partial I_A}{\partial [\mathbf{A}]} = [\mathbf{I}]. \tag{5.103}$$

Exercise 5.4.2   Show that

$$\frac{\partial II_A}{\partial [\mathbf{A}]} = (\text{trace } \mathbf{A})[\mathbf{I}] - [\mathbf{A}]. \tag{5.104}$$

To obtain a working expression for the determinant of a matrix in terms of its components, we introduce the *permutation symbol* ε_{ijk} defined as

$$\varepsilon_{ijk} = \begin{cases} 1 & \text{if } (ijk) = (123), \ (231), \text{ or } (312) \\ -1 & \text{if } (ijk) = (321), \ (132), \text{ or } (213) \\ 0 & \text{otherwise, i.e., if an index value is repeated.} \end{cases} \tag{5.105}$$

The determinant can be expressed as follows:

$$\det [\mathbf{A}] = \sum_{i=1}^{3}\sum_{j=1}^{3}\sum_{k=1}^{3} \varepsilon_{ijk} a_1^i a_2^j a_3^k. \tag{5.106}$$

This expression corresponds exactly to the statement that ternary products are taken without repeating a row or a column, and that the sign of each product is chosen according to whether the permutation of the rows is even or odd. An equivalent way to express the determinant is the following:

$$\varepsilon_{mnp} \det [\mathbf{A}] = \sum_{i=1}^{3}\sum_{j=1}^{3}\sum_{k=1}^{3} \varepsilon_{ijk} a_m^i a_n^j a_p^k. \tag{5.107}$$

Exercise 5.4.3   The determinant of a product. Derive equation (5.100). Use expressions (5.106) and/or (5.107).

Exercise 5.4.4   The derivative of a determinant. Prove that for a non-singular matrix $\mathbf{A}$,

$$\frac{\partial III_A}{\partial [\mathbf{A}]} = \frac{\partial \det [\mathbf{A}]}{\partial [\mathbf{A}]} = (\det[\mathbf{A}]) \, [\mathbf{A}]^{-T}, \tag{5.108}$$

where the superscript $-T$ denotes the inverse transpose.

5.4.3   The Symmetric Case

A case of particular interest in applications is that for which the matrix $[\mathbf{A}]$ is *symmetric*:

$$[\mathbf{A}]^T = [\mathbf{A}]. \tag{5.109}$$

The following theorems summarize the most important results.

Theorem 5.4.5   Reality of eigenvalues. *The eigenvalues of a real symmetric matrix are real.*

Proof. Assume that the characteristic polynomial of a real symmetric matrix has a complex root $\lambda = \alpha + \beta i$, where α and β are real and i is the imaginary unit. Let $\{\mathbf{u}\}$ be an eigenvector associated with λ. Since we have opened up the possibility of complex numbers for the eigenvectors, we should, in principle, allow $\{\mathbf{u}\}$ to have complex entries. By equation (5.88), we have

$$[\mathbf{A}]\{\mathbf{u}\} = \lambda\{\mathbf{u}\}. \tag{5.110}$$

Recall that the *complex conjugate* of a complex number is obtained by changing the sign of its imaginary part. Moreover, the conjugate of the product of two complex numbers is equal to the product of the conjugates of the factors. Denoting the complex conjugate by means of an overbar, and recalling that the matrix $[\mathbf{A}]$ is real (and, therefore, equal to its conjugate), we obtain from equation (5.110) the following result:

$$[\mathbf{A}]\{\bar{\mathbf{u}}\} = \bar{\lambda}\{\bar{\mathbf{u}}\}. \tag{5.111}$$

Multiplying (5.110) on the left by $\{\bar{\mathbf{u}}\}^T$ yields the scalar equation

$$\{\bar{\mathbf{u}}\}^T[\mathbf{A}]\{\mathbf{u}\} = \lambda\{\bar{\mathbf{u}}\}^T\{\mathbf{u}\}. \tag{5.112}$$

On the other hand, multiplying the transpose of equation (5.111) on the right by $\{\mathbf{u}\}$, we obtain

$$\{\bar{\mathbf{u}}\}^T[\mathbf{A}]\{\mathbf{u}\} = \bar{\lambda}\{\bar{\mathbf{u}}\}^T\{\mathbf{u}\}, \tag{5.113}$$

where we have taken into consideration the fact that $[\mathbf{A}]$ is symmetric. Clearly, this has been the crucial step in the proof so far. Subtracting (5.112) from (5.113), we can write

$$0 = (\bar{\lambda} - \lambda)\{\bar{\mathbf{u}}\}^T\{\mathbf{u}\}. \tag{5.114}$$

Since $\{\mathbf{u}\}$ is non-zero, the product $\{\bar{\mathbf{u}}\}^T\{\mathbf{u}\}$ is real and strictly positive, whence

$$\bar{\lambda} - \lambda = 0, \tag{5.115}$$

which means that λ must be real.   ■

Theorem 5.4.6   Orthogonality of eigenvectors. *The eigenvectors of a symmetric matrix associated with two different eigenvalues are mutually orthogonal.*

Proof. Let $\lambda_1 \neq \lambda_2$ be two eigenvalues of the symmetric matrix $[\mathbf{A}]$, and let $\{\mathbf{u}\}_1$ and $\{\mathbf{u}\}_2$ be corresponding eigenvectors. By equation (5.88) we have the following two relations:

$$[\mathbf{A}]\{\mathbf{u}\}_1 = \lambda_1\{\mathbf{u}\}_1 \tag{5.116}$$

and

$$[\mathbf{A}]\{\mathbf{u}\}_2 = \lambda_2\{\mathbf{u}\}_2. \tag{5.117}$$

Multiplying the first on the left by $\{\mathbf{u}\}_2^T$, transposing the second and multiplying it on the right by $\{\mathbf{u}\}_1$, and then subtracting the resulting scalar equations, we obtain

$$\{\mathbf{u}\}_2^T \left([\mathbf{A}] - [\mathbf{A}]^T\right)\{\mathbf{u}\}_1 = (\lambda_1 - \lambda_2)\{\mathbf{u}\}_2^T\{\mathbf{u}\}_1. \tag{5.118}$$

Just as in the case of the previous theorem, we observe that, due to the assumed symmetry of $[\mathbf{A}]$, the left-hand side vanishes. Finally, since the eigenvalues have been assumed to be different, it follows that

$$\{\mathbf{u}\}_2^T\{\mathbf{u}\}_1 = 0, \tag{5.119}$$

which is the desired result.   ∎

Theorem 5.4.7   Multiple eigenvalues. *The eigenvectors corresponding to an eigenvalue of a symmetric matrix span a subspace of dimension equal to the multiplicity of the eigenvalue.*

We omit the proof. The eigenvectors of a simple eigenvalue span a line. The eigenvectors of a double eigenvalue span a plane. By linearity, every vector on this plane is an eigenvector, and one can choose two orthogonal ones. The eigenvector of a triple eigenvalue (in the three-dimensional case) span the whole space, so that in this case every vector is an eigenvector. The subspace of $\mathbb{R}^3$ spanned by the eigenvectors of an eigenvalue is called the *eigenspace* associated with this eigenvalue. Every vector in an eigenspace is an eigenvector.

Exercise 5.4.8   Show that if $[\mathbf{A}]$ is a 3×3 symmetric matrix with a triple eigenvalue, then $[\mathbf{A}]$ must be *spherical*, that is, proportional to the unit matrix.

Combining Theorems 5.4.6 and 5.4.7 we obtain the following corollary:

Corollary 5.4.9   Orthonormal eigenbasis. *For a real symmetric matrix there exists a complete set of orthonormal eigenvectors.*

By 'complete set' we mean a basis of the underlying vector space ($\mathbb{R}^3$ in our case). If all the eigenvalues are distinct, the orthogonal directions are unique and all we need to do is choose a unit vector along each of these directions (which is thus determined uniquely except for its sense). In the case of a multiple eigenvalue, the orthogonal directions are not unique and can be chosen arbitrarily within the corresponding eigenspace.

Having chosen an orthonormal basis of eigenvectors $\{\mathbf{u}\}_1$, $\{\mathbf{u}\}_2$ and $\{\mathbf{u}\}_3$, we can form a matrix $[\mathbf{Q}]$ with these eigenvectors as columns, namely

$$[\mathbf{Q}] = \left[\; \{\mathbf{u}\}_1 \;\; \{\mathbf{u}\}_2 \;\; \{\mathbf{u}\}_3 \;\right]. \tag{5.120}$$

This matrix is called a *modal matrix*, a name that derives from the application to the theory of vibrating systems, where the eigenvalues are natural frequencies of vibration and the eigenvectors describe the natural modes of vibration of the system.

Exercise 5.4.10   Show that a modal matrix is, by construction, an orthogonal matrix.

Theorem 5.4.11   Diagonalization. *Every real symmetric matrix* [**A**] *can be expressed as*

$$[A] = [Q] [\Lambda] [Q]^T, \tag{5.121}$$

where [**Q**] *is a modal matrix of* **A**, *and* [**Λ**] *is the diagonal matrix of eigenvalues,*

$$[\Lambda] = \begin{bmatrix} \lambda_1 & 0 & 0 \\ 0 & \lambda_2 & 0 \\ 0 & 0 & \lambda_3 \end{bmatrix}, \tag{5.122}$$

the eigenvalues appearing in [**Λ**] *in the same order as the corresponding eigenvectors appear in* [**Q**]. *It is customary to order the eigenvalues in either ascending or descending order.*

Exercise 5.4.12   Prove Theorem 5.4.11. [Hint: read off [**Λ**] from equation (5.121) and carry out the matrix multiplications intelligently.]

A matrix [**A**] is *positive definite* if, for every non-zero input vector, the inner product with the output is strictly positive, namely

$$\{v\}^T [A]\{v\} > 0, \quad \text{for all } \{v\} \neq \{0\}. \tag{5.123}$$

In physical terms, the output vector points roughly in the same direction as the input (i.e., they form an acute angle). Positive definite symmetric matrices play an important role in the description of discrete and continuous systems in physics.

Exercise 5.4.13   Show that a real symmetric matrix is positive definite if and only if its eigenvalues are all positive.

5.4.4   Functions of Symmetric Matrices

Given a real-valued function of a real variable, $f(x)$, it is possible to define a corresponding symmetric-matrix-valued function $[f([A])]$ of real symmetric matrices [**A**] in a unique way by means of the following prescription:

$$[f([A])] = [Q] \begin{bmatrix} f(\lambda_1) & 0 & 0 \\ 0 & f(\lambda_2) & 0 \\ 0 & 0 & f(\lambda_3) \end{bmatrix} [Q]^T, \tag{5.124}$$

where [**Q**] is a modal matrix of [**A**].

Exercise 5.4.14   Prove that the eigenvectors of $[f([A])]$ are the same as those of [**A**]. Moreover, for each eigenvalue λ_i of [**A**], the corresponding eigenvalue of $[f([A])]$ is $f(\lambda_i)$. Conclude that the definition in equation (5.124) is consistently independent of the particular modal matrix adopted (even if there are multiple eigenvalues).

Notice that, just as the logarithm of real numbers is defined only on the open half-line $\mathbb{R}^+$, so too the logarithm of a symmetric matrix is defined only for positive definite matrices. A similar remark applies to the square root function. The square root of a symmetric positive definite matrix [**A**] is also a symmetric positive definite matrix denoted by $[A]^{\frac{1}{2}}$.

Exercise 5.4.15   The square root of a matrix. Verify that the matrix

$$[\mathbf{A}] = \begin{bmatrix} 2 & 1 & 0 \\ 1 & 2 & 0 \\ 0 & 0 & 1 \end{bmatrix} \tag{5.125}$$

is positive definite and find its square root $[\mathbf{A}]^{\frac{1}{2}}$. Verify your numerical result by checking that $\left([\mathbf{A}]^{\frac{1}{2}}\right)^2 = [\mathbf{A}]$.

5.5   Kinematics

5.5.1   Material Bodies

Having set up some of the mathematical scaffolding necessary to ascend from the one-dimensional to the three-dimensional formulation, we proceed to emulate the treatment presented in Chapter 1. Just as in Section 1.2 the underlying material universe $\mathcal{M}$ was a copy of the real line $\mathbb{R}$, we now declare $\mathcal{M}$ to be a copy of $\mathbb{R}^3$ and we introduce the following definition:

Definition 5.5.1   A *material body* $\mathcal{B}$ is a connected open subset of $\mathcal{M}$.

Remarks similar to those made in Section 1.2 pertaining to the exclusion of the boundary apply now as well.

5.5.2   Configurations, Deformations and Motions

Since the *physical space* of classical continuum mechanics is $\mathbb{E}^3$, it seems natural to define a *configuration* κ as a map

$$\kappa : \mathcal{B} \rightarrow \mathbb{E}^3. \tag{5.126}$$

Just as we did in Chapter 1, we will require that this map be continuous (preserving the body integrity) and injective (preventing matter penetrability).

For practical purposes, it proves convenient to introduce a fixed *reference configuration*,

$$\kappa_0 : \mathcal{B} \rightarrow \mathbb{R}^3. \tag{5.127}$$

From the technical standpoint, it is worthwhile pointing out that the reference configuration does not necessarily live in the Euclidean space $\mathbb{E}^3$, but rather in $\mathbb{R}^3$. Thus, a reference configuration fulfils the function of assigning labels (coordinates) to the particles of $\mathcal{B}$, but it is not subject to such physical ideas as changes of frame. In physical terms, all observers agree on the labels of the body particles.

Once a reference configuration has been chosen, we define the *deformation* from the reference configuration $\kappa_0(\mathcal{B})$ to the spatial configuration κ as the composition

$$\chi = \kappa \circ \kappa_0^{-1}, \tag{5.128}$$

as illustrated in Figure 5.6.

In terms of coordinates, if we denote by capital letters, X^I, $I = 1, 2, 3$, the natural coordinates of $\mathbb{R}^3$ in the reference configuration and by lower-case letters, x^i, $i = 1, 2, 3$,

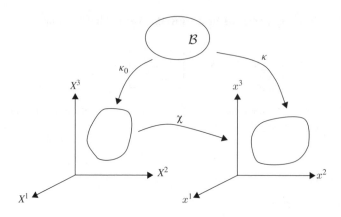

Figure 5.6   Configurations and deformation

the Cartesian coordinates induced by the choice of a spatial frame in $\mathbb{E}^3$, a deformation consists of three functions,

$$x^i = x^i(X^1, X^2, X^3), \quad i = 1, 2, 3. \tag{5.129}$$

Beyond continuity and injectivity, it is usually assumed that a deformation is smooth and has a smooth inverse over its range. Moreover, to satisfy the same physical conditions as were imposed in the one-dimensional case (see Section 1.3), we assume that the *Jacobian determinant*,

$$J = \left| \frac{\partial x^i}{\partial X^I} \right| = \begin{vmatrix} \dfrac{\partial x^1}{\partial X^1} & \dfrac{\partial x^1}{\partial X^2} & \dfrac{\partial x^1}{\partial X^3} \\[2mm] \dfrac{\partial x^2}{\partial X^1} & \dfrac{\partial x^2}{\partial X^2} & \dfrac{\partial x^2}{\partial X^3} \\[2mm] \dfrac{\partial x^3}{\partial X^1} & \dfrac{\partial x^3}{\partial X^2} & \dfrac{\partial x^3}{\partial X^3} \end{vmatrix}, \tag{5.130}$$

is everywhere bounded and strictly positive.

A *motion* is a (smooth) one-parameter family of configurations parametrized by time. In terms of a fixed reference configuration and a chosen observer, a motion is given by three smooth functions of four independent variables, namely

$$x^i = x^i(X^1, X^2, X^3, t), \quad i = 1, 2, 3. \tag{5.131}$$

5.5.3   The Deformation Gradient

The matrix with entries:

$$F_I^i = \left. \frac{\partial x^i}{\partial X^I} \right|_P, \quad i, I = 1, 2, 3, \tag{5.132}$$

is the coordinate representation of the *deformation gradient* at the point P with coordinates X^I. The deformation gradient $\mathbf{F}$ is a *linear transformation* or *tensor*, whose mathematical and physical meaning we now explore.

At each point P of the body in the reference configuration $\kappa_0(\mathcal{B})$, we can consider the collection of all vectors issuing from this point. These vectors form a *vector space* of dimension 3. The technical name of this entity is the *tangent space of* $\kappa_0(\mathcal{B})$ *at* P and its elements are called the *tangent vectors at* P or simply *vectors at* P. If we imagine a *strain gauge*[11] placed at P, we have a crude materialization of a tangent vector at that point.

Let the components of a tangent vector at P be denoted by dX^1, dX^2, dX^3. Premultiplying this vector by the matrix representing the deformation gradient we obtain a new vector with components given by

$$\left\{ \begin{matrix} dx^1 \\ dx^2 \\ dx^3 \end{matrix} \right\} = \begin{bmatrix} F^1_1 & F^1_2 & F^1_3 \\ F^2_1 & F^2_2 & F^2_3 \\ F^3_1 & F^3_2 & F^3_3 \end{bmatrix} \left\{ \begin{matrix} dX^1 \\ dX^2 \\ dX^3 \end{matrix} \right\}. \tag{5.133}$$

This image vector belongs to the tangent space of $\kappa(\mathcal{B})$ at the image point $p = \chi(P)$ in the current configuration, as shown in Figure 5.7

Equation (5.133) can be written as

$$dx^i = F^i_1 dX^1 + F^i_2 dX^2 + F^i_3 dX^3, \quad i = 1, 2, 3, \tag{5.134}$$

or, more compactly, as

$$dx^i = \sum_{I=1}^{3} F^i_I dX^I, \quad i = 1, 2, 3. \tag{5.135}$$

On the other hand, recalling equation (5.132), we can rewrite this equation as

$$dx^i = \sum_{I=1}^{3} \frac{\partial x^i}{\partial X^I} dX^I, \quad i = 1, 2, 3. \tag{5.136}$$

We conclude that, according to the definition of the differential of a function of several variables, the deformation gradient is nothing but the first-order (i.e., linear) approximation to the deformation (5.129) at the point P. It transforms the tangent vector space at P into the tangent vector space at the image point $p = \chi(P)$. In particular, if we consider a small brick (a unit cube, say) at P, the effect of the deformation gradient is to transform it into a deformed parallelepiped at p (see Figure 5.8). The smaller the brick of departure, the more accurate the representation of the deformation χ will be.

Remark 5.5.2 The condition following equation (5.130), stipulating the positive definiteness of the determinant of $\mathbf{F}$, can be now interpreted physically as demanding that the parallelepiped must not collapse to a two-dimensional object ($J \neq 0$) and that it must preserve (rather than reverse) its orientation ($J > 0$).

---

[11] A strain gauge is a small copper wire that is attached at a point (usually, but not necessarily) located on the surface of a body. The change in length of this wire during the process of deformation can be related to its change in electrical resistance.

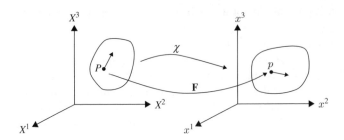

Figure 5.7 The deformation gradient maps a vector at P to a vector at $p = \chi(P)$

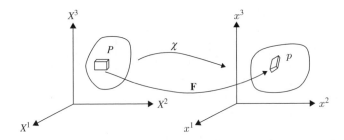

Figure 5.8 The deformation gradient maps a small parallelepiped at P to its counterpart at $p = \chi(P)$

5.5.4 Local Configurations

Given a configuration of a body, we may be interested in focusing attention on a particular body point P and a small neighbourhood only. In other words, we may identify two different configurations as long as they coincide in some open neighbourhood of the point. The collection of all configurations satisfying this condition is called a *configuration germ* or a *local configuration* at P. This concept turns out to be too strong for applications.

Instead, we can introduce a weaker requirement as follows. Given a configuration at which the image of P is denoted by p, we can adopt it as a reference configuration and generate an infinitude of other configurations by applying subsequent deformations ξ such that: (i) $\xi(p) = p$, (ii) $\mathbf{F}(p) = \mathbf{I}$. This collection of configurations, which coincide with each other at P 'up to first order', is called a *local first-order configuration* or, more commonly, just a *local configuration* at P.

More intuitively, we may define a local configuration at P by stating that it is characterized by a fixed deformation gradient $\mathbf{F}$ at P, measured with respect to some reference configuration. Why is this concept useful? Most materials encountered in practice are *local* in the sense that they react by producing internal forces only in response to their immediate environment. The material response at a point is sensitive only to the value of the local configuration at that point, while it is oblivious of what happens elsewhere. Materials of this kind are called *simple* or *first-grade* materials.

Exercise 5.5.3 A local configuration. Consider the following two configurations expressed in terms of deformations with respect to some fixed reference configuration with coordinates X, Y, Z:

$$\begin{cases} x = X \cos Z - Y \sin Z \\ y = X \sin Z + Y \cos Z \\ z = Z + \pi/2 \end{cases} \tag{5.137}$$

and

$$\begin{cases} x = -Y + \cos Z \\ y = X + \cos Z \\ z = 2Z + \cos Z \end{cases} \tag{5.138}$$

where x, y, z are the spatial coordinates. (a) Show that these two different configurations represent the same local configuration at the point P with coordinates $(X, Y, Z) = (1, 1, \pi/2)$ by checking that at this particular point the deformations and their gradients coincide. (b) Construct the explicit expression for the deformation of the second configuration with respect to the first and verify that the image point p remains fixed and that the deformation gradient thereat is the identity map.

5.5.5 A Word on Notation

In equation (5.133), we have represented the nine components of the deformation gradient as a 3×3 matrix. Clearly, if we were to change our coordinate system, these components would change according to some trigonometric expressions. But the entity represented, namely the deformation gradient itself (the map that transforms the brick), remains the same. For this reason, it is convenient to denote the deformation gradient not by the collection of its components but rather by a single (bold-faced) symbol $\mathbf{F}$, representing precisely the underlying linear operator or tensor. If $\mathbf{N}$ is a tangent vector at P, then the action of the deformation gradient at P is a new vector $\mathbf{n}$ at p. We indicate this action compactly as

$$\mathbf{n} = \mathbf{F} \, \mathbf{N}. \tag{5.139}$$

The advantage of this so-called *block notation*, which we have already used in Section 5.2.4, is that it is more closely related to the physical meaning, whereas an expression such as (5.133) is more readily amenable to numerical computations in particular coordinate systems. When the coordinate system has been fixed once and for all, as is the case in most applications, we may use the compact matrix notation, according to which equation (5.139) would read

$$\{\mathbf{n}\} = [\mathbf{F}] \, \{\mathbf{N}\}. \tag{5.140}$$

In this equation, $\{\mathbf{N}\}$ and $\{\mathbf{n}\}$ stand for the one-column matrices (or column vectors) whose entries are the components of the vectors $\mathbf{N}$ and $\mathbf{n}$, respectively, while $[\mathbf{F}]$ stands for the matrix whose entries are given in equation (5.132).

As far as the notation (5.135), called *index notation*, is concerned, it can be considered as intermediate between the two extremes. This is the kind of notation that Albert Einstein used in formulating his theory of general relativity. Einstein noticed that in all monomials that have a physical meaning independent of observer, the summation takes place on indices that are diagonally repeated once (i.e., a subscript and a superscript). Usually, if an index in a monomial is repeated more than once, or not diagonally, the expression contains a mistake. For this reason, the summation symbol $\sum$ becomes superfluous: it is tacitly understood to be in force whenever an index in a monomial appears once diagonally repeated. These are called *dummy indices*. The remaining indices should appear only once in every monomial. These are called *free indices*. Free indices should be *balanced*, that is, they should appear in every monomial making up the expression, and each one should appear at exactly the same level (as either a subscript or a superscript). Thus, equation (5.136) is equivalent to

$$dx^i = \frac{\partial x^i}{\partial X^I} dX^I. \tag{5.141}$$

All indices, in our case, are understood to vary in the range $1, 2, 3$. The collection of the rules just explained is known as the *Einstein summation convention*.[12]

5.5.6 *Decomposition of the Deformation Gradient*

Our intuitive interpretation of the deformation gradient as an operation that transforms a small parallelepiped into another, as shown in Figure 5.8, raises the following question: is the deformation gradient a measure of strain? The answer to this question is negative. Indeed, if the transformed parallelepiped happens to be congruent with the initial one, we must conclude that no elongation or contraction whatsoever has taken place in any direction and, whatever definition of strain one eventually adopts, this case must correspond to zero strain. On the other hand, the deformation gradient in this case is not trivial. Let us explore this question further by finding out what is the form of a deformation gradient $\mathbf{R}$ when the two objects (the two 'infinitesimal parallelepipeds') are congruent.

The tangent vector spaces at P and p are endowed with an *inner product* (or *dot product*). Otherwise, the concepts of the length of a vector and of the angle between two vectors would be meaningless and, accordingly, so would the notion of strain. Because of the assumed metric nature of the spaces of both the reference and the actual configuration, the inner product in terms of vector components has the simple form prescribed by Pythagoras' theorem, namely, if $\mathbf{M}$ and $\mathbf{N}$ are two vectors at P with components M^I and N^I, respectively, their inner product is given by

$$\mathbf{M} \cdot \mathbf{N} = \{\mathbf{M}\}^T \{\mathbf{N}\}. \tag{5.142}$$

According to equation (5.140), the corresponding deformed vectors $\mathbf{m}$ and $\mathbf{n}$ are given by

$$\{\mathbf{m}\} = [\mathbf{R}] \{\mathbf{M}\} \tag{5.143}$$

and

$$\{\mathbf{n}\} = [\mathbf{R}] \{\mathbf{N}\}, \tag{5.144}$$

respectively, and their inner product is

$$\mathbf{m} \cdot \mathbf{n} = \{\mathbf{m}\}^T \{\mathbf{n}\} = ([\mathbf{R}] \{\mathbf{M}\})^T [\mathbf{R}] \{\mathbf{N}\} = \{\mathbf{M}\}^T \left([\mathbf{R}]^T [\mathbf{R}]\right) \{\mathbf{N}\}, \tag{5.145}$$

---

[12] When restricting attention to Cartesian coordinate systems, the distinction between subscripts and superscripts can be dropped and, accordingly, the summation convention may be understood to apply to any once-repeated index in a monomial.

where we have used the well-known properties of matrix multiplication and transposition. For the linear transformation $\mathbf{R}$ to enjoy the property of preservation of the inner product *identically* for all pairs of vectors $\mathbf{M}, \mathbf{N}$, namely

$$\mathbf{m} \cdot \mathbf{n} = \mathbf{M} \cdot \mathbf{N}, \tag{5.146}$$

we must have, on comparing equation (5.142) with (5.145), that

$$[\mathbf{R}]^T \, [\mathbf{R}] = [\mathbf{I}], \tag{5.147}$$

where $[\mathbf{I}]$ is the identity matrix. By equation (5.51), $[\mathbf{R}]$ is an *orthogonal matrix*. An immediate consequence of this equation is that

$$(\det[\mathbf{R}])^2 = 1, \tag{5.148}$$

where we have used the fact that the determinant of a matrix is equal to that of its transpose. Since, according to equation (5.130), we have imposed the condition of strict positivity for the determinant of any deformation gradient, we conclude that, for our purposes, an orthogonal matrix satisfies

$$\det[\mathbf{R}] = 1. \tag{5.149}$$

Orthogonal transformations with this property are called *proper orthogonal transformations*. They can be viewed as rigid *rotations*, a terminology that we will use freely, although this interpretation requires the virtual identification of the reference ambient space with the actual configuration ambient space. In this context, orthogonal transformations with a negative determinant (equal to -1) would correspond to rotations followed or preceded by *reflections* about a plane.

Remark 5.5.4   Proper orthogonal transformations as rotations. Strictly speaking, the interpretation of a proper orthogonal transformation as a rotation is somewhat sloppy, since a deformation gradient is a map between two *different* vector spaces. Nevertheless, because of the special nature of $\mathbb{E}^3$, the following identifications can be made naturally. The first identification that can be made without sacrifice of rigour consists of viewing the referential coordinate axes X^I and the spatial coordinate axes x^i as coincident. This artifice is particularly natural whenever one chooses as reference a configuration actually occupied by the body at some initial time. There is, moreover, a second kind of identification that can be made in an affine space, namely, a natural *equipollence* (or *distant parallelism*) between the tangent spaces at different points, as shown in Figure 5.9 by the ability to translate the original brick from the source point P to the target point p. Under these conditions, a proper orthogonal transformation can be legitimately regarded as a rotation, an interpretation that is obviously very suggestive in most applications.

Having demonstrated that non-trivial deformation gradients exist that preserve the length of all vectors, namely rotations, it has become abundantly clear that the deformation gradient *per se* cannot be used as a measure of strain. In general, a deformation gradient will produce both rotations and changes of length, and there seems to be no clear-cut way to separate the former from the latter. It is a remarkable fact, however, that, given any deformation gradient, it is possible to extract from it *in a unique way* the rotational component as distinct from the part that causes strain. From the mathematical standpoint,

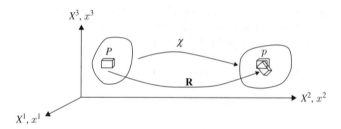

Figure 5.9 Identifying referential and spatial coordinates and exploiting the natural distant parallelism of Euclidean space

this is the content of a theorem in linear algebra known as the *polar decomposition theorem*, whose proof we will presently provide.

Theorem 5.5.5 **Polar decomposition.** *Every non-singular matrix* $[\mathbf{F}]$ *can be uniquely decomposed as a product*

$$[F] = [R][U], \tag{5.150}$$

where $[\mathbf{R}]$ *is orthogonal and* $[\mathbf{U}]$ *is symmetric and positive definite.*

   Proof. The matrix

$$[C] = [F]^T[F] \tag{5.151}$$

is symmetric and positive definite. Indeed,

$$[C]^T = \left([F]^T[F]\right)^T = [F]^T \left([F]^T\right)^T = [F]^T[F] = [C], \tag{5.152}$$

which shows that $[\mathbf{C}]$ is symmetric. Moreover, for any vector $\{\mathbf{N}\} \neq \{\mathbf{0}\}$ we have

$$\{N\}^T[C]\,\{N\} = \{N\}^T \left([F]^T[F]\right)\,\{N\} = ([F]\{N\})^T \,([F]\{N\}) \geq 0, \tag{5.153}$$

where we have used the fact that, since $[\mathbf{F}]$ is not singular, the vector $([\mathbf{F}]\{\mathbf{N}\})$ cannot vanish. Equation (5.153) shows that $[\mathbf{C}]$ is positive definite. We now define

$$[U] = [C]^{\frac{1}{2}}, \tag{5.154}$$

which is a well-defined symmetric and positive definite matrix, according to Section 5.4.4. We define the matrix

$$[R] = [F][U]^{-1}. \tag{5.155}$$

This matrix is orthogonal. Indeed,

$$[R]^T[R] = \left([F][U]^{-1}\right)^T \left([F][U]^{-1}\right) = [U]^{-T}[F]^T[F][U]^{-1} \tag{5.156}$$

$$= [U]^{-1}[C][U]^{-1} = [U]^{-1}[U]^2[U]^{-1} = [I].$$

Finally, assume that two supposedly different polar decompositions are given, namely $[\mathbf{F}] = [\mathbf{R}][\mathbf{U}] = [\mathbf{R}]'[\mathbf{U}]'$. It follows that $[\mathbf{F}]^T[\mathbf{F}] = [\mathbf{U}]^2 = [\mathbf{U}]'^2$, whence $[\mathbf{U}] = [\mathbf{U}]'$ and, consequently, $[\mathbf{R}] = [\mathbf{R}]'$. This completes the proof. ∎

In the case of the deformation gradient, the matrix

$$[\mathbf{C}] = [\mathbf{F}]^T[\mathbf{F}] \tag{5.157}$$

is known as the (coordinate representation of the) *right Cauchy–Green tensor*. Moreover, since the determinant of $[\mathbf{F}]$ is always positive, so is the determinant of $[\mathbf{R}]$, which means that $[\mathbf{R}]$ represents a pure rotation. The matrix $[\mathbf{U}]$ is the coordinate representation of the *right stretch tensor*. This terminology arises from the fact that if $\mathbf{u}$ is an eigenvector of $\mathbf{U}$ whose associated eigenvalue is λ_u, then λ_u clearly measures the following *stretch ratio*:

$$\lambda_u = \frac{ds}{dS}, \tag{5.158}$$

where ds is the length of the output vector along $\mathbf{u}$ that is obtained by applying $\mathbf{U}$ to a vector of length dS along $\mathbf{u}$. The eigenvectors and eigenvalues of $\mathbf{U}$ are also called *principal directions* and *principal stretches*, respectively. The *principal relative elongation* ε_u along the principal direction $\mathbf{u}$ is measured, accordingly, by

$$\varepsilon_u = \lambda_u - 1. \tag{5.159}$$

To disclose more fully the physical meaning of the polar decomposition theorem applied to the deformation gradient $[\mathbf{F}]$, consider the action of $[\mathbf{F}]$ on one of the (unit) eigenvectors of $[\mathbf{U}]$, say $\{\mathbf{u}_1\}$:

$$[\mathbf{F}]\{\mathbf{u}_1\} = [\mathbf{R}][\mathbf{U}]\{\mathbf{u}_1\} = [\mathbf{R}]\lambda_1\{\mathbf{u}_1\}. \tag{5.160}$$

This action, as shown by the above equation, involves the multiplication of the unit eigenvector by its corresponding eigenvalue λ_1 and the subsequent rotation of this vector according to the matrix $[\mathbf{R}]$. If the eigenvalue is larger (smaller) than 1, the vector is lengthened (shortened) and then rotated. Now, let us choose, as we always can (according to Corollary 5.4.9), an orthonormal basis of eigenvectors of $[\mathbf{U}]$ or, more graphically, a unit cube aligned with three eigenvectors. The action of the deformation gradient on this unit cube consists of pure elongations or contractions without any angular distortion, followed by a rigid rotation, as illustrated in Figure 5.10. The eigenvectors of $[\mathbf{U}]$ are also called the *right principal directions* of the deformation at P.

Exercise 5.5.6   Components of the right Cauchy–Green tensor. Show that the components of the right Cauchy–Green tensor, that is, the entries of the matrix $[\mathbf{C}]$, are given by the formula

$$C_{IJ} = \sum_{k=1}^{3} F_I^k F_J^k = \sum_{k=1}^{3} \frac{\partial x^k}{\partial X^I} \frac{\partial x^k}{\partial X^J}. \tag{5.161}$$

Observe that, as indicated by the free indices, this tensor transforms vectors in the reference configuration to vectors that are also in the reference configuration.

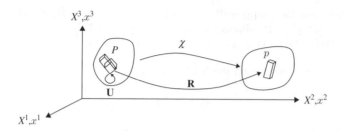

Figure 5.10   Polar decomposition of the deformation gradient $\mathbf{F} = \mathbf{RU}$

Exercise 5.5.7   Let $d\mathbf{X}$ be a (small) vector at a point P in the reference configuration. Show that the length ds of its 'deformed' counterpart $d\mathbf{x}$ is given by

$$ds^2 = \sum_{I=1}^{3}\sum_{J=1}^{3} C_{IJ}\, dX^I\, dX^J. \tag{5.162}$$

Exercise 5.5.8   Left polar decomposition. Prove the left polar decomposition theorem, which states that every non-singular matrix $[\mathbf{F}]$ can be uniquely decomposed as a product $[\mathbf{V}][\mathbf{R}]$, where $[\mathbf{V}] = [\mathbf{B}]^{\frac{1}{2}}$ is known as the *left stretch tensor*. The matrix $[\mathbf{B}] = [\mathbf{F}][\mathbf{F}]^T$ is the coordinate expression for the *left Cauchy–Green tensor*. Show that the orthogonal matrix $[\mathbf{R}]$ coincides with its counterpart in the usual (right) polar decomposition theorem. Show that the eigenvalues of $[\mathbf{V}]$ are equal to those of $[\mathbf{U}]$. Establish the relation between $[\mathbf{V}]$ and $[\mathbf{U}]$. How are the eigenvectors of $[\mathbf{V}]$ related to the eigenvectors of $[\mathbf{U}]$? Draw the counterpart of Figure 5.10 for the left polar decomposition theorem.

Exercise 5.5.9   Components of the left Cauchy–Green tensor. Show that the components of the left Cauchy–Green tensor, that is, the entries of the matrix $[\mathbf{B}]$, are given by the formula

$$B^{ij} = \sum_{K=1}^{3} F_K^i F_K^j = \sum_{K=1}^{3} \frac{\partial x^i}{\partial X^K}\frac{\partial x^j}{\partial X^K}. \tag{5.163}$$

Observe that, as indicated by the free indices, this tensor transforms vectors in the spatial configuration to vectors also in the spatial configuration.

Exercise 5.5.10   Show that the components of the inverse $\mathbf{B}^{-1}$ of the left Cauchy–Green tensor are given by the formula

$$B_{ij}^{-1} = \sum_{K=1}^{3} \frac{\partial X^K}{\partial x^i}\frac{\partial X^K}{\partial x^j}. \tag{5.164}$$

Conclude that the length dS of a reference vector $d\mathbf{X}$ mapped by the deformation into the spatial vector $d\mathbf{x}$ is given by

$$dS^2 = \sum_{i=1}^{3}\sum_{j=1}^{3} B_{ij}^{-1} dx^i\, dx^j. \tag{5.165}$$

Exercise 5.5.11 Obtain the right and left polar decompositions for the following deformation gradient at a point:

$$[\mathbf{F}] = \frac{1}{2} \begin{bmatrix} 1 + \sqrt{3} & -1 + \sqrt{3} & 0 \\ (3 - \sqrt{3})/2 & (3 + \sqrt{3})/2 & \sqrt{2} \\ (1 - \sqrt{3})/2 & -(1 + \sqrt{3})/2 & \sqrt{6} \end{bmatrix}. \tag{5.166}$$

5.5.7 Measures of Strain

Intuitively speaking, the *strain* at a point should be a measure of the changes in size
and shape of a small neighbourhood of the point as it is mapped from the reference to
the actual configuration. Whatever is meant by terms such as 'size and shape' or 'small
neighbourhood of a point', the following condition will be imposed on any proposed
measure of strain:

Condition 5.5.12 *Any proposed strain measure must vanish for rigid-body motions.*

The standard measure of strain at a point P in the reference configuration is the
Lagrange strain tensor $\mathbf{E}$. defined as

$$\mathbf{E} = \frac{1}{2}(\mathbf{C} - \mathbf{I}). \tag{5.167}$$

Let $d\mathbf{X}$ be a vector at a point P in the reference configuration, and let dS denote
its length. Denoting by ds the length of its deformed counterpart, equations (5.162) and
(5.167) imply that

$$ds^2 - dS^2 = 2 \sum_{I=1}^{3} \sum_{J=1}^{3} E_{IJ} dX^I dX^J. \tag{5.168}$$

Thus, the Lagrange strain tensor provides a measure of the difference between the squares
of spatial and referential lengths. The presence of the factor $\frac{1}{2}$ in equation (5.167) is
explained by the desirability of the eigenvalues of $\mathbf{E}$ approaching the principal relative
elongations (5.159) for *small* (or *infinitesimal*) *strain* regimes (see Window 5.3). Indeed,
it follows directly from equation (5.167) that the eigenvectors of $\mathbf{E}$ coincide with those of
$\mathbf{C}$ (and $\mathbf{U}$). Moreover, if $\mathbf{u}$ is an eigenvector of $\mathbf{U}$ with eigenvalue λ_u, the corresponding
eigenvalue of $\mathbf{E}$ is

$$\lambda_u^E = \frac{1}{2}\left(\lambda_u^2 - 1\right) = \frac{1}{2}\left(\lambda_u + 1\right)\left(\lambda_u - 1\right) = \frac{1}{2}(2 + \varepsilon_u)\varepsilon_u. \tag{5.169}$$

For infinitesimal strains ($\varepsilon_u \ll 1$), we obtain

$$\lambda_u^E \approx \varepsilon_u, \tag{5.170}$$

as intended.

The *Eulerian strain tensor* $\mathbf{e}$ at a point p in the spatial configuration is defined as

$$\mathbf{e} = \frac{1}{2}(\mathbf{I} - \mathbf{B}^{-1}), \tag{5.171}$$

where, in a abuse of notation, we use the same symbol $(\mathbf{I})$ to denote the identity at p (in the spatial configuration) as we have used to denote the identity at P (in the reference configuration). If $d\mathbf{x}$ is a vector of length ds at p, we can write

$$ds^2 - dS^2 = 2\sum_{i=1}^{3}\sum_{j=1}^{3} e_{ij}\,dx^I\,dx^J. \tag{5.172}$$

Exercise 5.5.13   Logarithmic strain. Show that the *logarithmic strain*, defined as

$$\mathbf{E}_L = \frac{1}{2}\ln\mathbf{C}, \tag{5.173}$$

satisfies Condition 5.5.12 and reduces to the Lagrangian strain $\mathbf{E}$ for infinitesimal strains.

Window 5.3   Infinitesimal strains

By an *infinitesimal strain regime* we mean a situation such that the change of length of any vector (as brought about by the deformation gradient) is very small compared with the length of the vector itself. Denoting, as we have already done, by dS and ds the lengths of corresponding vectors in the reference and spatial configurations, respectively, an infinitesimal strain regime at a given material point P implies that

$$\varepsilon = \frac{ds - dS}{dS} \ll 1,$$

for all directions at P. Small strains do not imply, nor are they implied by, small rotations. For example, the fuselage of an airplane undergoes (hopefully) very small strains during flight, in spite of the fact that huge rotations can take place. Conversely, a rubber band can be stretched to more than double its original length without undergoing any rotation. In the case of infinitesimal strains, therefore, while the rotation $\mathbf{R}$ can be arbitrary, the tensors $\mathbf{U}$ and $\mathbf{C}$ (or, for that matter, $\mathbf{V}$ and $\mathbf{B}$) of the polar decomposition must be very close to the unit tensor $\mathbf{I}$. The eigenvalues of these tensors must be very close to 1.

5.5.8   *The Displacement field and its Gradient*

The reference and spatial configurations inhabit essentially different worlds, and the formulation of continuum mechanics does not require that these worlds be identified with each other. But they can be, and in many applications it is convenient to do so. The most common instance is provided by the identification of the reference configuration with a configuration actually occupied by the body at some putative initial time. In solid mechanics applications, this initial configuration is sometimes called *undeformed*, while the subsequent spatial configurations are called *deformed*.

The identification of the two worlds alluded to above is obtained, as we have already suggested in Remark 5.5.4, by identifying the referential coordinate axes with their respective spatial counterparts. This device is, therefore, tantamount to converting the whole kinematic stage into a single copy of $\mathbb{R}^3$. Under these conditions, upper-case

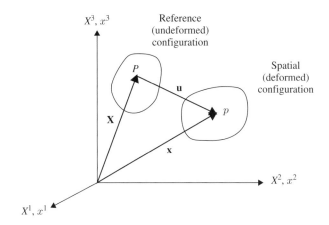

Figure 5.11   Displacement vector

indices $(I, J, \ldots)$ are interchangeable with lower-case indices $(i, j, \ldots)$. Moreover, denoting respectively by $\{X\}$ and $\{x\}$ the coordinate column vectors of a material point in the reference configuration and in its spatial counterpart, one can define the *displacement vector* $\{u\}$ via the difference

$$\{u\} = \{x\} - \{X\}, \tag{5.174}$$

as shown in Figure 5.11. It is important to bear in mind that this definition would be meaningless were it not that the aforementioned identification has been assumed.

Given a deformation and attaching to each material point the corresponding displacement vector, one obtains a *displacement field*. Given a motion, this construction leads to a time-dependent vector field, described in coordinates by three smooth functions:

$$u^i = u^i(X^1, X^2, X^3, t), \quad i = 1, 2, 3. \tag{5.175}$$

The *displacement gradient* at a material point P is represented by the matrix $[\nabla u]$ of partial derivatives of $\{u\}$ with respect to the referential coordinates (X^1, X^2, X^3). The relation between the displacement gradient and the deformation gradient is, therefore, given by

$$[\nabla u] = [F] - [I]. \tag{5.176}$$

Exercise 5.5.14   Show that the Lagrangian strain (5.167) can be expressed as

$$E = \frac{1}{2} \left(\nabla u^T \nabla u + \nabla u + \nabla u^T \right), \tag{5.177}$$

or, in components,

$$E_{ij} = \frac{1}{2} \left(\sum_{k=1}^{3} \frac{\partial u_k}{\partial x^i} \frac{\partial u_k}{\partial x^j} + \frac{\partial u_i}{\partial x^j} + \frac{\partial u_j}{\partial x^i} \right). \tag{5.178}$$

5.5.9   The Geometrically Linearized Theory

The *geometrically linearized theory* is the result of assuming that, when identifying the referential and spatial axes, *both* the rotation [**R**] and the right stretch tensor [**U**] are very close to the unit matrix [**I**]. Consequently, the deformation gradient [**F**] itself is also close to [**I**]. Notice that in the linearized theory the strains are necessarily small (see Window 5.3), but small strains do not imply the linearized theory, since the rotations may be large. In the linearized theory, *both* the strains and the rotations are assumed to be small. Writing

$$[\mathbf{F}] = [\mathbf{I}] + [\Delta\mathbf{F}], \tag{5.179}$$

$$[\mathbf{R}] = [\mathbf{I}] + [\Delta\mathbf{R}] \tag{5.180}$$

and

$$[\mathbf{U}] = [\mathbf{I}] + [\Delta\mathbf{U}], \tag{5.181}$$

the polar decomposition theorem yields

$$[\mathbf{I}] + [\Delta\mathbf{F}] = ([\mathbf{I}] + [\Delta\mathbf{R}])([\mathbf{I}] + [\Delta\mathbf{U}]) = [\mathbf{I}] + [\Delta\mathbf{R}] + [\Delta\mathbf{U}] + [\Delta\mathbf{R}][\Delta\mathbf{U}]. \tag{5.182}$$

In the linearized theory, since the entries of both [$\Delta\mathbf{R}$] and [$\Delta\mathbf{U}$] are small compared with the number 1, the quadratic term on the right-hand side of equation (5.182) may be neglected, thus yielding the following infinitesimal version of the polar decomposition theorem:

$$[\Delta\mathbf{F}] = [\Delta\mathbf{R}] + [\Delta\mathbf{U}]. \tag{5.183}$$

By virtue of equations (5.176) and (5.179), equation (5.183) can also be written as

$$[\nabla\mathbf{u}] = [\Delta\mathbf{R}] + [\Delta\mathbf{U}]. \tag{5.184}$$

The result of the linearization is that the *multiplicative* decomposition of the deformation gradient is converted into an *additive* decomposition of its increment. This considerable simplification is responsible for the historical success of this and other linearized field theories. To further investigate the consequences of the assumed smallness of the increments, we recall that [**R**] is an orthogonal matrix, namely

$$[\mathbf{R}]^T[\mathbf{R}] = [\mathbf{I}], \tag{5.185}$$

or

$$([\mathbf{I}] + [\Delta\mathbf{R}]^T)([\mathbf{I}] + [\Delta\mathbf{R}]) = [\mathbf{I}]. \tag{5.186}$$

Neglecting the quadratic term, we obtain

$$[\Delta\mathbf{R}]^T + [\Delta\mathbf{R}] = [\mathbf{0}]. \tag{5.187}$$

In other words, the incremental (small) rotation matrix is skew-symmetric. On the other hand, since both [**I**] and [**U**] are symmetric, so is [$\Delta\mathbf{U}$]. Thus, the decomposition (5.184) expresses the resolution of the displacement gradient [$\nabla\mathbf{u}$] as the sum of a symmetric part and a skew-symmetric part. Since such a decomposition is unique (as can be easily shown by assuming the contrary), we obtain the following important results:

$$[\Delta U] = \frac{1}{2} \left([\nabla u] + [\nabla u]^T \right) \tag{5.188}$$

and

$$[\Delta R] = \frac{1}{2} \left([\nabla u] - [\nabla u]^T \right). \tag{5.189}$$

Thus, the linearized theory is equivalent to the assumption that the entries of the displacement gradient are small. A large translational motion does not affect the validity of the theory.

In the linearized theory it is common to rename $[\Delta U]$ the *infinitesimal strain* $[\varepsilon]$ and to call $[\Delta R]$ the *infinitesimal rotation* $[\omega]$.

Exercise 5.5.15   Show that the components of the infinitesimal strain and of the infinitesimal rotation are given, respectively, by the formulas

$$\varepsilon_{ij} = \frac{1}{2} \left(\frac{\partial u_i}{\partial x^j} + \frac{\partial u_j}{\partial x^i} \right) \tag{5.190}$$

and

$$\omega_{ij} = \frac{1}{2} \left(\frac{\partial u_i}{\partial x^j} - \frac{\partial u_j}{\partial x^i} \right). \tag{5.191}$$

Compare equation (5.190) with (5.178) and conclude that the infinitesimal strain $[\varepsilon]$ is indeed the linearized version of the Lagrangian strain $[E]$.

Remark 5.5.16   In spite of their obvious limitations, formulas (5.190) and (5.191) are usually all that most undergraduate engineering students ever learn about the deformation of a continuous medium. In engineering textbooks, the off-diagonal terms of the infinitesimal strain matrix $[\varepsilon]$ appear multiplied by the factor 2 and are indicated with the symbol γ, as opposed to the diagonal terms, which preserve the original symbol ε. The former are the so-called *shear strain components*, while the latter are known as the *normal strain components*. These components can be respectively obtained by a careful calculation of the change of shape and the change of size of an infinitesimal die aligned with the coordinate system.

Exercise 5.5.17   A body is subjected to a rotation of magnitude α about the x^3-axis. Write the components of the displacement as functions of α. Verify that the Lagrangian strain vanishes identically, independently of α. Check, on the other hand, that using the infinitesimal strain formula (5.190) would lead to non-vanishing strains. Find the numerical values corresponding to $\alpha = 0.01\pi$ and $\alpha = 0.5\pi$. Comment on the results.

5.5.10   Volume and Area

The deformation gradient F at a point P contains, as we have learned in equation (5.162), all the information required to calculate the length ds in the spatial configuration of a material segment with components dX^I at P in the reference configuration. We would now like to extract from the deformation gradient the corresponding formulas for the calculation of area and volume elements as they accompany the process of deformation.

Material Volume Element

We have already established in Section 5.5.6 that the effect of the deformation gradient on a unit cube aligned with the principal directions of **U** consists of pure elongations or contractions of its edges without any angular distortion, followed by the rotation **R**, prescribed by the polar decomposition theorem, as depicted in Figure 5.10. Moreover, the new lengths of the edges of the unit cube are measured precisely by the eigenvalues $\lambda_1, \lambda_2, \lambda_3$ of the right stretch tensor **U**. On the other hand, since **U** is symmetric, according to Theorem 5.4.11 it can be diagonalized. It follows, therefore, from equations (5.121) and (5.122) that

$$\det \mathbf{U} = \lambda_1 \lambda_2 \lambda_3. \tag{5.192}$$

In other words, the determinant of **U** measures precisely the new volume of the unit cube. More generally, denoting by dV and dv, respectively, the values of a referential material volume and of its deformed counterpart, we can establish that

$$\frac{dv}{dV} = \det \mathbf{U}. \tag{5.193}$$

According to the polar decomposition theorem, however, since **R** is properly orthogonal (see equation (5.149)), we have

$$\det \mathbf{F} = \det \mathbf{U}. \tag{5.194}$$

Combining this result with equation (5.193), we conclude that

$$\frac{dv}{dV} = \det \mathbf{F} = J, \tag{5.195}$$

which is the desired formula for volume elements.

Material Area Element

Let dX^J and dY^K be the components in the reference configuration of two vectors issuing from a material point P. These vectors subtend a material area element (a parallelogram) of extent dA and a right-handed unit normal with components N_I. All these quantities are related via the cross product of the two given vectors. More precisely,

$$N_I \, dA = \sum_{J=1}^{3} \sum_{K=1}^{3} \varepsilon_{IJK} \, dX^J \, dX^K, \tag{5.196}$$

where ε_{IJK} is the permutation symbol defined in equation (5.105).

Exercise 5.5.18 Cross product. Verify that the right-hand side of equation (5.196) is consistent with the standard formula for the components of the *cross product* of two vectors, **a** and **b**, in a Cartesian coordinate system (x, y, z), namely

$$\mathbf{a} \times \mathbf{b} = \begin{vmatrix} \mathbf{i} & \mathbf{j} & \mathbf{k} \\ a_x & a_y & a_z \\ b_x & b_y & b_z \end{vmatrix}, \tag{5.197}$$

where $\mathbf{i}, \mathbf{j}, \mathbf{k}$ are unit vectors along the respective coordinate axes.

The deformed versions of dX^J and dY^K are given respectively by

$$dx^j = \sum_{J=1}^{3} F^j_J \, dX^J \tag{5.198}$$

and

$$dy^k = \sum_{K=1}^{3} F^k_K \, dY^K, \tag{5.199}$$

which subtend an area of extent da and unit normal n_i interrelated by

$$n_i \, da = \sum_{j=1}^{3} \sum_{k=1}^{3} \varepsilon_{ijk} dx^j dy^k. \tag{5.200}$$

Combining the preceding three equations yields

$$n_i \, da = \sum_{j=1}^{3} \sum_{k=1}^{3} \varepsilon_{ijk} \sum_{J=1}^{3} F^j_J \, dX^J \sum_{K=1}^{3} F^k_K \, dY^K. \tag{5.201}$$

At this point, we observe that our formula (5.201) contains only two contributions of **F**, whereas we would like to include a third one so as to be able to make use of the formula for the determinant of a matrix as given in equation (5.107). To this end, we can make use of the identity

$$\sum_{L=1}^{3} F^m_L \, F^{-L}_{\ i} = \delta^m_i, \tag{5.202}$$

where δ^m_i is the Kronecker symbol introduced in equation (5.31). Equation (5.202) is the component expression of the fact that the multiplication of a non-singular matrix by its inverse is equal to the unit matrix. We have indicated the components of the inverse of **F** by $F^{-L}_{\ i}$, a shorthand for $\left(\mathbf{F}^{-1}\right)^L_i$. On the other hand, by the filtering property of the Kronecker symbol, as described by equation (5.59), we may write

$$\varepsilon_{ijk} = \sum_{m=1}^{3} \varepsilon_{mjk} \, \delta^m_i = \sum_{m=1}^{3} \varepsilon_{mjk} \sum_{L=1}^{3} F^m_L \, F^{-L}_{\ i}, \tag{5.203}$$

where we have used equation (5.202). Introducing this result in equation (5.201), we obtain

$$n_i \, da = \sum_{j=1}^{3} \sum_{k=1}^{3} \sum_{m=1}^{3} \varepsilon_{mjk} \sum_{L=1}^{3} F^m_L \, F^{-L}_{\ i} \sum_{J=1}^{3} F^j_J \, dX^J \sum_{K=1}^{3} F^k_K \, dY^K \tag{5.204}$$

or, collecting terms,

$$n_i \, da = \sum_{L=1}^{3} \sum_{J=1}^{3} \sum_{K=1}^{3} \left(\sum_{m=1}^{3} \sum_{j=1}^{3} \sum_{k=1}^{3} \varepsilon_{mjk} \, F^m_L \, F^j_J \, F^k_K \right) F^{-L}_{\ i} \, dX^J \, dY^K. \tag{5.205}$$

Using equation (5.107) to simplify the bracketed expression, we obtain

$$n_i \, da = \sum_{L=1}^{3} \sum_{J=1}^{3} \sum_{K=1}^{3} \left(\varepsilon_{LJK} \det \mathbf{F}\right) F_{\,i}^{-L} \, dX^J \, dY^K. \tag{5.206}$$

Finally, by virtue of equation (5.196) we can write

$$n_i \, da = \det \mathbf{F} \left(\sum_{L=1}^{3} F_{\,i}^{-L} N_L\right) dA. \tag{5.207}$$

In terms of matrix notation, this equation can be written as

$$\{\mathbf{n}\} \, da = (\det \mathbf{F}) \, [\mathbf{F}]^{-T} \{\mathbf{N}\} \, dA, \tag{5.208}$$

where the superscript $-T$ denotes the transpose of the inverse. Equation (5.208) is known as *Nanson's formula*.[13]

Exercise 5.5.19   The volume element revisited. Given *any* three non-coplanar vectors at a point P in the reference configuration, verify that the volume of the parallelepiped they determine transforms according to the formula (5.195). Perform the calculations by recalling that the volume of the parallelepiped is given by the (absolute value of the) determinant of the matrix having the components of the given vectors as rows.

Exercise 5.5.20   Using the summation convention. Check that all the calculations in this section could have been carried out using Einstein's summation convention, described in Section 5.5.5. Thus, the summation symbols '$\sum$' would be eliminated. It is sufficient to check that all dummy (i.e., summed up) indices in a monomial appear exactly once diagonally repeated and that the free indices are balanced.

5.5.11   The Material Derivative

In continuum mechanics one works with scalar, vector and tensor *fields*. These fields can be defined and evaluated (by calculation or experiment) either in a reference configuration or in the present configuration that the body happens to occupy in space at a certain time t. Consider, for example, the *velocity field* $\mathbf{v}$. If the motion is given by an equation such as (5.131), we define the components of the velocity field as

$$v^i = \frac{\partial x^i (X^I, t)}{\partial t} = v^i (X^1, X^2, X^3, t). \tag{5.209}$$

It is opportune to make several remarks. Firstly, we stress the fact that these are *spatial* components, since, naturally, the velocity is a spatial (rather than referential) physical entity. Nevertheless, by the very essence of the notion of partial derivative, these components are evaluated at a *fixed material point* with coordinates X^1, X^2, X^3 in the reference configuration. We conclude, accordingly, that the velocity field just introduced

---

[13] This formula was obtained originally in 1878 by Edward J. Nanson (1850–1936), an English-born mathematician and renowned professor at the University of Melbourne, Australia. An advocate of proportional representation, he is best remembered for his contributions to the mathematical theory of political elections.

is defined over the reference configuration and we call it a *referential field* or a *Lagrangian field*.

On the other hand, we may not have an explicit knowledge of the whole motion, as represented by equation (5.131), but rather we may have placed some (perhaps radioactive) markers in a fluid (blood, say) and taken a visual record of their motion over a small period of time dt. In this way, we can obtain an experimental approximation to the velocity field. Each vector of this vector field, however, is attached to a *spatial location* with coordinates x^1, x^2, x^3, regardless of which material particle happens to occupy this position at the time t of the experiment. Thus, we have obtained the velocity field at time t as

$$v^i = \tilde{v}^i(x^1, x^2, x^3, t).\tag{5.210}$$

This is called a *spatial* or *Eulerian field*. In the case of the velocity, therefore, we have obtained two different representations of the same physical field (the velocity field). We may say that the Eulerian representation is obtained when one adopts the present configuration (at time t) as reference configuration, in which case both representations would coincide at time t. Notice that on the right-hand side of equation (5.210) we have used a tilde over the function to indicate that the functional dependence thereat is different from that in equation (5.209). This notational distinction is seldom enforced, since the context and/or the indication of the independent variables are sufficient to deduce which representation (Lagrangian or Eulerian) is being used.

In general, given some extensive physical quantity Ψ, we have at our disposal either the Lagrangian field

$$\Psi = \Psi(X^1, X^2, X^3, t),\tag{5.211}$$

or the corresponding Eulerian field

$$\Psi = \tilde{\Psi}(x^1, x^2, x^3, t).\tag{5.212}$$

The motion itself can be used to establish the relation between these two representations, via the following composition identity:

$$\Psi(X^1, X^2, X^3, t) = \tilde{\Psi}\left(x^1(X^1, X^2, X^3, t), x^2(X^1, X^2, X^3, t), x^3(X^1, X^2, X^3, t), t\right),\tag{5.213}$$

or, more compactly,

$$\Psi(X^I, t) = \tilde{\Psi}\left(x^i(X^I, t), t\right).\tag{5.214}$$

Now assume that one needs to calculate the time derivative of the quantity Ψ. Will this be the time derivative of the function $\Psi(X^I, t)$ or that of the function $\tilde{\Psi}(x^i, t)$? They both make sense. The first one describes the rate of change of the quantity Ψ *at a fixed material particle*, while the second provides the rate of change of Ψ *at a fixed position in space*. To visualize this distinction, imagine that Ψ is the temperature field. To repeat the argument of Section 1.10, the derivative $\frac{\partial \tilde{\Psi}}{\partial t}$ is what an observer would measure and calculate while sitting with a thermometer in a laboratory chair, whereas the derivative $\frac{\partial \Psi}{\partial t}$ is what a thermocouple glued to the particle would allow us to determine. Because the Eulerian representation is closer to the experimental setting, but the time derivative at a fixed particle has the more intrinsic meaning, it is important to obtain a formula that allows us to calculate the latter in terms of the former. The result is called the *material deriva-tive* (or *material time derivative*). It is the partial time derivative of Ψ, but expressed in terms of partial derivatives of $\tilde{\Psi}$. By the chain rule applied to equation (5.214), we obtain

$$\frac{\partial \Psi}{\partial t} = \frac{\partial \tilde{\Psi}}{\partial t} + \sum_{i=1}^{3} \frac{\partial \tilde{\Psi}}{\partial x^i} \frac{\partial x^i}{\partial t}. \tag{5.215}$$

To avoid the cumbersome tilde, it is customary to denote the material derivative by D/Dt, or with a superimposed dot, while retaining the symbol Ψ without a tilde. By virtue of equation (5.209), the material derivative can be written as

$$\frac{D\Psi}{Dt} = \dot{\psi} = \frac{\partial \Psi}{\partial t} + \sum_{i=1}^{3} \frac{\partial \Psi}{\partial x^i} v^i. \tag{5.216}$$

In other words, the material derivative corrects the partial time derivative by means of an additive term equal to the inner (or dot) product of the spatial gradient of the field with the velocity. A moment's reflection reveals that this correction is exactly what one should expect. Indeed, imagine water flowing through a narrow tube. As time goes on, two fixed spatial observers are recording the water temperature at two windows separated by an axial distance Δx. Imagine that each one of them happens to observe a constant (but different) temperature (stationary temperature). Is the temperature of the particles passing through the windows constant in time? Evidently not. If the instantaneous speed of the fluid is v, the time it takes for a particle to go from the first window to the next is $\Delta t = \Delta x/v$. If the difference of the temperatures recorded by the observers is $\Delta \Psi$ it is clear that the rate of change of the temperature *as would be measured by an observer moving with the particle* is $\Delta \Psi/(\Delta x/v) = (\Delta \Psi/\Delta x) v$, which is the one-dimensional version of our correction term.

Exercise 5.5.21   The acceleration. The acceleration of a particle is the second time derivative of the motion or, equivalently, the material time derivative of the velocity field. Assuming that the velocity field is known in its Eulerian representation, find the expression for the acceleration field. [Hint: apply the formula (5.216) to each component]. Notice that the result contains a ('convected') term non-linear in the velocity.

5.5.12   Change of Reference Configuration

A reference configuration, as we have already established, is a convenient device that allows us to express fields and their derivatives and to eventually formulate the governing equations in terms of coordinates. It should be clear, on the other hand, that, whereas the spatial configurations are real (in the sense that they are amenable to experiment and measurement), the reference configuration is, in principle, an arbitrary fiction. It is important, therefore, to be aware of the explicit dependence of a given physical quantity on the reference configuration chosen. More precisely, we want to determine in which way the components of a physical quantity change upon a change of reference configuration.

Let a motion be referred to two different reference configurations, κ_0 and κ_1, as shown in the Figure 5.12. They are related by composition

$$\kappa_1 = \xi \circ \kappa_0, \tag{5.217}$$

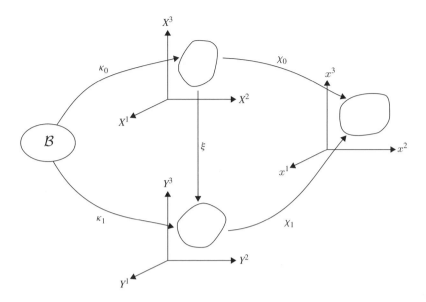

Figure 5.12   Change of reference configuration

where ξ is the deformation of $\kappa_1(\mathcal{B})$ with respect to κ_0. Denoting by $X^I(I = 1, 2, 3)$ and $Y^A(A = 1, 2, 3)$ the natural coordinates of material points in the configurations $\kappa_0(\mathcal{B})$ and $\kappa_1(\mathcal{B})$ respectively, the transition function ξ is given by three time-independent smooth functions

$$Y^A = Y^A(X^1, X^2, X^3), \quad A = 1, 2, 3. \tag{5.218}$$

By the chain rule of differentiation, we obtain for the relation between the corresponding deformation gradients $\mathbf{F}_0$ and $\mathbf{F}_1$ the expression

$$\mathbf{F}_0 = \mathbf{F}_1 \, \nabla \xi, \tag{5.219}$$

where ∇ stands for the gradient operator. This is the desired rule for change of reference configuration. In coordinates, it reads:

$$\frac{\partial x^i}{\partial X^I} = \sum_{A=1}^{3} \frac{\partial x^i}{\partial Y^A} \frac{\partial Y^A}{\partial X^I}, \quad I, i = 1, 2, 3. \tag{5.220}$$

Remark 5.5.22   Reference configurations and coordinates. Our notion of reference configuration consists of a map from the body $\mathcal{B}$ into $\mathbb{R}^3$. Since $\mathbb{R}^3$ is endowed with a natural coordinate system, X^I say, this map is indeed an assignment of coordinate labels to each material body point. Moreover, $\mathbb{R}^3$ is also endowed with a natural inner product, so that in the given reference configuration lengths and angles can be calculated accordingly. Without changing this configuration, on the other hand, it is quite possible to consider other coordinate systems in $\mathbb{R}^3$ which need not even be Cartesian. A common example is provided by the use of cylindrical or spherical coordinates in problems exhibiting geometric symmetries. The particle labels in these new coordinate systems will, of course, change. Nevertheless, since the configuration is fixed, the lengths and angles should be calculated in a manner consistent with the original coordinates. Thus, Pythagoras' theorem in the new coordinates will attain in general a more involved expression than in the natural coordinates. In contradistinction to this situation, the adoption of a new reference configuration entails a map from the body to a new copy of $\mathbb{R}^3$ with natural coordinates Y^A. By assigning new numerical coordinate labels to the same particles, this looks very much like a coordinate transformation, but the lengths and angles in the new reference configuration are not necessarily consistent with those in the old one. In the new natural coordinate system Y^A, the usual form of Pythagoras' theorem holds. Some, but not all, of the transition formulas for coordinate transformations within the same reference configuration, on the one hand, and for changes of reference configuration, on the other hand, are the same. Those that are not involve precisely the metric properties of the surrounding space.

5.5.13   The Velocity Gradient

If we compute the material derivative of equation (5.219) we obtain

$$\dot{\mathbf{F}}_0 = \dot{\mathbf{F}}_1 \, \nabla \xi, \tag{5.221}$$

since ξ is independent of time. For economy, we use a dot over a symbol to denote its material derivative. Using equation (5.219) itself to read off $\nabla \xi$, it turns out that the combination

$$\mathbf{L} = \dot{\mathbf{F}}_0 \, \mathbf{F}_0^{-1} = \dot{\mathbf{F}}_1 \, \mathbf{F}_1^{-1} = \dot{\mathbf{F}} \, \mathbf{F}^{-1} \tag{5.222}$$

is independent of the reference configuration chosen. It is a purely spatial tensor called the *velocity gradient*. The name arises from the fact that, since the quantity is independent of reference configuration, we may choose the present configuration as reference, whereby the deformation gradient is instantaneously equal to $\mathbf{I}$, immediately leading (by the equality of mixed partial derivatives) to the interpretation of $\mathbf{L}$ as the spatial gradient of the velocity field. In terms of components, we have

$$L^i_j = \frac{\partial v^i}{\partial x^j}. \tag{5.223}$$

Remark 5.5.23   Covariant derivatives. If non-Cartesian coordinates are used in space, the partial derivatives should be replaced with the *covariant derivatives* associated with that coordinate system, a topic beyond our scope. It is customary to indicate covariant derivatives with a comma preceding the coordinate index, so that equation (5.223) would read

$$L^i_j = v^i{}_{,j}. \tag{5.224}$$

Naturally, this notation can also be used for Cartesian coordinates, in which case the covariant derivative is equal to the partial derivative.

Remark 5.5.24 Using the comma notation introduced in Remark 5.5.23 and the summation convention presented in Section 5.5.5, equation (5.77) for the divergence of a vector field can be written more compactly as

$$\operatorname{div} \mathbf{f} = f^i{}_{,i}. \tag{5.225}$$

Interpreting commas as indicating covariant derivatives, this formula is invariant under arbitrary changes of coordinates, whether Cartesian or not.

Adopting the present configuration (at time t) as an instantaneous reference, and indicating corresponding quantities with the subscript t, the polar decomposition $\mathbf{F}_t = \mathbf{R}_t \mathbf{U}_t$ leads by differentiation to

$$\mathbf{L} = \dot{\mathbf{F}}_t = \dot{\mathbf{R}}_t + \dot{\mathbf{U}}_t, \tag{5.226}$$

where we have used the fact that $\mathbf{F}_t = \mathbf{R}_t = \mathbf{U}_t = \mathbf{I}$ at time t. We observe that $\mathbf{D} = \dot{\mathbf{U}}_t$ is symmetric. On the other hand, since $\mathbf{R}$ is always orthogonal, it follows (by taking the material derivative of $\mathbf{R}\mathbf{R}^T = \mathbf{I}$) that $\dot{\mathbf{R}}_t$ (at time t) is skew-symmetric. By the uniqueness of the additive decomposition of a matrix into symmetric and skew-symmetric parts we obtain that

$$\mathbf{D} = \dot{\mathbf{U}}_t = \frac{1}{2}(\mathbf{L} + \mathbf{L}^T) \tag{5.227}$$

and

$$\mathbf{W} = \dot{\mathbf{R}}_t = \frac{1}{2}(\mathbf{L} - \mathbf{L}^T) \tag{5.228}$$

The tensors $\mathbf{D}$ and $\mathbf{W}$ are called *stretching* (or rate of deformation) and *spin* (or vorticity), respectively.

Exercise 5.5.25 **Rates and the linearized theory.** Compare the expressions (5.227) and (5.228) with the strain and rotation of the geometrically linearized theory (Section 5.5.9) and draw your own conclusions.

6

Balance Laws and Constitutive Equations

6.1 Preliminary Notions

6.1.1 Extensive Properties

An *extensive property* in continuum mechanics is a physical quantity assigned to each *subset* of a material body $\mathcal{B}$. For example, the mass, the electric charge, the momentum and the internal energy are extensive properties. The temperature, the strain, the stress, the heat flux vector, on the other hand, are not extensive properties, but rather quantities, known as *intensive properties*, assigned to each *point* of the body.

Let Ψ be an extensive property. As just described, an extensive property is characterized by a *set function*, namely, a function Ψ whose argument is a variable subset of the total set $\mathcal{B}$. It is usually assumed, on physical grounds, that extensive properties are *additive*, namely, if $\mathcal{R}_1$ and $\mathcal{R}_2$ are two *disjoint* subsets of $\mathcal{B}$, then

$$\Psi(\mathcal{R}_1 \cup \mathcal{R}_2) = \Psi(\mathcal{R}_1) + \Psi(\mathcal{R}_2). \tag{6.1}$$

Under appropriate continuity assumptions, an additive set function Ψ can be shown to be completely characterized by a *density*, namely, an ordinary function g, via an integral

$$\Psi(\mathcal{R}) = \int_{\mathcal{R}} g \, d\mathcal{R}. \tag{6.2}$$

In this expression, $\mathcal{R} \subset \mathcal{B}$ is an arbitrary measurable subset of $\mathcal{B}$, while the volume element $d\mathcal{R}$ depends on whether Ψ is evaluated in the reference or the spatial configuration. It may be said that the extensive properties under consideration are such that they can ultimately be expressed in terms of an intensive property, namely, the appropriate density function.[1]

Remark 6.1.1 Strictly speaking, the integration in equation (6.2) should be understood in the sense of Lebesgue. What this means is that the function Ψ is well defined and

---

[1] The additivity assumption may be interpreted physically as ruling out scale, or non-local, effects.

The Elements of Continuum Biomechanics, First Edition. Marcelo Epstein.
© 2012 John Wiley & Sons, Ltd. Published 2012 by John Wiley & Sons, Ltd.

vanishes when evaluated on sets of measure zero. In particular, the extensive property Ψ cannot be concentrated on subsets of dimension lower than 3.

6.1.2 Transport Theorem

In the formulation of balance laws in continuum mechanics one is confronted with the evaluation of the rate of change of the content of an extensive physical quantity (such as momentum, energy or entropy) in the body. Let such content Ψ be given by the (Riemann) integral of some density g over a *spatial* volume of interest v, namely

$$\Psi(t) = \int_{v(t)} g(x^i; t)\, dv. \tag{6.3}$$

If we were to follow this volume as it is dragged by the motion of the body (i.e., if we were to ensure that we are always enclosing the same material particles), we would perceive that the function Ψ depends on time by virtue of two independent factors: (i) the dependence of the integrand itself on time; (ii) the dependence of the (moving) domain of integration on time. Our objective is to calculate the derivative of this function of time.

Although it is mathematically possible to carry out a careful passage to the limit in the form

$$\frac{d\Psi}{dt} = \lim_{h \to 0} \frac{\Psi(t+h) - \Psi(t)}{h}, \tag{6.4}$$

we will opt for a more physically meaningful derivation. Since we have agreed to let our volume be dragged by the motion χ of the body, we can exploit the motion itself to transform (or *pull back*) the integration to the reference configuration, where the corresponding domain $V = \chi^{-1}(v)$ is independent of time. This pullback is

$$\Psi(t) = \int_{V} g[x^i(X^I; t); t]\, J\, dV. \tag{6.5}$$

The appearance of the determinant $J = \det \mathbf{F}$ is explained by equation (5.195) or, equivalently, by the rule of change of variables of integration. Since the new domain of integration is fixed in time, the order of integration and differentiation is immaterial.[2] We obtain, therefore,

$$\frac{d\Psi}{dt} = \int_{V} \left[\frac{Dg}{Dt}\, J + g\, \frac{DJ}{Dt} \right] dV. \tag{6.6}$$

To proceed, we need an explicit formula for the derivative of the determinant (J) of a matrix $(\mathbf{F})$ depending on a parameter (t). This formula is

$$\dot{J} = J\, \text{trace}\, (\mathbf{F}^{-1}\dot{\mathbf{F}}), \tag{6.7}$$

as obtained in Exercise 6.1.2.

---

[2] Integration and differentiation are both defined as limits. As long as two successive passages to the limit are independent of each other, they are commutative. On the other hand, if the process of passage to the limit in one variable is intertwined with the passage to the limit in the other, the limits no longer commute.

Exercise 6.1.2 The derivative of a determinant. Deduce equation (6.7) for the derivative of the determinant of a matrix depending on a parameter. Use equation (5.108).

In view of equations (5.222), (5.223) and (5.77), we obtain

$$\dot{J} = J \text{ trace } (\mathbf{F}^{-1}\dot{\mathbf{F}}) = J \text{ trace } (\dot{\mathbf{F}}\mathbf{F}^{-1}) = J \text{ trace } \mathbf{L} = J \text{ div } \mathbf{v}, \qquad (6.8)$$

where div stands for the spatial divergence. Plugging these results back into equation (6.6), we obtain

$$\frac{d\Psi}{dt} = \int_V \left(\frac{Dg}{dt} + g \text{ div } \mathbf{v} \right) J \, dV, \qquad (6.9)$$

which allows us to change the domain of integration back to the original v:

$$\frac{d\Psi}{dt} = \int_v \left[\frac{Dg}{Dt} + g \text{ div } \mathbf{v} \right] dv. \qquad (6.10)$$

This result is known as *Reynolds' first transport theorem*. Similar theorems can be proven for integrals along areas and lines, but will not be needed here.

Remark 6.1.3 An apparent paradox. A *control volume* is an arbitrary bounded spatial region $v(t)$ which varies smoothly in time. As a moving volume, a control volume may have nothing to do with the motion of a material body. The only velocity that we can intrinsically assign to a control volume is the normal speed at each point of its boundary ∂v, since there is no other natural way to associate points at two 'successive' positions (at times t and $t + dt$) of this moving entity but to draw the normal to the boundary at a point, measure the length dn of the segment between the two surfaces and then divide by dt. Assume that an extensive property $g(x, t)$ (such as electromagnetic energy) exists in space. It makes sense to ask how the content captured by the control volume changes in time. This question has nothing to do with the motion of a material, and yet we could imagine any material motion we wish (such that the particle velocities at the boundary happen to have a normal component equal to that of the surface, so that no particles escape) and then apply Reynolds' theorem. The result would obviously be correct, regardless of the velocity field of the assumed motion within the body, as long as the normal velocity on the boundary equals that of the given control volume. On the other hand, according to Reynolds' transport theorem, the integral on the right-hand side of equation (6.10) does depend on the values of the velocity field within the region! To solve this apparent paradox, we need only invoke our formula for the material derivative (5.216) to rewrite (6.10) as

$$\frac{d\Psi}{dt} = \int_v \left[\frac{\partial g}{\partial t} + \frac{\partial g}{\partial x^i} v^i + g \text{ div } \mathbf{v} \right] dv = \int_v \left[\frac{\partial g}{\partial t} + \text{div}(g\mathbf{v}) \right] dv. \qquad (6.11)$$

By the divergence theorem, we may change the volume integral of a divergence into the total flux through the boundary, so that

$$\frac{d\Psi}{dt} = \int_v \frac{\partial g}{\partial t} dv + \int_{\partial v} (g\mathbf{v} \cdot \mathbf{n}) \, da, \qquad (6.12)$$

where $\mathbf{n}$ is the unit normal to the element of area da of the boundary ∂v. As expected, the result depends only on the normal speed of the boundary.

Exercise 6.1.4   A direct 'derivation' of Reynolds' theorem. Starting from equation (6.4), provide a direct derivation of Reynolds' transport theorem in the form (6.12). Notice that the volume per unit time swept by the boundary is given by the integral over the boundary of the normal velocity.

Exercise 6.1.5   The transport theorem and the space-time divergence theorem. Provide an alternative proof of Reynolds' transport theorem by applying the divergence theorem in $\mathbb{R}^4$ to the vector field $\psi\,\mathbf{w}$, where $\mathbf{w}$ is the 4-velocity. The 4-velocity results from introducing a material time coordinate T in the body numerically equal to the physical time t. Its first three components provide the ordinary velocity vector $\mathbf{v}$, while the fourth component is always equal to 1. The trace of the body motion in space-time has a tubular shape. Applying the divergence theorem to the field $\psi\,\mathbf{w}$ within the four-dimensional tube contained between two nearby times, the theorem follows directly.

Exercise 6.1.6   Leibniz's formula. Show that, in a one-dimensional context, Reynolds' transport theorem reduces to the *Leibniz integral rule* for the derivative of an integral with variable limits, namely

$$
\frac{d}{dt}\int_{a(t)}^{b(t)} f(x,t)dx = \int_{a(t)}^{b(t)} \frac{\partial f(x,t)}{\partial t}dx + \frac{db(t)}{dt}f(b(t),t) - \frac{da(t)}{dt}f(a(t),t). \tag{6.13}
$$

This result was mentioned in Section 2.3.

6.2   Balance Equations

6.2.1   The General Balance Equation

A *balance equation* in continuum physics is a rigorous accounting of the causes for the change in content, over a certain domain, of an extensive physical quantity.[3] In continuum mechanics, in particular, we attribute physical content only to a material body or a part thereof (as opposed to a field theory such as electromagnetism, where a vacuum can be a carrier of fields). Just like in balancing a cheque book, we want to account for the change in content of the quantity Ψ in the body as time goes on.

   We will postulate that the time rate of change of the content of a physical quantity Ψ in the body is entirely due to only two causes: a *production* (or source) Π within the body and a *flux* Φ through its boundary:

$$
\frac{d\Psi}{dt} = \Pi + \Phi. \tag{6.14}
$$

Naturally, both the production and the flux are rates (per unit time). This equation is the generic statement of an equation of balance.

   We want, perhaps at a price, to convert this general statement into a more specific form suitable for computations. We can do so in either the referential (Lagrangian) or the spatial (Eulerian) setting. A configuration is assumed to inherit the Cartesian volume element of the ambient space, so that, according to the assumptions made in Section 6.1.1, we can

---

[3] Sections 6.2 and 6.3 parallel the one-dimensional treatment of Chapters 2 and 3 and follow closely the condensed presentation in the Appendix of Epstein (2010).

represent the total content in terms of the referential volume element dV or its spatial counterpart dv as follows:

$$\Psi = \int_V G\, dV = \int_v g\, dv,\tag{6.15}$$

where V and v are the corresponding material volumes in reference and space, and where G and g represent, respectively, the content of Ψ per unit referential and spatial volume. These two densities are related by

$$G = Jg.\tag{6.16}$$

Similar assumptions apply to the production, thus yielding:

$$\Pi = \int_V P\, dV = \int_v p\, dv,\tag{6.17}$$

with $P = Jp$.

The flux term needs some more finessing. Firstly, we will assume that a flux is defined not only through the boundary of the body, but also through any internal boundary between sub-bodies. Moreover, we will assume that the flux is given by an integral of a flux density over the area of interest. But, unlike the case of the volume integrals, now we are not allowed to claim that this density is a mere function of position on the surface. A simple example will clarify this situation. If you are sunbathing on the beach, it makes a difference whether you are lying horizontally or standing up. The heat energy flows at a maximum rate when the exposed surface is perpendicular to the rays. This is as true for the outer boundary as for an internal point. When you consider an interior point of the endodermis, the flux at that point will vary depending on the orientation of the imagined surface element you consider at the point. Inspired by this example, we will assume that the flux density depends on the local normal to the boundary.[4] Having made these assumptions, we can write

$$\Phi = \int_{\partial V} H(X^I, N^I; t)\, dA = \int_{\partial v} h(x^i, n^i; t)\, da.\tag{6.18}$$

where N^I and n^i are the exterior unit normals to the area elements dA and da of the boundaries ∂V and ∂v, and where H and h represent, respectively, the Lagrangian and Eulerian flux densities per unit area and per unit time.

To obtain the relation between the Lagrangian and Eulerian flux densities, we would like to invoke equation (5.207). But there is a snag, since in this equation the dependence on the normal is explicitly linear, whereas so far our assumed dependence of the flux densities on the normal is arbitrary. It would appear that the only way out of the impasse would be to *assume* such linearity (a logical thing to do, particularly considering the example of the sun's rays, whose effect on the skin is governed by a simple projection on the normal). It is a rather remarkable fact, however, that such an assumption is superfluous, for it can be *derived* from the general statement of the balance law. This result is justly called *Cauchy's theorem*, since it was Cauchy who, in a series of papers published in the 1820s, first established its validity in the context of his theory of stress.

---

[4] In principle, the dependence on the boundary could be of higher order, involving, for example, the local curvature of the boundary.

To see how this works, we will use, for example, the Lagrangian setting. The generic equation of balance so far reads:

$$\frac{d}{dt} \int_V G \, dV = \int_V P \, dV + \int_{\partial V} H \, dA.$$

(6.19)

The only extra assumption we need to make (actually, already implicitly made) is that this statement of the general balance law is valid, with the same densities, for any sub-body. This assumption essentially eliminates scale effects involved in phenomena such as surface tension. This being the case, the theorem follows from the *Cauchy tetrahedron argument*.

Theorem 6.2.1   Cauchy's theorem. *If the balance equation (6.19) is valid for any sub-body V, the dependence of the flux density H on the local normal to the boundary is necessarily linear, namely, there exists a linear operator $\mathbf{H}$ such that*

$$H(X^I, N_I; t) = \mathbf{H}(X^I; t) \cdot \mathbf{N}.$$

(6.20)

Proof. Let P be a point lying on ∂V and let $\mathbf{N}$ be the exterior unit normal at P, as illustrated in Figure 6.1. To sketch the proof, we begin by considering a tetrahedron within the body with three faces aligned with the coordinate system and the fourth face having its vertices lying on a portion of ∂V containing P. We apply the balance equation (6.19) to this tetrahedron. According to the mean value theorem, we can approximate the two volume integrals appearing in equation (6.19) as the product of the volume of the tetrahedron times the value of the integrand at an interior point. As far as the surface integral is concerned, the mean value theorem delivers a sum of four terms, one for each face of the tetrahedron. Each one of these four terms consists of the product of the area of the face times the value of the flux density at some interior point of that face. We now let the tetrahedron shrink by moving the far vertex towards the surface, along the line joining it with P. In this limiting process, as the volume of the tetrahedron becomes smaller and smaller, the two volume terms approach zero with the cube of the length k

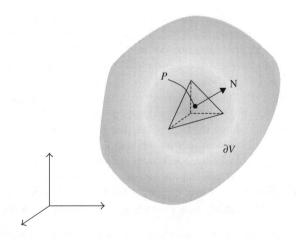

Figure 6.1   The Cauchy tetrahedron

of a side of the tetrahedron. The four surface terms, on the other hand, approach zero as the square of k. Therefore, dividing the whole expression by k^2, we see that the volume terms cancel in the limit and the only survivors are the four surface terms. Recall that each of these terms consists of a product of the value of H at an interior point times the area of a triangular face. Accordingly, the surviving expression has the following form:

$$-H^1 \, \Delta A_1 - H^2 \, \Delta A_2 - H^3 \, \Delta A_3 + H \, \Delta A \to 0. \tag{6.21}$$

In this expression, we have called $-H^I$ and ΔA_I, respectively, the flux density and the area of the face whose normal is aligned with the coordinate axis X^I, while H and ΔA represent the flux and the area of the inclined face of the tetrahedron. But in a tetrahedron, the areas of the faces are not independent. In fact, the area of the incline multiplied by each component of the unit normal delivers precisely the area of the corresponding face, namely

$$\Delta A_I = \Delta A \, N_I. \tag{6.22}$$

The result follows suit. In other words, there exists a linear operator **H** (function of position alone) which, when acting on the normal **N**, delivers the flux H per unit area. ∎

We observe that if the quantity being balanced is a scalar, as we have assumed, then **H** is a vector and the linear operation is given by the dot product $H = \mathbf{H} \cdot \mathbf{N}$, or

$$H = H^I N_I. \tag{6.23}$$

If, on the other hand, the quantity being balanced is vectorial, the theorem can be applied component by component and **H** becomes a tensor acting on the normal. Thus, for example, for a spatial vector density we would write

$$H^i = H^{iI} \, N_I. \tag{6.24}$$

In equations (6.23) and (6.24) we have used Einstein's summation convention, described in Section 5.5.5. From now on, we will strictly adhere to this summation convention.

In view of Cauchy's theorem, we can write the flux term in equation (6.19) as

$$\int_{\partial V} H \, dA = \int_{\partial V} \mathbf{H} \cdot \mathbf{N} \, dA, \tag{6.25}$$

and similarly for the Eulerian formulation,

$$\Phi = \int_{\partial v} \mathbf{h} \cdot \mathbf{n} \, da, \tag{6.26}$$

with an obvious notation. Invoking now equation (5.208), we obtain the following relation between the Lagrangian and Eulerian flux densities:

$$\mathbf{H} = J \, \mathbf{F}^{-1} \, \mathbf{h}. \tag{6.27}$$

Finally, the integral statement of the generic law of balance in Lagrangian form is

$$\frac{d}{dt} \int_V G \, dV = \int_V P \, dV + \int_{\partial V} \mathbf{H} \cdot \mathbf{N} \, dA, \tag{6.28}$$

and in Eulerian form

$$\frac{d}{dt} \int_v g \, dv = \int_v p \, dv + \int_{\partial v} \mathbf{h} \cdot \mathbf{n} \, da. \tag{6.29}$$

These integral statements of the balance law may suffice for numerical computations (such as with the finite-volume method). Nevertheless, under suitable assumptions of smoothness, we can obtain their local (differential) forms by using the divergence theorem to convert the flux integral into a volume integral and by invoking the fact that the resulting three integrals must be identically equal regardless of the domain of integration within the body. The results are, therefore,

$$\frac{\partial G}{\partial t} = P + \text{Div } \mathbf{H} \tag{6.30}$$

and

$$\frac{Dg}{Dt} + g \, \text{div } \mathbf{v} = p + \text{div } \mathbf{h}, \tag{6.31}$$

where the latter equation necessitated the application of Reynolds' transport theorem. The operators Div and div stand for the referential and spatial divergence, respectively. To avoid confusion, we list the component versions of these equations:

$$\frac{\partial G}{\partial t} = P + H^I_{,I} \tag{6.32}$$

and

$$\frac{Dg}{Dt} + g \, v^i_{,i} = p + h^i_{,i}. \tag{6.33}$$

If the quantity of departure were vectorial (whether referential or spatial), we would simply apply these equations component by component (taking advantage of the Euclidean structure), the result being an added index (referential or spatial) to each quantity. Appropriate boundary conditions are needed, but we will not deal with them at this stage.

6.2.2 The Balance Equations of Continuum Mechanics

We have devoted perhaps too much detail to the justification of the general form of an equation of balance in both the Lagrangian and the Eulerian forms. In this section we hope to reap the reward when applying the general prescription just obtained to the five fundamental quantities to be balanced in a traditional continuum mechanics treatment.

Conservation of Mass

A balance law is said to be a *conservation law* if the production and the flux vanish identically. This is the case for the mass of a continuum.[5] Denoting by ρ_0 and ρ the

---

[5] In modern theories of biological growth, however, or in theories of chemically reacting mixtures (when looking at each component of the mixture), conservation of mass does not hold, and specific mass production and/or flux terms are to be included.

referential and spatial mass densities, respectively, we obtain, by a direct application of equations (6.30) and (6.31), the Lagrangian and Eulerian differential versions as

$$\frac{\partial \rho_0}{\partial t} = 0 \tag{6.34}$$

and

$$\frac{D\rho}{Dt} + \rho \operatorname{div} \mathbf{v} = 0. \tag{6.35}$$

The latter (Eulerian) version is known in hydrodynamics as the *continuity equation*. Expanding the material time derivative, it can also be written as

$$\frac{\partial \rho}{\partial t} + \operatorname{div}(\rho \mathbf{v}) = 0, \tag{6.36}$$

or in other equivalent forms.

Balance of Linear Momentum

This balance is a statement of Newton's second law as applied to a deformable continuum. It is, therefore, important to recall that we must assume that our frame of reference (which we have identified with a Cartesian coordinate system) is actually *inertial*. The quantity to be balanced is the (vectorial) linear momentum, whose Lagrangian and Eulerian densities are, respectively, $\rho_0 \mathbf{v}$ and $\rho \mathbf{v}$. Note that in both cases we have a *spatial* vector to balance, whether the statement of the law is referential or spatial.

In broad terms, Newton's second law of motion establishes that the sources of change of the linear momentum of a body are the external forces acting on the body. Accordingly, the production term is given by the (distributed) forces per unit volume, or *body forces*, with densities $\mathbf{B}$ and $\mathbf{b}$, respectively, in the Lagrangian and Eulerian formulations. The flux term is represented by the *surface tractions* (i.e., forces per unit referential or spatial area) $\hat{\mathbf{T}}$ and $\hat{\mathbf{t}}$, respectively, acting on the boundary. We again emphasize that these forces, even when measured per unit referential area, are *spatial* vectors. The integral Lagrangian and Eulerian forms of the equation of balance of linear momentum are, respectively,

$$\frac{d}{dt} \int_V \rho_0 v^i \, dV = \int_V B^i \, dV + \int_{\partial V} \hat{T}^i \, dA \tag{6.37}$$

and

$$\frac{d}{dt} \int_v \rho v^i \, dV = \int_v b^i \, dV + \int_{\partial v} \hat{t}^i \, da. \tag{6.38}$$

By Cauchy's theorem, we know that the surface tractions are governed by a (tensorial) flux, which we will denote by $\mathbf{T}$ and $\mathbf{t}$ for the Lagrangian and Eulerian settings, respectively. To avoid confusion, we express them in components as follows:

$$\hat{T}^i = T^{iI} N_I, \tag{6.39}$$

and

$$\hat{t}^i = t^{ij} n_j, \tag{6.40}$$

Identifying the various terms with their counterparts in the corresponding generic law of balance, and invoking the conservation of mass already obtained, we obtain the following Lagrangian and Eulerian forms of the balance of linear momentum:

$$T^{iI}_{,I} + B^i = \rho_0 \frac{Dv^i}{Dt} \tag{6.41}$$

and

$$t^{ij}_{,j} + b^i = \rho \frac{Dv^i}{Dt}. \tag{6.42}$$

The tensorial fluxes **T** and **t** are called, respectively, the *first Piola–Kirchhoff stress* (or just the *Piola stress*) and the *Cauchy stress*. The physical meaning of the Cartesian components of the Cauchy stress **t** at a spatial point P is that of components of forces per unit spatial area on the coordinate planes passing through P, as shown in Figure 6.2. Note that, whereas the Cauchy stress is a purely spatial tensor (a linear map of the tangent space at a point in the current configuration into itself), the first Piola–Kirchhoff stress is a mixed tensor (a linear map between the tangent space at the reference point and its counterpart in space). The Piola and Cauchy stresses produce linearly out of unit vectors with components N_I and n_i, respectively, the spatial forces per unit area acting on the elements of area to which they are normal. The relation between these tensors follows directly from equation (6.27):

$$T^{iI} = J \, (F^{-1})^I_j \, t^{ij}. \tag{6.43}$$

The physical meaning of the Piola stress can be gathered from equation (6.39). It is a linear operator that, when applied to a unit vector **N** in the reference configuration, produces a force $\hat{\mathbf{T}}$ in the spatial configuration measured per unit referential area perpendicular to **N**. If we represent the Piola stress tensor by means of a matrix [**T**] and the

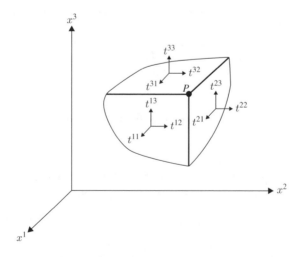

Figure 6.2   The Cartesian components of the Cauchy stress tensor at a point P as components of forces per unit spatial area acting on this side of the coordinate planes through P

vector $\mathbf{N}$ by means of a column $\{\mathbf{N}\}$, then the column vector representing the force $\hat{\mathbf{T}}$ is given by the matrix product $\{\hat{\mathbf{T}}\} = [\mathbf{T}]\{\mathbf{N}\}$. The Cauchy stress, on the other hand, acts on a unit vector $\mathbf{n}$ in the spatial configuration to produce a force $\hat{\mathbf{t}}$ measured per unit spatial area perpendicular to $\mathbf{n}$. The forces per unit area produced by these two tensors differ only in magnitude, but not in direction, the difference being due to the different surfaces of reference. If dA is an area perpendicular to $\mathbf{N}$, the deformation transforms it into a different area da perpendicular to $\mathbf{n}$. The relation between the forces per unit area just discussed is given by $\hat{T}^i \, dA = \hat{t}^i \, da$.

In addition to these two stress tensors, which arise naturally from the Lagrangian and Eulerian formulations of the balance of linear momentum, it is possible to define other stress tensors that are useful in applications. One of the most widely used is the *second Piola–Kirchhoff stress tensor* $\mathbf{S}$ defined as

$$\mathbf{S} = \mathbf{F}^{-1} \, \mathbf{T} \tag{6.44}$$

or, in components,

$$S^{IJ} = \left(F^{-1}\right)^I_i \, T^{iJ}. \tag{6.45}$$

The second Piola–Kirchhoff stress is purely referential. When acting on the normal vector $\mathbf{N}$, this tensor does not produce the actual physical force per unit referential area, but rather the pullback of this force to the reference configuration. For this reason, the formulation of the law of balance of momentum in terms of the second Piola–Kirchhoff stress leads to an awkward outcome. The main usefulness of $\mathbf{S}$ is its convenient application to the formulation of constitutive equations.

Remark 6.2.2   The two sides of a plane. Given a plane (or, more generally, a surface element), we have the option of choosing either of the two unit normals $\mathbf{n}$ and $-\mathbf{n}$. The physical meaning of this choice is the specification of one or the other side of this plane. The corresponding forces per unit area are $\hat{\mathbf{t}}^+ = \mathbf{tn}$ and $\hat{\mathbf{t}}^- = -\mathbf{tn}$, so that these forces are equal in magnitude and opposite in direction, as they should be. To provide a more graphical representation, imagine (in a two-dimensional picture) a surgeon making a small incision in the skin, thus opening a wound with two originally matching sides. The wound opens as a result of the release of the internal forces present in the skin. At the end of the operation, the surgeon sutures the wound by applying (with a thread, say) equal and opposite forces so as to bring the two sides of the wound again into coincidence and restore the original state of stress.

Remark 6.2.3   Normal and shear components. Given a point x on a spatial plane with unit normal $\mathbf{n}$, equation (6.39) provides the force per unit area acting on this plane as $\hat{t}^i = t^{ij} n_j$. This vector $\hat{\mathbf{t}}$ (with components $\hat{t}^i$) can be projected on the normal itself to provide the (scalar) result

$$\sigma = \hat{\mathbf{t}} \cdot \mathbf{n} = \hat{t}^i n_i = t^{ij} n_i n_j. \tag{6.46}$$

In the classical literature, this notation (σ) is used extensively to denote the *normal stress* acting on the given plane at the point x. To obtain the tangential part of the stress, we subtract the normal part as follows:

$$\tau = \hat{\mathbf{t}} - \sigma \mathbf{n}, \tag{6.47}$$

or, in components:

$$\tau^k = \hat{t}^k - \sigma n^k = t^{kj} n_j - t^{ij} n_i n_j n^k. \tag{6.48}$$

The magnitude $|\tau|$ of the vector τ constitutes the *shear stress* acting on the given plane at x.

Exercise 6.2.4 Verify that

$$\tau^k n_k = 0, \tag{6.49}$$

so that the vector with components τ^k indeed lies on the plane with normal **n**.

Exercise 6.2.5   The engineering notation. Apply equation (6.46) to each of the three coordinate planes and show that σ coincides, in each case, with t^{11}, t^{22} and t^{33}, respectively. The usual engineering notation (going back to Cauchy himself) for these 'normal stresses' is σ^1, σ^2 and σ^3, or σ_x, σ_y and σ_z, respectively. The remaining components, namely t^{ij} (with $i \neq j$), are the 'tangential or shear stresses', denoted usually by τ^{ij} (with $i \neq j$) or, more commonly, by $\tau_{xy}, \tau_{xz}, \tau_{yx}, \tau_{yz}, \tau_{zx}, \tau_{zy}$.

Remark 6.2.6   On the acceleration term. Note that in the Lagrangian version (6.41), the material time derivative of the velocity (appearing on the right-hand side of the equation) reduces to a partial derivative, since the velocity field is expressed in terms of X^I and t. In the Eulerian version (6.42), on the other hand, the velocity field is expressed in terms of x^i and t, and the material time derivative includes the non-linear convected term $\frac{\partial v^i}{\partial x^j} v^j$.

Exercise 6.2.7   Linear momentum. Work out the details of the derivation of the differential equations of balance of linear momentum.

Exercise 6.2.8 Write out explicitly, term by term, the three scalar equations of balance of linear momentum implied in equation (6.42). Express the result in terms of the engineering notation described in Exercise 6.2.5.

Exercise 6.2.9 Draw an infinitesimal die or brick with faces aligned with a Cartesian coordinate system. Indicate with arrows on each of the six faces of this infinitesimal brick the corresponding three components of the Cauchy stress, with due attention to the fact that, on parallel faces, they point in opposite directions and they differ by a differential amount. On the basis of this drawing, obtain the Eulerian equation of balance (6.42) by enforcing Newton's second law of motion on this 'free-body diagram'.

Exercise 6.2.10 Starting from equation (6.42), use relation (6.43) to derive the Lagrangian equation (6.41).

Balance of Angular Momentum

In the Newtonian mechanics of systems of particles, the law of balance of angular momentum follows from Newton's second law under the assumption that the particles of the system interact by means of forces abiding by Newton's third law ('action and reaction'). In the case of a continuum, the analogue of such internal forces is the stress tensor, but the analogy is not easy to pursue rigorously. In fact, we will soon discover that in continuum mechanics the postulation of a law of balance of momentum leads to a *new result*, without a clear analogue in discrete systems.

The law of balance of angular momentum states that the rate of change of the total angular momentum with respect to a fixed point (the origin, say) of an inertial frame is equal to the moment of all the external forces with respect to the same point. The density of the angular momentum (or moment of momentum) is given by the cross product $\mathbf{r} \times (\rho \mathbf{v})$, where $\mathbf{r}$ is the spatial position vector. The production density is $\mathbf{r} \times \mathbf{b}$, and the flux density is $\mathbf{r} \times (\mathbf{tn})$. Expressing these cross products by means of the permutation symbol (5.105), using the general prescription of the Eulerian balance law, and invoking conservation of mass and balance of linear momentum, the somewhat surprising final result is the survival of just one term, namely

$$\varepsilon_{ijk}\, t^{jk} = 0. \tag{6.50}$$

Exercise 6.2.11 Angular momentum. Work out the details of the derivation of the local law of balance of angular momentum to obtain the final result just stated. [Hint: recall that the cross product of a vector with itself vanishes.]

By the complete skew symmetry of the permutation symbol ε_{ijk}, an equivalent way to express equation (6.50) is

$$t^{ij} = t^{ji}, \tag{6.51}$$

that is to say, *the Cauchy stress is symmetric*. At this point it is appropriate to go back to the discrete analogy. There one assumes that the internal forces between the particles of the system abide by the principle of action and reaction. In the continuum case, what we have implicitly assumed is that the surface interactions are merely forces (stresses) and that there are no extra contributions of surface couples. A similar assumption was made regarding the external body forces. In other words, the only contribution to the moment equation is that arising from the moments of forces (no couple interactions). This assumption may have to be abandoned when dealing with electrically or magnetically polarizable materials. In those cases, the antisymmetric part of the stress may not vanish, but is balanced by the external body-couple.

Now that we have equation (6.51), we obtain the Lagrangian version by a direct use of equation (6.43):

$$\mathbf{FT}^{\mathrm{T}} = \mathbf{TF}^{\mathrm{T}}. \tag{6.52}$$

Note that the first Piola–Kirchhoff tensor is not symmetric in the ordinary sense, nor could it be, since it is a two-point tensor.

Exercise 6.2.12 Symmetry of the second Piola–Kirchhoff stress. Prove that the law of balance of angular momentum implies that the second Piola–Kirchhoff stress is symmetric.

Exercise 6.2.13 Given two mutually perpendicular spatial planes, show that the components of the tangential forces per unit area on each of the two planes at a point on the common edge in the direction normal to the edge are numerically equal. Moreover, they both either converge to or diverge from this edge.

Balance of Energy (first law of thermodynamics)

In the case of a single particle or a rigid body, a direct application of Newton's laws yields the result that the rate of change in kinetic energy K is exactly balanced by the mechanical power W_{ext} of the external forces acting on the system. In other words, the application of an external force over time along a trajectory produces (or extracts) work, and this work is entirely expended in increasing (or decreasing) the kinetic energy of the system.

In a discrete non-rigid system of interacting particles, on the other hand, one can obtain, by applying Newton's equations of motion to each particle and then adding over all particles of the system, that the rate of change of the total kinetic energy of the system is equal to the power of the external forces plus the power of the internal forces. For a system consisting of two particles, a detailed treatment was presented in Section 2.8. By the law of action and reaction, the power of the internal forces between any two particles is equal to the force exerted by the first on the second dot-multiplied by the relative velocity of the second with respect to the first (so that, for example, if the mutual force is one of attraction, the internal power will be positive if the particles are moving towards each other). The (positive, say) power of the external forces is now spent in part on increasing the kinetic energy and in part on overpowering the particle interactions. This purely mechanical result has an exact analogue in continuum mechanics.

If we start from the local statement of the law of balance of linear momentum (whether in Lagrangian or in Eulerian form), dot-multiply it by the velocity, integrate over the body and apply the divergence theorem to shift a term to the boundary, we indeed obtain the result

$$\frac{dK}{dt} = W_{ext} + W_{int}, \tag{6.53}$$

where K is the *kinetic energy*, given by

$$K = \int_V \frac{1}{2}\rho_0 \mathbf{v} \cdot \mathbf{v} dV = \int_v \frac{1}{2}\rho \mathbf{v} \cdot \mathbf{v} dv, \tag{6.54}$$

and W_{ext} is the power of the external forces

$$W_{ext} = \int_V \mathbf{B} \cdot \mathbf{v} dV + \int_{\partial V} (\mathbf{TN}) \cdot \mathbf{v} dA = \int_v \mathbf{b} \cdot \mathbf{v} dv + \int_{\partial v} (\mathbf{tn}) \cdot \mathbf{v} da. \tag{6.55}$$

We have suggestively denoted the remaining term in (6.53) by W_{int}, to signify that this is the term that corresponds to the power of the internal forces in the discrete-system analogy. Its exact expression is

$$W_{int} = -\int_V T^{il} v_{i,I} dV = -\int_v t^{ij} v_{i,j} dv, \tag{6.56}$$

or, recalling that $v_{i,j}$ are the components of the velocity gradient $\mathbf{L} = \mathbf{W} + \mathbf{D}$, and that the Cauchy stress is symmetric,

$$W_{\text{int}} = -\int_v \text{trace} \, (\mathbf{t} \, \mathbf{D}) \, dv, \qquad (6.57)$$

where the minus sign is only indicative of the fact that, when defining the flux operator through Cauchy's tetrahedron argument, we made a choice of sign that (in the case of the stress) corresponds to the reaction (rather than the action) in the discrete analogue. We will call the term W_{int} the *internal power* (or stress power) in a continuum.

Exercise 6.2.14 Kinetic energy. Follow the steps indicated and derive the law of balance of kinetic energy (6.53).

We have devoted some attention to the law of balance of kinetic energy only to show that it is not an independent law, but just a consequence of the previous, purely mechanical, balance laws. On the other hand, our treatment has served to bring into evidence the presence and nature of the extra power term W_{int}, attributable to the intrinsic deformability of the continuum. Continuous media, however, react as well to other types of energy input, particularly to what one would refer in everyday life as *thermal energy* or *heating*. Indeed, it is a matter of daily experience that continuous media deform under applied heat and, conversely, deformation may lead to the emission of heat (bending or twisting a metal paper-clip repeatedly until it breaks is a good experiment to reveal this common effect). There are other occurrences of non-mechanical energy sources (chemical reactions, electromagnetic fields, etc.) We will lump all the non-mechanical power input together and denote it by W_{heat}. As we are wont to do in all balance laws, we will recognize two heating inputs: one corresponding to distributed volumetric sources (*radiation*), and the other to input across the boundaries (*conduction*). The non-mechanical power input is, therefore, given by

$$W_{\text{heat}} = \int_V R \, dV + \int_{\partial V} Q \, dA = \int_v r \, dv + \int_{\partial v} q \, da, \qquad (6.58)$$

where R and r denote radiation, and where Q and q denote conduction in the Lagrangian and Eulerian versions, respectively.

The law of balance of energy (first law of thermodynamics) asserts that, in addition to the kinetic energy K, there exists another kind of energy content, called *internal energy*, U, such that the rate of change of the total energy content $K + U$ exactly balances the combined external mechanical and heating powers, namely

$$\frac{d(K + U)}{dt} = W_{\text{ext}} + W_{\text{heat}}. \qquad (6.59)$$

Comparing this result with the balance of kinetic energy (6.53), we obtain

$$\frac{dU}{dt} = W_{\text{heat}} - W_{\text{int}}. \qquad (6.60)$$

This equation implies roughly that an increase in internal energy can be achieved by either an input of heating power (with no strain), or by working against the internal forces of the system (with no heating), or by a combination of both.

To obtain the local versions of this balance law, we assume that the internal energy is given by an integral of a density u, which is usually assumed to be given per unit mass (rather than unit volume),[6] namely

$$U = \int_V \rho_0 u \, dV = \int_v \rho u \, dv. \tag{6.61}$$

Moreover, using the tetrahedron argument, one can show that the flux terms in the heating input in equation (6.58) are given by referential and spatial *heat flux vectors* as $Q = -\mathbf{Q} \cdot \mathbf{N}$ and $q = -\mathbf{q} \cdot \mathbf{n}$, the minus signs being chosen so that the heat flux vectors $\mathbf{Q}$ or $\mathbf{q}$ point in the direction of the flow of heat (if the normals are the *exterior* normals to the boundary of the domain of interest). The standard procedure yields the following Lagrangian and Eulerian forms of the local equations of energy balance:

$$\rho_0 \dot{u} = R - Q^I_{,I} + T^{iI} v_{i,I} \tag{6.62}$$

and

$$\rho \dot{u} = r - q^i_{,i} + t^{ij} v_{i,j} = r - \text{div } \mathbf{q} + \text{trace}(\mathbf{tD}). \tag{6.63}$$

Naturally, in this last (Eulerian) version, the material time derivative $\dot{u}$ includes the convected term $\nabla u \cdot \mathbf{v}$.

Exercise 6.2.15 Energy. Provide the missing steps in the derivation of the local equation of energy balance.

Entropy Inequality (second law of thermodynamics)

An important conceptual element in continuum mechanics is the presence of an arrow of time, that is, the natural irrevocable direction of phenomena prescribed by the second law of thermodynamics. There are different ways to deal with this delicate issue, but here we will only present the formulation based on the Clausius–Duhem inequality. There are two new elements that need to be added to the picture. The first is a new extensive quantity, the *entropy S*, whose content will be measured in terms of the integral of a density s per unit mass:

$$S = \int_V \rho_0 s \, dV = \int_v \rho s \, dv. \tag{6.64}$$

The second element to be introduced is a new field θ, called the *absolute temperature*. It is assumed that θ is strictly positive and measurable instantaneously and locally by an appropriate instrument (such as a thermocouple). This temperature scale is consistent with the temperature appearing naturally in the theory of ideal gases. It is assumed, moreover, that there are two universal sources of entropy production, one volumetric and the other through the boundary. These sources are obtained, respectively, by dividing the corresponding (volume or surface) heating source by the local value of the absolute temperature. The Clausius–Duhem inequality asserts that the rate of entropy production

---

[6] In the one-dimensional formulation of Section 2.9.2 we used the notation $\bar{u}$ to avoid confusion with the displacement. This distinction is now unnecessary.

is never less than what can be accounted for by these universal sources, namely

$$\frac{d}{dt} \int_V \rho_0 s \, dV \geq \int_V \frac{R}{\theta} dV - \int_{\partial V} \frac{\mathbf{Q} \cdot \mathbf{N}}{\theta} dA, \tag{6.65}$$

or, in the Eulerian version,

$$\frac{d}{dt} \int_v \rho s \, dv \geq \int_v \frac{r}{\theta} dv - \int_{\partial v} \frac{\mathbf{q} \cdot \mathbf{n}}{\theta} da. \tag{6.66}$$

The equality corresponds to *reversibility* of a physical process, while all other processes (for which the strict inequality holds) are *irreversible*. We will later exploit this inequality to derive restrictions on the possible material responses: a material cannot exist in nature for which, for any conceivable process, the Clausius–Duhem inequality might be violated.

The local forms of the inequality are obtained by the standard procedure as

$$\rho_0 \dot{s} \geq \frac{R}{\theta} - \text{Div}\left(\frac{\mathbf{Q}}{\theta}\right) \tag{6.67}$$

and

$$\rho \dot{s} \geq \frac{r}{\theta} - \text{div}\left(\frac{\mathbf{q}}{\theta}\right). \tag{6.68}$$

It is often convenient to replace, as we may, the Clausius–Duhem inequality by a linear combination with the balance of energy. In particular, subtracting from the (Lagrangian) equation of energy balance (6.62) the (Lagrangian) entropy inequality (6.67) multiplied by the absolute temperature yields

$$\rho_0(\dot{u} - \theta\dot{s}) \leq T^{il} v_{i,I} - \frac{1}{\theta} Q^I \theta_{,I}, \tag{6.69}$$

which has the nice feature of not involving the radiation term R. Defining now the *(Helmholtz) free-energy density* per unit mass as

$$\psi = u - \theta s, \tag{6.70}$$

we can write (6.69) in the form

$$\rho_0 \dot{\psi} + \rho_0 \dot{\theta} s - T^{il} v_{i,I} + \frac{1}{\theta} Q^I \theta_{,I} \leq 0. \tag{6.71}$$

Similarly, the Eulerian form of the modified Clausius–Duhem inequality is obtained as

$$\rho_0 \dot{\psi} + \rho_0 \dot{\theta} s - t^{ij} D_{ij} + \frac{1}{\theta} q^i \theta_{,i} \leq 0. \tag{6.72}$$

6.3 Constitutive Theory

6.3.1 Introduction and Scope

We have completed the formulation of all the universal laws of continuum mechanics, that is, those laws that apply to all material bodies, regardless of their physical constitution. They are equally valid for solids, liquids and gases of any kind. Even a cursory count of equations reveals that these laws are not sufficient to solve for all the fields involved. Nor should they be. Any science must stop at some level of description, and continuum mechanics stops at what might be regarded as a rather high level, where the description of material response loses the character of what the philosopher of science Mario Bunge would call an *interpretive* or *mechanismic* explanation. This is hardly a deficiency: it is simply a definition of the point of view already conveyed by the very name of the discipline. It is this particular point of view, namely the consideration of matter as if it were a continuum, that has paved the way for the great successes of continuum mechanics, both theoretical and practical. Much of engineering is still based on its tenets, and is likely to remain so for as long as it strives to be guided by intellectually comprehensible models rather than by computer brute force. The price to pay for this luxury is that the representation of the material response is not universal but must be tailored to each material or class of materials. This tailoring is far from arbitrary. It must respect certain principles, the formulation of which is the job of the constitutive theory.

Emulating the treatment in Chapter 3, if we define the *history* of a body as the pair of its motion $\chi^i = \chi^i(X^i; t)$ and the absolute temperature function $\theta = \theta(X^I; t)$ for all points of the body and for all times in the interval $(-\infty, t]$, t being the present instant, then the principles of *causality* and *determinism* assert that the history completely determines the present values of the stress tensor ($\mathbf{T}$ or $\mathbf{t}$), the heat-flux vector ($\mathbf{Q}$ or $\mathbf{q}$), the internal energy density (u) and the entropy density (s). We also assume (*principle of equipresence*) that the list of independent variables appearing in each of the functionals just mentioned is, *a priori*, the same in all functionals. For example, if the temperature is a determining factor for the heat flux, then it should *a priori* be considered determining also for the stress.

The theory of materials with fading memory was presented in Section 3.8 within the one-dimensional context, and will not be repeated here. Instead, in the three-dimensional context, we will present only a few examples of *classes of materials*, each of these classes being characterized by a dependence not on the whole history of the motion and the temperature but just on the present values of the deformation gradient, the temperature, its gradient and possibly their time derivatives. Moreover, we will assume that these materials are *local*, so that the fluxes and densities at a point depend only on the values of the independent variables (just listed) at that point.

Even with these very restrictive assumptions (which, nevertheless, are general enough to encompass many material models in widespread use), we will see that the remaining two tenets of the constitutive theory, namely the *principle of material frame indifference* and the *principle of thermodynamic consistency*, are strong enough to impose severe restrictions upon the possible constitutive laws that one might measure in the laboratory. For experimentalists (unless they have grounds to challenge the established ideas) these principles constitute a bonanza, since they drastically reduce the type and number of experiments to perform. The general validity of the principle of material frame indifference and the particular methodology to implement thermodynamic consistency that we will present, have both been challenged on various grounds, which we shall not discuss.

6.3.2 The Principle of Material Frame Indifference and Its Applications

In Section 5.3.5 we discussed the effect that a change of frame, or change of observer, has upon the reckoning of a motion. It is clear that a change of frame will also have an effect on all observable quantities, such as deformation gradients, vorticities and temperature gradients, the exact effect depending on the intrinsic or assumed nature of the quantity at hand. The principle of material frame indifference asserts that, although the independent and dependent variables of a given constitutive equation may be affected by a change of frame, the *constitutive functions themselves are not affected*, regardless of whether or not the frames are inertially related. In plain words, what the principle states is that material properties such as the stiffness of a spring, the heat conductivity of a substance or the coefficient of thermal expansion can be determined in any laboratory frame.[7]

Before we can proceed to apply this important principle to particular cases, we must establish once and for all how the measurements of some of the most common physical quantities change under a change of frame, namely, a transformation of the form (5.62). The most primitive quantity is the spatial distance between two simultaneous events. By construction, as shown in Section 5.3, the most general change of frame involves just an orthogonal spatial transformation, whence it follows that *all observers agree on the distance between two simultaneous events*. A scalar quantity, the result of whose measurement is independent of the frame, is called a *frame-indifferent scalar*. On physical grounds (e.g., by claiming that the length of the mercury line of a thermometer is the distance between two simultaneous events, or perhaps through a more sophisticated argument or assumption) we establish that the *absolute temperature is a frame-indifferent scalar*. Since observers agree on length, they must surely agree on volume, and they certainly should agree on the counting of particles, so that it is reasonable to assume that *mass density is a frame-indifferent scalar*. On similar grounds we will agree that *internal energy density and entropy density are frame-indifferent scalars*.

Moving on now to vector quantities, we start with the oriented segment **d** joining two simultaneous events, two flashlights blinking together in the dark, as it were. Denoting by x^i and y^i the spatial coordinates of the events, applying equation (5.62) to each point and then subtracting, we obtain

$$\hat{y}^i - \hat{x}^i = Q_j^i \ (y^j - x^j) \tag{6.73}$$

or

$$\hat{\mathbf{d}} = \mathbf{Q}\,\mathbf{d}. \tag{6.74}$$

A vector, such as **d**, that transforms in this manner, is called a *frame-indifferent vector*. The unit normal **n** to a *spatial* element of area is frame-indifferent, since it can be thought of as an arrow of the type just described. The unit normal **N** to a *referential* element of area, however, is not frame-indifferent, since as far as the reference configuration is concerned, a change of frame has no consequence whatsoever.[8] There is nothing sacred about being frame-indifferent. In particular, the principle of material frame indifference *does not claim that the quantities involved in a constitutive law must be frame-indifferent*. Quite the contrary, it affirms that *even though the quantities involved are in general not frame-indifferent, the constitutive functionals themselves are invariant under a change of frame*.

Next we consider the velocity vector **v**. The observed motions are related by

$$\hat{x}^i(X^I;\hat{t}) \ = \ c^i(t) \ + \ Q_j^i(t) \ x^i(X^I;t), \quad \hat{t} = t + a, \tag{6.75}$$

---

[7] In fact, part of the criticism levelled against the principle of material frame indifference stems from the conceptual possibility of having materials with microscopic gyroscopic effects at the molecular level, materials which would then be sensitive to the non-inertiality of the frame.

[8] Some authors, such as Ogden (1984), adopt a different point of view and distinguish between Eulerian and Lagrangian objectivity or frame indifference. According to this viewpoint, the referential normal **N** would be called 'objective'. Ultimately, this is just a matter of terminology.

where we have repeated equation (5.62) to emphasize the fact that the referential variables X^I remain intact. Taking the partial derivative with respect to $\hat{t}$, we obtain

$$\hat{v}^i = \dot{c}^i + \dot{Q}^i_j x^j + Q^i_j v^j \tag{6.76}$$

or

$$\hat{\mathbf{v}} = \dot{\mathbf{c}} + \dot{\mathbf{Q}}\mathbf{x} + \mathbf{Q}\,\mathbf{v}, \tag{6.77}$$

so that, as expected, the velocity is not frame-indifferent, not even if the observers are inertially related (an important clue in Galilean mechanics). Taking time derivatives once more, we obtain the following rule for the change of acceleration:

$$\hat{\mathbf{a}} = \ddot{\mathbf{c}} + \ddot{\mathbf{Q}}\mathbf{x} + 2\dot{\mathbf{Q}}\mathbf{v} + \mathbf{Q}\,\mathbf{a}. \tag{6.78}$$

Note that the acceleration, although not frame-indifferent, will appear to be so if the observers happen to be inertially related by equations (5.63) and (5.64).

Exercise 6.3.1   Coriolis and centripetal terms. Express equation (6.78) in the standard way found in textbooks of classical mechanics, namely, in terms of an angular velocity vector. Identify the Coriolis and centripetal terms. [Hint: since $\mathbf{Q}$ is orthogonal, its time derivative entails a skew-symmetric matrix. In an oriented three-dimensional Euclidean space, a skew-symmetric matrix can be replaced by a vector acting on other vectors via the cross product].

Moving on, finally, to tensors, we are bound to define a frame-indifferent tensor as a linear transformation that takes frame-indifferent vectors into frame-indifferent vectors. It follows from this criterion that a tensor $\mathbf{A}$ is frame-indifferent if it transforms according to the formula

$$\hat{\mathbf{A}} = \mathbf{Q}\,\mathbf{A}\,\mathbf{Q}^T. \tag{6.79}$$

Exercise 6.3.2   Transformation of tensors. Prove the above formula for the transformation of frame-indifferent tensors under a change of frame.

Assuming that *spatial forces are frame-indifferent*, it follows, according to the definition above (since spatial normals are also frame-indifferent), that *the Cauchy stress is a frame-indifferent tensor*. We check now the deformation gradient. By differentiation of (6.75) with respect to X^I, we obtain

$$\hat{\mathbf{F}} = \mathbf{Q}\,\mathbf{F}, \tag{6.80}$$

which shows that the deformation gradient is not frame-indifferent. Its determinant J, however, is a frame-indifferent scalar. The first Piola–Kirchhoff stress tensor $\mathbf{T}$ is not frame-indifferent. It transforms according to

$$\hat{\mathbf{T}} = \mathbf{Q}\,\mathbf{T}. \tag{6.81}$$

Exercise 6.3.3   The velocity gradient. Show that the velocity gradient L is not frame-indifferent and that it transforms according to the formula

$$\hat{\mathbf{L}} = \mathbf{Q}\,\mathbf{L}\,\mathbf{Q}^T + \dot{\mathbf{Q}}\mathbf{Q}^T. \tag{6.82}$$

[Hint: notice that to obtain $\hat{\mathbf{L}}$, one must take derivatives of $\hat{\mathbf{v}}$ with respect to $\hat{\mathbf{x}}$.]

We are now in a position to apply the principle of material frame indifference to reduce the possible forms of constitutive equations. We will consider two examples:

1. **Elasticity**. A material is said to be *elastic* if the stress at a point is a function of just the present value of the deformation gradient at that point, namely

$$\mathbf{t} = \mathbf{f}(\mathbf{F}),\tag{6.83}$$

where $\mathbf{f}$ is a tensor-valued function. Leaving aside the other constitutive functions (such as the internal energy), we will find what restrictions, if any, the principle of material frame indifference imposes on this constitutive law. According to this principle, in another frame we must have

$$\hat{\mathbf{t}} = \mathbf{f}(\hat{\mathbf{F}})\tag{6.84}$$

identically, for all non-singular tensors $\mathbf{F}$. Note the conspicuous absence of a hat over the function $\mathbf{f}$, which is the whole point of the principle of frame indifference. Using equation (6.80) and the frame-indifferent nature of the Cauchy stress, we can write (6.84) as

$$\mathbf{Q}\,\mathbf{f}(\mathbf{F})\,\mathbf{Q}^T = \mathbf{f}(\mathbf{Q}\,\mathbf{F}),\tag{6.85}$$

which is an equation that the function $\mathbf{f}$ must satisfy identically for all non-singular $\mathbf{F}$ and all orthogonal $\mathbf{Q}$, certainly a severe restriction. To make this restriction more explicit, we use the polar decomposition theorem to write

$$\mathbf{Q}\,\mathbf{f}(\mathbf{RU})\,\mathbf{Q}^T = \mathbf{f}(\mathbf{Q}\,\mathbf{RU}).\tag{6.86}$$

Since this is an identity, we can choose (at each instant) $\mathbf{Q} = \mathbf{R}^T$ and, rearranging some terms, we obtain

$$\mathbf{t} = \mathbf{f}(\mathbf{RU}) = \mathbf{R}\,\mathbf{f}(\mathbf{U})\,\mathbf{R}^T.\tag{6.87}$$

What this restriction means is that the dependence of the Cauchy stress can be arbitrary as far as the strain part ($\mathbf{U}$) of the deformation gradient is concerned, but the dependence on the rotational part is canonical. Notice that, since the tensor $\mathbf{U}$ is symmetric, the number of independent variables has reduced from nine to six.

2. **A viscous fluid**. In elementary textbooks on fluid mechanics one sometimes finds a statement to the effect that the shearing stress in some fluids is proportional to the velocity gradient. To substantiate this conclusion, an experiment is described whereby two parallel solid plates, separated by a layer of fluid, are subjected to opposite parallel in-plane forces and the resulting relative velocity is measured. Dividing this velocity by the distance between the plates, the statement appears to be validated, at least as a first approximation, by the behaviour of many fluids. Without contesting the experimental result itself, we will now show that the conclusion, even if true for that particular experiment, cannot be elevated directly to the status of a general constitutive law.

We start from an assumed law of the form

$$\mathbf{t} = \mathbf{f}(J,\,\mathbf{L}),\tag{6.88}$$

so that the stress may arise as a consequence of a change of volume (compressible fluid) as well as from a non-vanishing velocity gradient. Applying the principle of frame indifference and invoking (6.82), we obtain

$$\mathbf{t} = \mathbf{Q}^T\,\mathbf{f}(J,\,\mathbf{Q}\mathbf{L}\mathbf{Q}^T + \dot{\mathbf{Q}}\mathbf{Q}^T).\tag{6.89}$$

Choosing $\mathbf{Q} = \mathbf{I}$ and $\dot{\mathbf{Q}} \neq 0$, we obtain

$$\mathbf{t} = \mathbf{f}(J \, , \, \mathbf{L} + \boldsymbol{\Omega}), \qquad (6.90)$$

where $\boldsymbol{\Omega}$ is an arbitrary skew-symmetric tensor, as already encountered in Section 5.5.13. We may, therefore, adjust instantaneously $\boldsymbol{\Omega} = -\mathbf{W}$, thereby obtaining

$$\mathbf{t} = \mathbf{f}(J \, , \, \mathbf{D}). \qquad (6.91)$$

Further restrictions arise in this case, but we will not consider them here (see, however, Exercise 6.4.9 below). The conclusion is that the stress cannot depend on the whole velocity gradient, but only on its symmetric part (to which, for example, it may be proportional).

Exercise 6.3.4 Homogeneity of space. Show that neither the spatial position nor the velocity can appear as independent variables in a constitutive equation.

6.3.3 The Principle of Thermodynamic Consistency and Its Applications

The second law of thermodynamics is a restriction that nature imposes on observable phenomena: certain things simply cannot happen. We will adopt the following point of view to ensure that those things that should not happen never come out as a solution of the equations of continuum mechanics: we will exclude any constitutive law for which, under any conceivable process, the Clausius–Duhem inequality might be violated, even instantaneously. In the classes of materials we have been dealing with, this statement implies that the Clausius–Duhem inequality must hold true identically for any instantaneous combination of the independent variables and their space or time derivatives. We will explore the restrictions that are obtained from the principle of thermodynamic consistency in two cases:

   1. **Thermoelastic heat conductors**. In this class of materials we assume, by definition, that the constitutive variables are functions of the deformation gradient, the temperature and the temperature gradient, namely:

$$T^{il} = T^{il}(F^j_J, \theta, \theta_{,J}), \qquad (6.92)$$

$$Q^I = Q^I(F^j_J, \theta, \theta_{,J}), \qquad (6.93)$$

$$s = s(F^j_J, \theta, \theta_{,J}), \qquad (6.94)$$

$$\psi = \psi(F^j_J, \theta, \theta_{,J}), \qquad (6.95)$$

where, for convenience, we have substituted the free energy ψ for the internal energy u. Notice, by the way, an application of the principle of equipresence: we have assumed exactly the same list of arguments for all constitutive variables, letting thermodynamics tell us eventually whether or not an argument should be excluded from a particular constitutive law. We plug equations 6.92–6.95 into (6.71), use the chain rule and group terms together to obtain the following result:

$$\left[\rho_0 \frac{\partial \psi}{\partial F^j_J} - T^J_j\right]\dot{F}^j_J + \rho_0\left[\frac{\partial \psi}{\partial \theta} + s\right]\dot{\theta} + \rho_0\left[\frac{\partial \psi}{\partial \theta_{,J}}\right]\dot{\theta}_{,J} + \frac{1}{\theta}Q^J\theta_{,J} \leq 0, \qquad (6.96)$$

where we have used the fact that $v^j_{,J} = \dot{F}^j_J$, by the symmetry of mixed partial derivatives.

The inequality we have just obtained should be valid identically for all combinations of $\dot{F}^j_J, \dot{\theta}, \theta_{,J}$ and $\dot{\theta}_{,J}$. But this inequality is *linear* in $\dot{F}^j_J, \dot{\theta}$ and $\dot{\theta}_{,J}$, since these variables do not appear anywhere except as multipliers of other expressions. This is not the case for the variable $\theta_{,J}$, since the heat-flux vector which it multiplies may depend on it, according to our constitutive assumptions (6.92)–(6.95). Given that a linear function cannot have a constant sign over the whole domain, we conclude that the identical satisfaction of the inequality demands the satisfaction of the following equations:

$$T^J_j = \rho_0 \frac{\partial \psi}{\partial F^j_J}, \tag{6.97}$$

$$s = -\frac{\partial \psi}{\partial \theta}, \tag{6.98}$$

$$\frac{\partial \psi}{\partial \theta_{,J}} = 0, \tag{6.99}$$

and the *residual inequality*

$$Q^J \theta_{,J} \leq 0. \tag{6.100}$$

These restrictions can be summarized as follows: the free-energy density is independent of the temperature gradient (equation (6.99)) and acts as a *potential* for the stress (equation (6.97)) and the entropy density (equation (6.98)), both of which are, consequently, also independent of the temperature gradient. The ten constitutive functions F^j_J and s boil down, therefore, to a single scalar function ψ. Moreover, according to the residual inequality (6.100), heat cannot flow from lower to higher temperatures, since the heat-flux vector cannot form an acute angle with the temperature gradient. If, for example, we postulate Fourier's law of conduction which establishes that the heat-flux vector is proportional to the temperature gradient, we conclude that the constant of proportionality must be negative. The coefficient of heat conduction for real materials is in fact defined as the negative of this constant, so as to be positive.

Remark 6.3.5 Thermoelasticity and hyperelasticity. If we were to limit the processes that a thermoelastic material undergoes to isothermally homogeneous processes (i.e., with a constant temperature in time and in space), we would arrive at the conclusion that the resulting elastic material is *hyperelastic*, namely, it is completely characterized by a stress potential, which is obtained as the restriction of the free-energy density to the pre-established temperature. This argument has sometimes been adduced to claim that all elastic materials are actually hyperelastic (see Section 6.5.2).

Exercise 6.3.6 Symmetry of the stress. Show that the fact that the first Piola–Kirchhoff stress derives from a scalar potential is consistent with the symmetry of the Cauchy stress.

Naturally, the thermodynamic restrictions just derived should be combined with those arising from the principle of material frame indifference.

Exercise 6.3.7 Apply the principle of material frame indifference to the Helmholtz free-energy density $\psi = \psi(\mathbf{F}, \theta)$ of a thermoelastic heat conductor to obtain the restricted form

$$\psi = \psi(\mathbf{C}, \theta). \tag{6.101}$$

Use equation (6.97) to prove that

$$\mathbf{S} = 2\rho_0 \frac{\partial \psi(\mathbf{C}, \theta)}{\partial \mathbf{C}}, \tag{6.102}$$

where $\mathbf{S}$ is the second Piola–Kirchhoff stress tensor defined by equation (6.44).

2. **A viscous fluid**. We now revisit the viscous fluid, whose reduced constitutive equation (6.91) we have obtained by means of the principle of material frame indifference. To derive further restrictions dictated by the principle of thermodynamic consistency, we will add the temperature to the list of independent variables and we will supplement equation (6.91) with equations for the heat flux, entropy density and free-energy density. The proposed constitutive laws are, therefore, of the form

$$t^{ij} = t^{ij}(J, \mathbf{D}, \theta), \tag{6.103}$$

$$q^i = q^i(J, \mathbf{D}, \theta), \tag{6.104}$$

$$s = s(J, \mathbf{D}, \theta), \tag{6.105}$$

$$\psi = \psi(J, \mathbf{D}, \theta), \tag{6.106}$$

where we are using the Eulerian formulation for a change. Substitution of these equations into (6.72) yields

$$\left[\rho \frac{\partial \psi}{\partial J} \delta^{ij} - t^{ij} \right] D_{ij} + \rho \left[\frac{\partial \psi}{\partial \theta} + s \right] \dot{\theta} + \left[\rho \frac{\partial \psi}{\partial D_{ij}} \right] \dot{D}_{ij} + \left[\frac{1}{\theta} q^i \right] \theta_{,i} \leq 0, \tag{6.107}$$

where the formula (6.7) for the derivative of a determinant has been used. Notice that the content within the first square brackets cannot be claimed to vanish now, since the stress may depend on $\mathbf{D}$. On the other hand, the term within the last square brackets is now zero, since we have not assumed any dependence on the temperature gradient. This is, therefore, a fluid heat insulator. To summarize, we obtain the equations

$$\frac{\partial \psi}{\partial D_{ij}} = 0, \tag{6.108}$$

$$s = -\frac{\partial \psi}{\partial \theta} \tag{6.109}$$

and

$$q^i = 0, \tag{6.110}$$

plus the residual inequality

$$\text{trace} \, (\mathbf{t}_{\text{irr}} \, \mathbf{D}) \geq 0 \tag{6.111}$$

where we have defined the *irreversible part of the stress* as $\mathbf{t}_{\text{irr}} := \mathbf{t} - \mathbf{t}_{\text{rev}}$. The reversible part $\mathbf{t}_{\text{rev}}$ is the hydrostatic (spherical) expression

$$\mathbf{t}_{\text{rev}} = \rho \frac{\partial \psi}{\partial J} J \, \mathbf{I}. \tag{6.112}$$

If, for example, the irreversible, or dissipative, part of the stress is assumed to be proportional to the rate of deformation $\mathbf{D}$, then the principle of thermodynamic consistency prescribes that the constant of proportionality, or *viscosity*, must be positive.

6.4 Material Symmetries

6.4.1 Symmetries and Groups

Symmetries of all kinds play a fundamental role in classical and modern physics. From the mathematical point of view, the concept of symmetry is closely associated with the notion of *group*.

Beyond being a mere collection of objects, a set need not display any additional structure. On the other hand, many applications arising from mathematics itself and from science in general have resulted in the consideration of certain classes of sets that possess some type of extra structure. We have encountered at least two of these classes in Chapter 5, where *vector spaces* and *affine spaces* emerged naturally as unifying concepts for sets endowed with certain operations satisfying meticulously defined properties. One of the great paradigmatic successes of modern mathematics has been precisely its power to prove theorems and to establish features common to all sets belonging to the same class, regardless of the particular application at hand, thus achieving an unprecedented degree of synthesis.

One of the most fundamental of these classes is embodied in the notion of group. A *group* is a set $\mathcal{G}$ with an internal binary operation called *group multiplication*. By *binary* we mean that the operation involves an ordered pair of elements (a, b) of the set. By *internal* we mean that the result of the operation is an element of the set, which we denote by ab. More formally, group multiplication is a map $\mathcal{G} \times \mathcal{G} \to \mathcal{G}$ represented by

$$(a, b) \mapsto ab, \quad a, b, ab \in \mathcal{G}. \tag{6.113}$$

This operation satisfies just three properties:

1. *Associativity.* $(ab)c = a(bc)$ for all $a, b, c \in \mathcal{G}$.
2. *Unit element.* There exists $e \in \mathcal{G}$ such that $ea = ae = a$ for all $a \in \mathcal{G}$.
3. *Inverse.* For each $a \in \mathcal{G}$ there exists $a^{-1} \in \mathcal{G}$ such that $aa^{-1} = a^{-1}a = e$.

The *unit element* e is also called the *neutral element*. The notation a^{-1} for the *inverse* of a is only suggestive.

Exercise 6.4.1 Uniqueness. Prove that in a group: (a) the unit element is unique; (b) for each a the inverse element is unique.

Two group elements, a and b, are said to *commute* if it so happens that $ab = ba$. If all the elements of a group commute pair by pair, the group is said to be *commutative* or *Abelian*. In an Abelian group, the group operation is usually denoted by a $+$ sign and the inverse of a by $-a$. The unit element is called the *zero element* and is denoted by 0. Thus, in a commutative group, we write $a + (-a) = 0$.

Exercise 6.4.2 Verify that every vector space is a commutative group with respect to vector addition.

Exercise 6.4.3 Use your inventiveness to come up with examples of multiplicative (i.e., non-commutative) and additive (i.e., commutative) groups. Provide an example of a set with a non-associative binary internal operation.

Exercise 6.4.4 Symmetries of a square. Consider a square in $\mathbb{R}^2$ with its centre at the origin. (a) Describe the collection of all rotations of $\mathbb{R}^2$ that leave the square congruent with itself. If two rotations differ by a full turn, count them as one. Verify that this collection is a group under the operation of composition. Is this group Abelian? (b) Describe the collection of all reflections of $\mathbb{R}^2$ that leave the square congruent with itself. Do they form a group? (c) Consider now the combination of all rotations and reflections that keep the square congruent with itself. Verify that this larger collection is a group under composition. Is this group Abelian?

Exercise 6.4.5 Symmetries of a cube. Consider a cube in $\mathbb{R}^3$ with its centre at the origin. Verify that the collection of all rotations that leave the cube congruent with itself is a group. Is this group Abelian?

The examples that we have seen so far are all *discrete groups*. If the underlying set $\mathcal{G}$ is a continuum and if the operations of multiplication and inverse are continuous (or smooth) in some precisely defined sense, then we have a *topological group* (or a *Lie group*).

Example 6.4.6 The general linear group. The *(real) general linear group* in n dimensions, denoted by GL$(n; \mathbb{R})$, is the set of all $n \times n$ non-singular matrices (with real entries) under the operation of matrix multiplication. This is a Lie group.

A *subgroup* of a group is a subset closed under the operations of multiplication and inverse. Thus, a subgroup is itself a group.

Example 6.4.7 The matrix groups. The subgroups of the general linear group GL$(n; \mathbb{R})$ are known as the *matrix groups*. They consist of square matrices of order n that satisfy certain definite properties. Some important examples are: (1) the subgroup GL$^+(n; \mathbb{R})$ of matrices with strictly positive determinant; (2) the *unimodular group* $\mathcal{U} = $ SL$(n; \mathbb{R})$ consisting of matrices whose determinant has a unit absolute value; (3) the *special linear group* SL$^+(n; \mathbb{R})$ of matrices with unit determinant; (4) the *orthogonal group* $\mathcal{O}(n; \mathbb{R})$ of orthogonal matrices; (5) the *proper or special orthogonal group* $\mathcal{O}^+(n; \mathbb{R})$ of proper orthogonal matrices. From the geometric point of view, when seen as transformations of $\mathbb{R}^n$, the elements of the GL$^+(n; \mathbb{R})$ preserve orientation, those of $\mathcal{U}$ preserve volume, those of SL$^+(n; \mathbb{R})$ preserve volume and orientation, those of $\mathcal{O}(n; \mathbb{R})$, consisting of rotations and reflections, preserve the inner product and, finally, those of $\mathcal{O}^+(n; \mathbb{R})$ are pure rotations. Notice the inclusions $\mathcal{O}^+(n; \mathbb{R}) \subset $ SL$^+(n; \mathbb{R}) \subset $ GL$^+(n; \mathbb{R})$ and $\mathcal{O}(n; \mathbb{R}) \subset $ SL$(n; \mathbb{R}) \subset $ GL$(n; \mathbb{R})$.

6.4.2 The Material Symmetry Group

The principle of material frame indifference establishes a *universal symmetry* of all constitutive laws. Indeed, it postulates that all constitutive laws are *invariant* under the left multiplication of the deformation gradient by an arbitrary orthogonal matrix **Q**. We express this by saying that constitutive equations are invariant under the *left action* of the orthogonal group.

Quite apart from this universal *spatial* symmetry, constitutive laws may have additional symmetries that arise from their *material* constitution. Although some of these symmetries, in the case of solid materials, may ultimately be due to the geometric symmetries of the underlying crystal lattices, the two concepts should not be confused, for

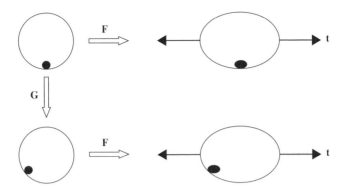

Figure 6.3   Material symmetry

a number of reasons which it is not our intention to discuss at this point. Suffice it to say that, in particular, constitutive equations may enjoy continuous symmetries, even if the underlying molecular structure is a regular lattice with discrete symmetries. From the point of view of a given constitutive law the only question that matters is the following: which material transformations, applied before each possible deformation, do not affect the response due to this deformation? Once the constitutive law is given, this is a purely mathematical question.

For the sake of brevity, we will consider the case of an elastic material, whose constitutive equation consists of a single tensor function $\mathbf{t}$ (the Cauchy stress) of a single tensor variable $\mathbf{F}$ (the deformation gradient). A somewhat simpler case would be a hyperelastic material, characterized by a single scalar function of $\mathbf{F}$. We are focusing our attention on a single point of the body, leaving for Chapter 7 the fascinating, geometrically rich, idea of comparing the responses of different points. A symmetry will, therefore, consist of a material deformation gradient[9] $\mathbf{G}$ that, when pre-applied to any further deformation gradient $\mathbf{F}$, does not affect the value of $\mathbf{t}$, that is,

$$\mathbf{t}(\mathbf{FG}) = \mathbf{t}(\mathbf{F}), \tag{6.114}$$

for all non-singular tensors $\mathbf{F}$. More pictorially, as suggested in Figure 6.3, if an experimenter has measured some material property on, say, a small spherical sample and, about to repeat the experiment, leaves the laboratory for a moment, during which interval an evil genie decides to rotate the sample through some angle, this rotation will be a symmetry if the repetition of the experiment with the rotated sample cannot lead the experimenter to detect or even suspect that the sample has been rotated.

It is not difficult to check that the collection of all material symmetries of a given constitutive law forms a group, called the *material symmetry group* $\mathcal{G}$ of the given law. We can say, therefore, that the constitutive law is invariant under the (right) action of its symmetry group. The symmetry group is never empty, since the identity transformation is always a symmetry.

Exercise 6.4.8   The symmetry group. Verify that the material symmetries (6.114) indeed form a group, whose operation is matrix multiplication. Recall that all you need to check is associativity, existence of a unit and existence of inverses.

---

[9] A material symmetry need not be orientation preserving, so that its determinant may be negative.

To speak of symmetries as matrices, we need to choose a reference configuration, at least locally, and the symmetry group $\mathcal{G}$ will depend on the reference chosen. No essential information is gained or lost, however, by changing the reference configuration, and the various symmetry groups thus obtained are mutually *conjugate*.[10] In principle, any non-singular matrix could act as a symmetry of a constitutive law, so that the symmetry group could be as large as the general linear group GL(3; $\mathbb{R}$), namely, the collection of all non-singular 3×3 matrices with the operation of multiplication. It is usually assumed, however, that symmetries are volume preserving, which implies that all possible symmetry groups are subgroups of the *unimodular group* $\mathcal{U}$ (understood as the multiplicative group of all matrices whose determinant is equal to ± 1).

An elastic material is a *solid* if its symmetry group, with respect to some reference configuration, is a subgroup of the *orthogonal group* $\mathcal{O}$. A material is called *isotropic* if its symmetry group, for some reference configuration, contains the orthogonal group. It follows that an isotropic solid is a material whose symmetry group, for some (local) reference configuration (called an *undistorted state*), is exactly the orthogonal group. An elastic material for which the symmetry group is the whole unimodular group $\mathcal{U}$ is called a *fluid*. Naturally, in this case the group is independent of the reference configuration chosen. This notion of fluidity corresponds to the intuitive idea that a fluid (at rest) adapts itself to the shape of the receptacle without any difficulty. Similarly, the idea of solidity is that, starting from a natural stress-free state of the material, one may perhaps find rotations that leave the response unaffected, but certainly any strain ($\mathbf{U} \neq \mathbf{I}$) will be felt. A theorem in group theory[11] allows us to establish that an isotropic material (of the type we are discussing) must either be an isotropic solid or a fluid. There is nothing in between. A material, therefore, is either a solid ($\mathcal{G} \subseteq \mathcal{O}$), a fluid ($\mathcal{G} = \mathcal{U}$), or a *crystal fluid*[12] ($\mathcal{G} \cap \mathcal{O} \neq \mathcal{O}$ and $\mathcal{G} \cap (\mathcal{U} - \mathcal{O}) \neq \emptyset$). As just remarked, crystal fluids cannot be isotropic.

To illustrate how the existence of a non-trivial symmetry group results in a further reduction of a constitutive law, beyond the reductions imposed by the principles of material frame indifference and of thermodynamic consistency, we will now exhibit the two classical examples of the representation of the most general constitutive law of an elastic fluid and an isotropic elastic solid.

For the elastic fluid, the symmetry condition reads:

$$\mathbf{t}(\mathbf{F}) = \mathbf{t}(\mathbf{F}\,\mathbf{G}), \quad \text{for all } \mathbf{F} \in \text{GL}(3; \mathbb{R}),\ \mathbf{G} \in \mathcal{U}. \tag{6.115}$$

Choosing, as we may, $\mathbf{G} = (\det\mathbf{F})^{1/3}\,(\mathbf{F}^{-1})$, we conclude that the most general constitutive equation of an elastic fluid must be of the form

$$\mathbf{t} = \mathbf{t}(\det\mathbf{F}), \tag{6.116}$$

which completely agrees with the intuitive idea that an elastic fluid (such as an ideal gas) is, mechanically speaking, only sensitive to changes in volume.

As far as the isotropic elastic solid is concerned, we start by stating the symmetry condition in terms of the Cauchy stress, namely

$$\mathbf{t}(\mathbf{F}) = \mathbf{t}(\mathbf{F}\,\mathbf{G}), \quad \text{for all } \mathbf{F} \in \text{GL}(3; \mathbb{R}),\ \mathbf{G} \in \mathcal{O}. \tag{6.117}$$

---

[10] For a definition of conjugacy, see Exercise 7.6.1.

[11] Noll (1965).

[12] Not to be confused with a liquid crystal, which is a different kind of material.

Invoking the left polar decomposition theorem (see Exercise 5.5.8), we can set $\mathbf{F} = \mathbf{VR}$ and then choose $\mathbf{G} = \mathbf{R}^T$, thus obtaining a first reduction,

$$\mathbf{t} = \mathbf{t}(\mathbf{V}) \tag{6.118}$$

or, equivalently, since $\mathbf{B} = \mathbf{V}^2$,

$$\mathbf{t} = \mathbf{t}(\mathbf{B}), \tag{6.119}$$

which states that, for an isotropic elastic solid, the Cauchy stress is a function of the left Cauchy–Green tensor $\mathbf{B} = \mathbf{FF}^T$. We may now exploit the principle of material frame indifference to write

$$\mathbf{Q}\ \mathbf{t}(\mathbf{B})\ \mathbf{Q}^T = \mathbf{t}(\mathbf{QBQ}^T), \tag{6.120}$$

for all orthogonal matrices $\mathbf{Q}$. In other words, the stress is an *isotropic function* of its matrix argument. Notice that both quantities are represented by symmetric matrices. A well-known representation theorem[13] establishes that the most general form that this function can attain is

$$\mathbf{t} = \phi_0\mathbf{I} + \phi_1\mathbf{B} + \phi_2\mathbf{B}^2, \tag{6.121}$$

where ϕ_0, ϕ_1 and ϕ_2 are arbitrary functions of the (positive) eigenvalues of $\mathbf{B}$. This representation is, of course, valid only if the tensor $\mathbf{B}$ is measured from an undistorted state.

Exercise 6.4.9   Revisiting the viscous fluid once more. What further reductions does the principle of material frame indifference impose on the constitutive equation (6.91)? [Hint: use the representation theorem just mentioned.]

Exercise 6.4.10   Principal stresses. Verify that in an isotropic elastic material every eigenvector of $\mathbf{B}$ is necessarily also an eigenvector of the Cauchy stress tensor. In the language of engineering, the principal directions of strain are also principal directions of stress.

The representation of the general constitutive law for an isotropic elastic solid[14] was one of the first results of the continuum mechanics renaissance that had its beginnings after the end of the Second World War.

6.5   Case Study: The Elasticity of Soft Tissue

6.5.1   Introduction

The structural complexity of soft biological tissues appears to defy any rational attempt at a macroscopic phenomenological modelling of their mechanical behaviour. Nevertheless, the use of continuum mechanics models has proved to be impressively successful in many applications, such as those related to the cardiovascular system.[15]

The mechanical properties of soft issue are largely determined by the *extracellular matrix* (ECM), rather than by the cells themselves. The ECM has a relatively high content of bound and unbound water. The former is responsible for the near *incompressibility*

---

[13] See, for example, Truesdell and Noll (1965).
[14] Reiner (1948).
[15] See Humphrey (2002).

of soft tissue, whereas the latter accounts for *viscous effects* of the kind that we have discussed in Section 4.5. Other components of the ECM are largely elastic. In most soft tissues, therefore, a good starting point is to consider the overall material approximately incompressible and elastic.

The presence of protein molecules (such as collagen and elastin) in the ECM confers on it preferred directions, thus accounting in many cases for *anisotropy* of the tissue at the macroscopic level. Moreover, the orientation and proportion of these molecules is responsible for a lack of *material uniformity*, namely for varying material properties from point to point. Finally, even when the material properties are the same at different points, processes of growth and remodelling may be responsible for a lack of compatibility between neighbouring elements, thus resulting in *residual stresses* and, more generally, in a lack of *material homogeneity*.

For continuum mechanics, therefore, soft tissues provide an ideal opportunity to contribute to biomechanics with a theory of incompressible, anisotropic, inhomogeneous and non-uniform elasticity.

6.5.2 Elasticity and Hyperelasticity

A material point is said to be *elastic* if its constitutive equation stipulates that the stress is a function of the deformation gradient alone, namely

$$\mathbf{T} = \mathbf{T}(\mathbf{F}).\qquad(6.122)$$

This idea is the mathematical expression of what we have in mind for the term 'elasticity' in everyday life. A rubber ball, for example, will require exactly the same forces to produce a given deformation, no matter how many times we repeat the experiment. Elastic solids (as opposed to elastic fluids) will in general possess *natural states*, that is, local configurations for which the state of stress is zero.

Exercise 6.5.1 Natural states. Show that, as a direct consequence of the principle of material frame indifference, any local configuration obtained from a natural state by applying a further rotation is also a natural state. Notice, however, that, in principle, an elastic solid may not possess any natural state. At the other extreme, not all natural states differ necessarily by a rotation, although this is the case for most real elastic solids.

If we cling to the strict principle of equipresence, having assumed the constitutive equation (6.122) for the stress compels us to assume that also the heat flux, the Helmholtz free-energy density and the entropy density depend on the deformation gradient alone. A direct application of the principle of thermodynamic consistency, leads to the conclusion that

$$\mathbf{T} = \rho_0 \frac{\partial \psi(\mathbf{F})}{\partial \mathbf{F}}.\qquad(6.123)$$

In other words, although we start from a constitutive law for the stress tensor that, in principle, consists of nine scalar equations (one for each component), and although in fact only six of these equations can be independent (by virtue of the algebraic symmetry required by equation (6.52)), the principle of equipresence and the principle of thermodynamic consistency lead us to the conclusion that only *a single scalar function* needs to be determined. This function, the Helmholtz free-energy density, acts as a scalar potential for the Piola stress. A material with the constitutive equation (6.123) is called *hyperelastic*.

Clearly, both from the analytic and the experimental points of view, a hyperelastic material is a considerably simpler model than a more general elastic material. Our reasoning thus far can be summarized as follows: the principle of equipresence implies that every elastic material is hyperelastic.

This coarse application of the principle of equipresence may be questioned on physical grounds. Can the internal energy and the entropy of a material be independent of temperature? For this reason, it is more prudent either to regard hyperelasticity as an independent theory for the stress or, more logically, to place hyperelasticity within the wider context of thermoelasticity. If we adopt the second option and carry out the analysis of the restrictions imposed on the constitutive laws by the second law of thermodynamics, as we have already done in Section 6.3.3, we conclude that

$$\mathbf{T} = \rho_0 \frac{\partial \psi(\mathbf{F}, \theta)}{\partial \mathbf{F}}. \tag{6.124}$$

It follows from this equation that, for *isothermal processes* (i.e., processes at constant temperature), we recover the hyperelastic formula (6.123). It is in this sense that one may say that hyperelasticity is implied by thermoelasticy restricted to isothermal processes, with the free-energy density acting as the hyperelastic potential.

Remark 6.5.2   The strain-energy function. It can be shown that the hyperelastic paradigm is also recovered for *isentropic processes*, with the internal energy acting as the hyperelastic potential. It is customary, therefore, to refer to the hyperelastic stress potential by the common designation of *strain-energy function*.

Remark 6.5.3   Entropic elasticity. It follows from equation (6.70) that contributions to an increase in the Helmholtz free energy arise from either an increase in internal energy or a decrease in entropy. The former effect, typical of crystalline solids, can be traced back to the existence of interatomic potentials (such as the Lennard-Jones potential), which are affected by the application of macroscopic deformations. The latter effect is known as *entropic elasticity*. It is characteristic of rubber-like materials. The theory of entropic elasticity can be developed from a statistical mechanics viewpoint. A network of long-chain molecules is made up of ideal chains such that, under no external mechanical stimulus, they have both ends coincident on average over time. The application of a deformation imposes a geometrical constraint (namely, an uncoiling of the chains), thereby resulting in a higher degree of order with a consequent decrease in entropy and an increase in free energy, that is, recoverable work (see Treloar, 1958, 1973).

Exercise 6.5.4   Follow the lead of Remark 6.5.3, read the pertinent parts of Treloar's works cited therein and elaborate on possible applications to soft tissue mechanics.

6.5.3   *Incompressibility*

An *internal constraint* at a material point is defined as a scalar-valued function $\gamma(\mathbf{F})$ of the deformation gradient, such that all the possible motions are restricted to those for which

$$\gamma(\mathbf{F}) = 0 \tag{6.125}$$

at the given point. A γ-constrained body is one made up of points all of which are subject to the constraint defined by γ.

Remark 6.5.5   Notice that the constraint is defined not by the function itself but rather by the collection (or manifold) of its zeros. Thus, two different functions having exactly the same zeros define the same internal constraint.

The following examples are of practical significance:

1. *Inextensibility in one direction*. Consider a solid matrix in which a family of non-intersecting rigid fibres has been smoothly embedded. To each point of the reference configuration of the composite body we can, therefore, assign a direction, that is, a unit vector **N**, that cannot possibly undergo any elongation. Since the deformation gradient **F** transforms **N** into the vector **n** = **FN**, the constraint stipulates that **n** must remain a unit vector. Using matrix notation, we can write

$$1 = \{\mathbf{n}\}^T \{\mathbf{n}\} = ([\mathbf{F}]\{\mathbf{N}\})^T ([\mathbf{F}]\{\mathbf{N}\}) = \{\mathbf{N}\}^T [\mathbf{F}]^T [\mathbf{F}]\{\mathbf{N}\} = \{\mathbf{N}\}^T [\mathbf{C}]\{\mathbf{N}\}. \quad (6.126)$$

In this case, therefore, the internal constraint of inextensibility is given by the formula

$$\gamma(\mathbf{F}) = \mathbf{N} \cdot \mathbf{C}\,\mathbf{N} - 1 = 0. \quad (6.127)$$

2. *Incompressibility*. Materials such as rubber and many soft biological tissues are approximately incompressible, as already remarked. In other words, they preserve their local volume throughout the process of deformation. In view of the volume formula (5.195), the incompressibility constraint is given by

$$\gamma(\mathbf{F}) = \det\mathbf{C} - 1 = 0. \quad (6.128)$$

A useful geometrical way to visualize a constraint is to imagine it represented in a nine-dimensional space with coordinates $F_1^1, F_2^1, \ldots, F_3^3$, as shown in Figure 6.4. In that case, equation (6.125) describes a *hypersurface* of dimension 8. Assuming the constraint function γ to be smooth (or, at least, continuously differentiable), the constraint hypersurface will have a well-defined and continuous tangent plane. It proves convenient (although not strictly necessary) to regard the nine coordinates as constituting a Cartesian system, so

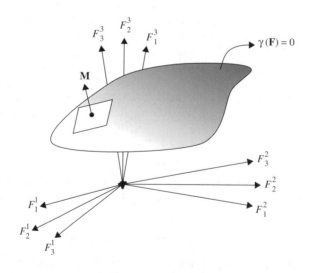

Figure 6.4   Constraint

that metric concepts can be easily invoked. In particular, given two 'vectors', **F** and **G**, their Cartesian dot product is defined by the formula

$$\mathbf{F} \cdot \mathbf{G} = F^1_1 G^1_1 + F^1_2 G^1_2 + \ldots + F^3_3 G^3_3 = \sum_{i=1}^{3} \sum_{I=1}^{3} F^i_I G^i_I, \tag{6.129}$$

or, in matrix notation,

$$\mathbf{F} \cdot \mathbf{G} = \text{trace} \left([\mathbf{F}]^T [\mathbf{G}] \right). \tag{6.130}$$

Referring to a fixed point of the hypersurface and taking the total differential of the constraint equation (6.125), we obtain

$$\frac{\partial \gamma}{\partial F^1_1} \, dF^1_1 + \frac{\partial \gamma}{\partial F^1_2} \, dF^1_2 + \ldots + \frac{\partial \gamma}{\partial F^3_3} \, dF^3_3 = 0, \tag{6.131}$$

which, according to equation (6.129), can also be written as

$$\frac{\partial \gamma}{\partial \mathbf{F}} \cdot d\mathbf{F} = 0. \tag{6.132}$$

Since $d\mathbf{F}$ is nothing but a vector tangent to the hypersurface at the chosen point and since equation (6.131) must hold for all possible increments $d\mathbf{F}$, we conclude that the vector

$$\mathbf{M} = \frac{\partial \gamma}{\partial \mathbf{F}} \tag{6.133}$$

is perpendicular to the hypersurface at the given point.[16]

Just like in the classical mechanics of particles, it makes sense in non-dissipative systems to require that constraints be *ideal* or *workless*, that is, that the forces necessary to maintain a constraint perform no mechanical power. For example, in the case of inextensibility, we may imagine that the constraint is achieved by the development within the material of a uniaxial stress in the direction of the fibre. Since the material is not supposed to elongate in this direction, there is no mechanical power done by the assumed forces. More generally, as we have learned in Section 6.2.2, the power of the stress per unit volume is given by

$$\sum_{i=1}^{3} \sum_{I=1}^{3} T^{iI} v_{i,I} = \sum_{i=1}^{3} \sum_{I=1}^{3} T^{iI} \dot{F}^i_{,I} = \mathbf{T} \cdot \dot{\mathbf{F}}. \tag{6.134}$$

Setting $d\mathbf{F} = \dot{\mathbf{F}} \, dt$ and comparing with equation (6.132), we conclude that the assumption that the constraint forces are workless is equivalent to stating that the corresponding stress $\mathbf{T}_{\text{constraint}}$ is in the direction of the normal **M** to the constraint hypersurface. According to equation (6.133), therefore, we must have

$$\mathbf{T}_{\text{constraint}} = \mu \, \frac{\partial \gamma}{\partial \mathbf{F}}, \tag{6.135}$$

where μ is an arbitrary scalar. The total stress **T** is not determined now by just a constitutive function of **F**. To see that this is the case, suppose that we prescribe a deformation gradient **F** (which, naturally, has to satisfy the given constraint). Let $\mathbf{T}_{\text{constitutive}} = \mathbf{T}_c(\mathbf{F})$ be a suggested constitutive equation. Clearly, for the same **F**, we can always add an arbitrary stress of constraint and obtain a new stress. Thus, we are led to the conclusion

---

[16] This argument is clearly a repetition of that in Section 5.3.6 dealing with the gradient of a general scalar field.

that the total stress $\mathbf{T}$ is made of two parts, as follows:

$$\mathbf{T} = \mathbf{T}_c(\mathbf{F}) + \mu \, \frac{\partial \gamma}{\partial \mathbf{F}}. \tag{6.136}$$

Because we are mainly interested in the constraint of incompressibility, we derive presently the particular form of the stress for this case. Starting from equation (6.128), we calculate

$$\frac{\partial \gamma(\mathbf{F})}{\partial \mathbf{F}} = \frac{\partial \det \mathbf{C}}{\partial \mathbf{F}} = \frac{\partial (\det \mathbf{F})^2}{\partial \mathbf{F}} = (2 \det \mathbf{F}) \, \mathbf{F}^{-T} = 2 \, \mathbf{F}^{-T}, \tag{6.137}$$

where we have used a result obtained in Exercise 6.1.2. To sum up, the total Piola stress in an incompressible material is given by

$$\mathbf{T} = \mathbf{T}_c(\mathbf{F}) + \mu \, \mathbf{F}^{-T}. \tag{6.138}$$

Exercise 6.5.6   Show that the total Cauchy stress in an incompressible material is given by

$$\mathbf{t} = \mathbf{t}_c(\mathbf{F}) - p \, \mathbf{I}, \tag{6.139}$$

where $p = -J^{-1}\mu$ is a new multiplier known as the *pressure*. From the physical point of view, we understand that in an incompressible material the stress is determined by the deformation only up to an arbitrary hydrostatic pressure.

Exercise 6.5.7   Derive an expression for the total stress in an inextensible material.

Exercise 6.5.8   Show that for a hyperelastic material subject to a constraint of the form (6.125), the total stress can be obtained from the modified strain-energy function:

$$\psi = \psi_c(\mathbf{F}) + \mu \gamma(\mathbf{F}), \tag{6.140}$$

where ψ_c acts as a potential for $\mathbf{T}_c$.

6.5.4   Isotropy

A *hyperelastic symmetry* at a point is a material symmetry of the strain-energy function ψ. In other words, a referential tensor $\mathbf{G}$ is a hyperelastic symmetry if

$$\psi(\mathbf{F}) = \psi(\mathbf{FG}) \tag{6.141}$$

identically, for all non-singular deformation gradients $\mathbf{F}$. Following the same reasoning as in Exercise 6.4.8, one can show that all hyperelastic symmetries of a given strain-energy function form a group $\mathcal{G}$. A hyperelastic material is a *hyperelastic solid* if, in some local reference configuration, this group is a subgroup of the orthogonal group. Any such local configuration is called an *undistorted state* of the material.[17] A hyperelastic solid is *isotropic* if its symmetry group in an undistorted state coincides with the orthogonal group. For an isotropic hyperelastic solid point, therefore, we have

$$\psi(\mathbf{F}) = \psi(\mathbf{FQ}) \tag{6.142}$$

---

[17] An undistorted state is not necessarily a natural state.

identically, for all non-singular $\mathbf{F}$ and for all orthogonal $\mathbf{Q}$. By the left polar decomposition theorem, we may write

$$\psi(\mathbf{F}) = \psi(\mathbf{FQ}) = \psi(\mathbf{V\,R\,Q}), \tag{6.143}$$

and, choosing $\mathbf{Q} = \mathbf{R}^T$, we conclude that

$$\psi(\mathbf{F}) = \psi(\mathbf{V}) = \psi(\mathbf{B}^{1/2}) = \hat{\psi}(\mathbf{B}), \tag{6.144}$$

where $\hat{\psi}$ is a new function. This result is independent of the principle of material frame indifference, which we now proceed to exploit, namely

$$\hat{\psi}(\mathbf{B}) = \hat{\psi}(\mathbf{F\,F}^T) = \hat{\psi}(\mathbf{QFF}^T\mathbf{Q}^T) = \hat{\psi}(\mathbf{Q\,B\,Q}^T) \tag{6.145}$$

identically, for all symmetric positive definite $\mathbf{B}$ and for all orthogonal $\mathbf{Q}$. We are now at liberty to choose $\mathbf{Q}$ as a modal matrix of $\mathbf{B}$ (see Section 5.4.3), whence it follows that $\hat{\psi}$ can be expressed as a function of the diagonalized form of $\mathbf{B}$ or, equivalently, as a function (which we rename ψ) of the eigenvalues of $\mathbf{B}$:

$$\psi = \psi(\lambda_1^2, \lambda_2^2, \lambda_3^2). \tag{6.146}$$

We note, however, that we can choose different modal matrices by interchanging columns. As a result, the order of the eigenvalues in the diagonalized form of $\mathbf{B}$ can be altered arbitrarily. Consequently, the function ψ must be indifferent to the order of its arguments. Noting that the principal invariants I_B, II_B, III_B already enjoy this property, we conclude that any function of the principal invariants is automatically a function with the desired property. The constitutive function of a hyperelastic isotropic solid can, therefore, be also represented as

$$\psi = \psi(I_B, II_B, III_B), \tag{6.147}$$

with an immaterial abuse of notation. Notice that the eigenvalues and invariants of $\mathbf{B}$ are the same as those of $\mathbf{C}$.

Remark 6.5.9 Given a constitutive equation in the form (6.147), it is a straightforward matter to write it in the form (6.146). The converse, however, although theoretically possible, is not always practicable.

Recalling that the third invariant is precisely the determinant of $\mathbf{B}$ and that, if the material happens to be incompressible, this determinant is fixed at the unit value, we conclude that the constitutive part of the strain-energy of a hyperelastic isotropic solid can be expressed as

$$\psi_c = \psi_c(I_B, II_B). \tag{6.148}$$

6.5.5 *Examples*

It is not our intention to provide a list of hyperelastic constitutive equations in current use, which can be found in the specialized literature, but only to illustrate their use in a few sample calculations. The *neo-Hookean* material model was used originally to describe the response of rubber. It was in fact obtained on the basis of statistical mechanics considerations assuming a purely entropic elastic response, as described in Remark 6.5.3. As a particular case of the general equation (6.148) for incompressible response, its constitutive

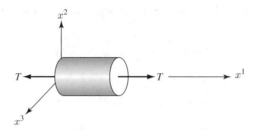

Figure 6.5   Uniaxially loaded specimen

function is given by

$$\rho_0 \psi_c = \frac{G}{2}(I_B - 3). \tag{6.149}$$

In this equation, G is a material constant.[18] Thus, the neo-Hookean material is completely characterized by a single constant G. The constant 3 is included only for convenience, so that in a stress-free reference configuration ψ_c vanishes. Our next task is to obtain an expression for the stress. According to equation (6.138),

$$\mathbf{T} = \mathbf{T}_c(\mathbf{F}) - Jp\ \mathbf{F}^{-T} = \rho_0 \frac{\partial \psi_c}{\partial \mathbf{F}} - Jp\ \mathbf{F}^{-T}. \tag{6.150}$$

Carrying out the operations indicated, and recalling that $J = 1$ because of the assumed incompressibility, we obtain

$$\mathbf{T} = G\mathbf{F} - p\mathbf{F}^{-T}, \tag{6.151}$$

or, in terms of the Cauchy stress,

$$\mathbf{t} = G\mathbf{B} - p\mathbf{I}. \tag{6.152}$$

Exercise 6.5.10   Derive equation (6.151) from (6.149) and (6.150).

Exercise 6.5.11   Uniaxial loading. A cylindrical specimen made of a neo-Hookean material is subjected to axial forces T uniformly distributed at the two ends, as shown in Figure 6.5. The lateral surface is free of force. Let ε denote the relative elongation of the cylinder (change in length over original length). Since the material is incompressible, the cross section will vary as the bar elongates or contracts. Show that (identifying the referential coordinate axes with their spatial counterparts) the deformation gradient at all points is given by the matrix

$$[\mathbf{F}] = \begin{bmatrix} 1+\varepsilon & 0 & 0 \\ 0 & 1-\alpha & 0 \\ 0 & 0 & 1-\alpha \end{bmatrix}, \tag{6.153}$$

with

$$(1-\alpha)^2 = \frac{1}{1+\varepsilon}. \tag{6.154}$$

---

[18] In a wider thermoelastic context, G is proportional to the absolute temperature.

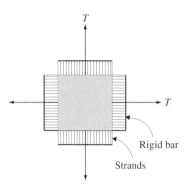

Figure 6.6   Biaxial loading

From the fact that the lateral surface is free of stress, deduce that

$$p = \frac{G}{1 + \varepsilon}.$$ (6.155)

Write T^{11} and t^{11} as functions of ε. Plot these functions (divided by G) over the range $-0.5 < \varepsilon < 2$. What are the relations between T^{11} and t^{11}, on the one hand, and the original cross-sectional area A and the applied force T, on the other hand?

Exercise 6.5.12   Biaxial loading. For thin-walled organs, such as the skin or the myocardium, it is important to study the response to in-plane biaxial loading, as shown in Figure 6.6.[19] A neo-Hookean square membrane of side L and thickness h is subjected to a total tensile force T on each side. Show that

$$p = \frac{G}{(1 + \varepsilon)^4},$$ (6.156)

where $\varepsilon = \Delta L/L$ is the relative elongation of each side of the square. Determine the relation between T and ΔL.

Exercise 6.5.13   Foam rubber. The neo-Hookean model is well adapted to nearly incompressible materials. As we have observed in Exercise 6.5.11, one of the consequences of incompressibility is that, upon extension (contraction) of a uniaxially loaded cylinder, the cross section shrinks (expands) so as to preserve volume. This effect, which is also present in not necessarily incompressible materials, is known as the *Poisson effect*. A *foam* is a natural or artificial solid or liquid material infused with a relatively large content of gas. In particular, a *foam rubber* consists of an air-filled rubber matrix. Such highly compressible materials tend not to exhibit an appreciable Poisson effect. Consider the following constitutive equation:

$$\rho_0 \psi = \frac{G}{2} \left(I_B - \ln III_B \right).$$ (6.157)

Obtain an expression for the Piola stress as a function of the deformation gradient. Show that in an experiment similar to that of Exercise 6.5.11 the cross section of the specimen remains unchanged. For the same experiment, plot the graph of stress against relative

---

[19] The experimental implementation of these conditions presents considerable challenges.

elongation, just as in Exercise 6.5.11. Is there a difference between Piola and Cauchy stresses in this particular experiment?

The modelling of anisotropic materials by constructing functions that are invariant under a proper subgroup of the orthogonal group is beyond the scope of our treatment. Nevertheless, it is possible to obtain anisotropic versions of originally isotropic constitutive equations by means of bias tensors that favour particular material directions. As an example, we will consider the case of *orthotropy*. In an orthotropic material there exists a local reference configuration (e.g., a natural state) in which there exist three mutually orthogonal material directions with respect to each of which the material response is invariant under 180° rotations. Roughly speaking, it may be said that the material properties in these three directions characterize the material response. A particular case of practical importance is that of *transverse isotropy*, for which the properties in two of these directions are identical and the material response is invariant under any rotation about the remaining axis.

Let $\mathbf{M}$ denote a positive definite symmetric material tensor that will play the role of a bias or fabric tensor. Note that, by construction, this tensor is diagonalizable in some Cartesian coordinate system, whose axes will be called the *principal directions of orthotropy*. If the three eigenvalues are different, we will have a legitimate case of orthotropy. If one eigenvalue is repeated, we have a case of transverse isotropy, while the case of three equal eigenvalues brings us back to isotropy. By the principle of material frame indifference, the constitutive equation of a general hyperelastic material point can be written as

$$\psi = \psi(\mathbf{C}). \tag{6.158}$$

Although not strictly necessary, from the physical standpoint it is convenient to normalize the bias matrix in the sense that

$$I_{M^2} = \text{trace } \mathbf{M}^2 = 1. \tag{6.159}$$

We now construct the modified constitutive equation:

$$\hat{\psi} = \hat{\psi}\left(\mathbf{M}(\mathbf{C} - \mathbf{I})\mathbf{M} + \mathbf{I}\right). \tag{6.160}$$

If the constitutive equation of departure, equation (6.158), is isotropic, the modified equation will represent an orthotropic material.

Exercise 6.5.14   Orthotropic material. Prove that, if equation (6.158) is isotropic, and if $\mathbf{M}$ is symmetric positive definite, then equation (6.160) is orthotropic. [Hint: by bringing $\mathbf{M}$ to its diagonal form, observe that the product of two symmetric matrices is commutative if, and only if, they have three common linearly independent eigenvectors.]

Example 6.5.15   A transversely isotropic example. A prismatic bar has its axis aligned with the X^1 coordinate axis. It is made of a transversely isotropic hyperelastic material with constitutive equation (6.160), modelled after equation (6.157), with a diagonalized form of the bias tensor given by the matrix

$$[\mathbf{M}_d] = \frac{1}{\sqrt{1.5^2 + 1^2 + 1^2}} \begin{bmatrix} 1.5 & 0 & 0 \\ 0 & 1 & 0 \\ 0 & 0 & 1 \end{bmatrix}. \tag{6.161}$$

The 'strong' direction $\mathbf{m}_1$ forms an angle $\theta = 30°$ with the main axis, as shown in Figure 6.7. The material constant G has a value of 10 MPa. The bar is subjected to an

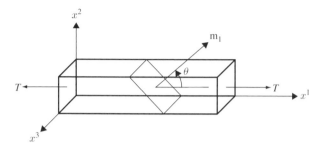

Figure 6.7   Example 6.5.15

axial tensile force $T = 50\,\text{N}$ applied uniformly over the original cross-sectional area of $A = 100\,\text{mm}^2$. Solve for the deformation.[20]

Solution. We identify the referential and spatial coordinate axes and, discarding an immaterial rigid-body rotation, we assume the deformation gradient to be of the form

$$[\mathbf{F}] = \begin{bmatrix} a & b & 0 \\ 0 & c & 0 \\ 0 & 0 & d \end{bmatrix}, \tag{6.162}$$

where a, b, c, d are constants to be determined. This assumed deformation gradient preserves the longitudinal direction of the bar and allows for shear deformations with respect to the x^1- and x^2-axes. The component form of the Piola stress is given by the matrix

$$[\mathbf{T}] = \begin{bmatrix} T/A & 0 & 0 \\ 0 & 0 & 0 \\ 0 & 0 & 0 \end{bmatrix}. \tag{6.163}$$

Notice in passing that the assumed form of the deformation gradient is consistent with the symmetry of the Cauchy stress, as expressed by equation (6.52). Since we are working in a specific coordinate system, we must express the bias tensor $\mathbf{M}$ in that system. Introducing the matrix $[\mathbf{P}]$ representing the rotation about x^3 that brings $\mathbf{m}_1$ into coincidence with x^1 as

$$[\mathbf{P}] = \begin{bmatrix} \cos\theta & -\sin\theta & 0 \\ \sin\theta & \cos\theta & 0 \\ 0 & 0 & 1 \end{bmatrix}, \tag{6.164}$$

we obtain

$$\mathbf{M} = \mathbf{P}\,\mathbf{M}\,\mathbf{P}^T. \tag{6.165}$$

The assumed constitutive equation has the following explicit form:

$$\rho_0\psi = \frac{G}{2}\,(\text{trace}\,(\mathbf{M}(\mathbf{C} - \mathbf{I})\mathbf{M} + \mathbf{I}) - \ln\,(\det\,(\mathbf{M}(\mathbf{C} - \mathbf{I})\mathbf{M} + \mathbf{I}))). \tag{6.166}$$

The Piola stress is

$$\mathbf{T} = \rho_0\frac{\partial\psi}{\partial\mathbf{F}} = G\,\mathbf{F}\,\mathbf{M}\left(\mathbf{I} - (\mathbf{M}(\mathbf{C} - \mathbf{I})\mathbf{M} + \mathbf{I})^{-1}\right)\mathbf{M}. \tag{6.167}$$

---

[20] This example is an adaptation of a similar one in the context of bone remodelling used in Epstein and Elżanowsi (2007).

```
G=10000000
theta=Pi/6
alpha=1.5
piola=500000
F={{a, b, 0},{0, c, 0}{0, 0, d}}
Cc=Transpose[F].F; MatrixForm[F]
P={{Cos[theta],-Sin[theta],0},{Sin[theta],Cos[theta],0}, {0, 0, 1}}; MatrixForm[P]
Md=(1/(alpha^2+1+1))*{{alpha, 0, 0},{0, 1, 0},{0, 0, 1}};MatrixForm[Md]
M=P.Md.Transpose[P]
one={{1, 0, 0}{0, 1, 0}{0, 0, 1}}
T={{piola, 0, 0}{0, 0, 0}{0, 0, 0}}
zeroeq=G*F.M.(one -Inverse[(M.(Cc-one).M + one)]).M-T
eq1=xeroeq[[1, 1]]
eq2=xeroeq[[1, 2]]
eq3=xeroeq[[2, 2]]
eq4=xeroeq[[3, 3]]
solving=FindRoot[{eq1==0, eq2==0, eq3==0, eq4==0},{{a, 1},{b, 0},{c, 1},{d, 1}}]
```

Figure 6.8 Mathematica® code for Example 6.5.15

The Mathematica® code shown in Figure 6.8 solves the resulting non-linear equations
numerically. The result is

$$a = 2.1184, \ b = -0.678\,943, \ c = 1.064\,04, \ d = 1. \tag{6.168}$$

The cross section tilts towards the plane of transverse isotropy, as schematically shown
in Figure 6.9.

More sophisticated models of anisotropic hyperelastic response can be found in the
literature. In particular, Gasser et al. (2006) have produced a realistic model based on the
anisotropy induced by the presence of collagen fibres with preferred orientations within
the various layers of the arterial wall.

All the examples of isotropic hyperelastic constitutive equations that we have discussed
so far have been expressed in terms of the principal invariants of the tensor **B**, which are
the same as those of **C** and which we have denoted by $\lambda_1^2, \lambda_2^2, \lambda_3^2$. The relation between
the invariants and the eigenvalues is easily obtained from the diagonal form of **B** or **C** as

$$I_B = \lambda_1^2 + \lambda_2^2 + \lambda_3^2, \tag{6.169}$$

$$II_B = \lambda_1^2\lambda_2^2 + \lambda_2^2\lambda_3^2 + \lambda_3^2\lambda_1^2 \tag{6.170}$$

and

$$III_B = \lambda_1^2 \, \lambda_2^2 \, \lambda_3^2. \tag{6.171}$$

It is, therefore, an elementary exercise to express any isotropic constitutive law given as
a function of the invariants as another function of the eigenvalues. As we have already

Figure 6.9 Deformation of transversely isotropic bar

pointed out, the transition from a function of the eigenvalues to the corresponding function of the invariants cannot in general be accomplished explicitly, thus opening the door to the formulation of novel isotropic hyperelastic constitutive equations. Valanis and Landel (1967) proposed the following *separable form* of the strain-energy function:

$$\psi = \phi(\lambda_1) + \phi(\lambda_2) + \phi(\lambda_3). \tag{6.172}$$

An important family of materials in this class is the family of *Ogden materials*,[21] whose constitutive equation is expressed as a sum of powers,

$$\psi = \sum_{i=1}^{n} \frac{G_i}{\alpha_i} \left(\lambda_1^{\alpha_i} + \lambda_2^{\alpha_i} + \lambda_3^{\alpha_i} - 3 \right). \tag{6.173}$$

The number of terms (as controlled by the integer n) and the chosen real exponents α_i afford a great flexibility in fitting experimental data while remaining amenable to analytical considerations.

6.6   Remarks on Initial and Boundary Value Problems

The mathematical apparatus presented in this and previous chapters is the embodiment of the physical principles and assumptions underlying processes of deformation, motion and heat in continuous media. As a reward for our labours we are entitled at the very least to expect that the combination of kinematic analysis, physical balance principles and constitutive laws will deliver a complete system of equations that could in principle be solved and result in useful descriptive and predictive models. Ultimately, we expect this system to be a mixture of algebraic, differential and possibly integral equations, as we have indeed found in the treatment of specific problems in one spatial dimension.

In addition to the equations themselves, static or steady-state situations require the specification of *boundary conditions*. A deformable body is usually supported at some part of its boundary and loaded elsewhere. Some parts of the boundary may be subjected to given temperatures and other parts to heat fluxes. In processes that evolve in time, considerations of causality lead us to expect that, over and above these boundary conditions, a process needs to get kick-started by the knowledge and specification of *initial conditions* that affect and determine the subsequent motion and heat transfer. The number and form of boundary and initial conditions to be specified in each case is a difficult mathematical question in the theory of partial differential equations.

A problem is said to be *well posed* when precise conditions have been met that guarantee the existence, uniqueness and stability of a solution. For example, the equilibrium of a transversely loaded membrane requires that its boundary be either fixed at each point (Dirichlet boundary condition) or loaded transversely (Neumann boundary condition), but not both simultaneously and independently. The Neumann condition is expressed in terms of the deflection derivative (or slope) in the direction normal to the boundary. Alternatively, a combination of a transverse displacement and loading (elastic support) is permitted (Robin boundary condition). Steady-state problems in continuum mechanics, in which these alternatives are crucial, tend to be of the so-called *elliptic type*. If we are interested in studying the vibrations of a membrane, we need to prescribe, in addition to the boundary conditions, the initial values of both the transverse deflection and the velocity

---

[21] For this and other related topics, see the classical treatise by Ogden (1984).

of each point of the membrane. This is a typical situation of the *hyperbolic type*, in which the highest time derivative that appears in the equations is of the second order. Problems of diffusion and heat are usually of the *parabolic type*, where the highest time derivative is the first. They require, in addition to appropriate boundary conditions, the specification of the initial state (concentration, temperature) only. The systems of equations arising in continuum mechanics are rarely solvable by exact analytic means. The combination of approximate numerical procedures and powerful computing devices is the preferred approach. Clearly, an ill-posed problem cannot lead to a reliable solution, no matter how sophisticated the numerical procedure or how powerful the computing device might be.

Example 6.6.1   The initial and boundary value problem in thermoelasticity. In solid mechanics, as opposed to fluid mechanics, the Lagrangian approach can be used advantageously in the formulation of initial and boundary value problems. For the sake of specificity, and to bring the equations into the standard forms found in treatises on partial differential equations, we will abandon the index notation and revert to the old notation, whereby the (Cartesian) referential coordinates X^1, X^2, X^3 are renamed X, Y, Z and the spatial coordinates x^1, x^2, x^3 are called x, y, z. We proceed to list the governing equations for a typical problem in thermoelasticity.

1. *Kinematic equations* . Identifying the referential and spatial coordinate axes, the components of the displacement vector field are given by

$$
\begin{cases}
u(X,Y,Z,t) = x - X \\
v(X,Y,Z,t) = y - Y \\
w(X,Y,Z,t) = z - Z.
\end{cases}
\tag{6.174}
$$

The components of the Lagrangian strain tensor are obtained from equation (5.178) as

$$
\begin{cases}
E_{xx} = \dfrac{\partial u}{\partial x} + \dfrac{1}{2}\left(\left(\dfrac{\partial u}{\partial x}\right)^2 + \left(\dfrac{\partial v}{\partial x}\right)^2 + \left(\dfrac{\partial w}{\partial x}\right)^2\right) \\[2mm]
E_{yy} = \dfrac{\partial v}{\partial y} + \dfrac{1}{2}\left(\left(\dfrac{\partial u}{\partial y}\right)^2 + \left(\dfrac{\partial v}{\partial y}\right)^2 + \left(\dfrac{\partial w}{\partial y}\right)^2\right) \\[2mm]
E_{xx} = \dfrac{\partial w}{\partial z} + \dfrac{1}{2}\left(\left(\dfrac{\partial u}{\partial z}\right)^2 + \left(\dfrac{\partial v}{\partial z}\right)^2 + \left(\dfrac{\partial w}{\partial z}\right)^2\right) \\[2mm]
E_{xy} = E_{yx} = \dfrac{1}{2}\left(\dfrac{\partial u}{\partial y} + \dfrac{\partial v}{\partial x} + \dfrac{\partial u}{\partial x}\dfrac{\partial u}{\partial y} + \dfrac{\partial v}{\partial x}\dfrac{\partial v}{\partial y} + \dfrac{\partial x}{\partial x}\dfrac{\partial w}{\partial y}\right) \\[2mm]
E_{yz} = E_{zy} = \dfrac{1}{2}\left(\dfrac{\partial v}{\partial z} + \dfrac{\partial w}{\partial y} + \dfrac{\partial u}{\partial y}\dfrac{\partial u}{\partial z} + \dfrac{\partial v}{\partial y}\dfrac{\partial v}{\partial z} + \dfrac{\partial x}{\partial y}\dfrac{\partial w}{\partial z}\right) \\[2mm]
E_{zx} = E_{xz} = \dfrac{1}{2}\left(\dfrac{\partial w}{\partial x} + \dfrac{\partial u}{\partial z} + \dfrac{\partial u}{\partial x}\dfrac{\partial u}{\partial z} + \dfrac{\partial v}{\partial x}\dfrac{\partial v}{\partial z} + \dfrac{\partial x}{\partial x}\dfrac{\partial w}{\partial z}\right).
\end{cases}
\tag{6.175}
$$

2. *Balance equations*
   (a) *Mass conservation*. In the Lagrangian formulation, mass conservation trivially becomes

$$
\rho_0 = \rho_0(X,Y,Z).
\tag{6.176}
$$

In other words, the density in the reference configuration is known and independent of time. Only in problems of growth, to be discussed in Chapter 7, will mass balance become an issue. In the Eulerian formulation, on the other hand, the continuity equation is a partial differential equation that needs to be enforced even in the absence of growth.

(b) *Balance of linear momentum* . In terms of the components of the Piola stress, the local equations of balance of linear momentum are, according to (6.41),

$$
\begin{cases}
\dfrac{\partial T_{xX}}{\partial X} + \dfrac{\partial T_{xY}}{\partial Y} + \dfrac{\partial T_{xZ}}{\partial Z} + B_x = \rho_0 \dfrac{\partial^2 u}{\partial t^2} \\[2mm]
\dfrac{\partial T_{yX}}{\partial X} + \dfrac{\partial T_{yY}}{\partial Y} + \dfrac{\partial T_{yZ}}{\partial Z} + B_y = \rho_0 \dfrac{\partial^2 v}{\partial t^2} \\[2mm]
\dfrac{\partial T_{zX}}{\partial X} + \dfrac{\partial T_{zY}}{\partial Y} + \dfrac{\partial T_{zZ}}{\partial Z} + B_z = \rho_0 \dfrac{\partial^2 w}{\partial t^2}.
\end{cases}
\tag{6.177}
$$

In these equations, B_x, B_y, B_z are the body forces per unit referential volume.

(c) *Balance of angular momentum* . The balance of angular momentum has been shown to be equivalent to the symmetry of the Cauchy stress. For the Piola stress, equation (6.52), we obtain the following three algebraic conditions:

$$
\begin{cases}
\left(1 + \dfrac{\partial u}{\partial X}\right) T_{yX} + \dfrac{\partial u}{\partial Y} T_{yY} + \dfrac{\partial u}{\partial Z} T_{yZ} = \dfrac{\partial v}{\partial X} T_{xX} + \left(1 + \dfrac{\partial v}{\partial Y}\right) T_{xY} + \dfrac{\partial v}{\partial Z} T_{xZ} \\[2mm]
\dfrac{\partial v}{\partial X} T_{zX} + \left(1 + \dfrac{\partial v}{\partial Y}\right) T_{zY} + \dfrac{\partial v}{\partial Z} T_{zZ} = \dfrac{\partial w}{\partial X} T_{yX} + \dfrac{\partial w}{\partial Y} T_{yY} + \left(1 + \dfrac{\partial w}{\partial Z}\right) T_{yZ} \\[2mm]
\dfrac{\partial w}{\partial X} T_{xX} + \dfrac{\partial w}{\partial Y} T_{xY} + \left(1 + \dfrac{\partial w}{\partial Z}\right) T_{xZ} = \left(1 + \dfrac{\partial u}{\partial X}\right) T_{zX} + \dfrac{\partial u}{\partial Y} T_{zY} + \dfrac{\partial u}{\partial Z} T_{zZ}.
\end{cases}
\tag{6.178}
$$

In practice, these algebraic conditions can be implemented directly in the formulation of the constitutive equations.

(d) *Balance of energy* . The first law of thermodynamics is given by equation (6.62) as follows:

$$
\begin{aligned}
\rho_0 \frac{\partial u_{\text{int}}}{\partial t} = R &- \frac{\partial Q_X}{\partial X} - \frac{\partial Q_Y}{\partial Y} - \frac{\partial Q_Z}{\partial Z} + T_{xX} \frac{\partial^2 u}{\partial X \partial t} + T_{yY} \frac{\partial^2 v}{\partial Y \partial t} + T_{zZ} \frac{\partial^2 w}{\partial Z \partial t} \\
&+ T_{xY} \frac{\partial^2 u}{\partial Y \partial t} + T_{xZ} \frac{\partial^2 u}{\partial Z \partial t} + T_{yX} \frac{\partial^2 v}{\partial X \partial t} + T_{yZ} \frac{\partial^2 v}{\partial Z \partial t} + T_{zX} \frac{\partial^2 w}{\partial X \partial t} \\
&+ T_{zY} \frac{\partial^2 w}{\partial Y \partial t},
\end{aligned}
\tag{6.179}
$$

where we have used the notation u_{int} for the internal energy per unit mass to avoid confusion with the first displacement component.

3. *Constitutive equations* . According to the restrictions placed by thermodynamic consistency, we need to specify constitutive equations only for the Helmholtz free-energy density ψ and for the heat-flux vector **Q**. The constitutive equations for the stress and for the entropy are obtained by differentiation of the free energy. Thus, we prescribe

$$
\psi = \psi(E_{XX}, E_{YY}, E_{ZZ}, E_{XY}, E_{YZ}, E_{ZX}, \theta; X, Y, Z).
\tag{6.180}
$$

Several remarks are in order. The explicit dependence on the body coordinates X, Y, Z indicates that we are contemplating either a possible material non-uniformity or, even if the body is homogeneous, a possible lack of homogeneity in the reference configuration. Notice that, by expressing the kinematic variables in terms of the Lagrangian strain components, we are satisfying the principle of material frame indifference. Moreover, it is not difficult to verify that the Piola stress components, given by

$$
\begin{cases}
T_{xX} = \rho_0 \, \dfrac{\partial \psi}{\partial (\partial u / \partial X)} \\[2mm]
T_{xY} = \rho_0 \, \dfrac{\partial \psi}{\partial (\partial u / \partial Y)} \\[2mm]
T_{xZ} = \rho_0 \, \dfrac{\partial \psi}{\partial (\partial u / \partial Z)} \\[2mm]
T_{yX} = \rho_0 \, \dfrac{\partial \psi}{\partial (\partial v / \partial X)} \\[2mm]
T_{yY} = \rho_0 \, \dfrac{\partial \psi}{\partial (\partial v / \partial Y)} \\[2mm]
T_{yZ} = \rho_0 \, \dfrac{\partial \psi}{\partial (\partial v / \partial Z)} \\[2mm]
T_{zX} = \rho_0 \, \dfrac{\partial \psi}{\partial (\partial w / \partial X)} \\[2mm]
T_{zY} = \rho_0 \, \dfrac{\partial \psi}{\partial (\partial w / \partial Y)} \\[2mm]
T_{zZ} = \rho_0 \, \dfrac{\partial \psi}{\partial (\partial w / \partial Z)},
\end{cases}
\tag{6.181}
$$

automatically satisfy the symmetries (6.178) imposed by the conservation of angular momentum, as already anticipated. The right-hand sides of (6.181) must be carefully calculated by means of the chain rule of differentiation via equation (6.175). This is left as an exercise. The internal energy density is obtained from the free energy by

$$
u_{\text{int}} = \psi - \theta \frac{\partial \psi}{\partial \theta},
\tag{6.182}
$$

where equations (6.42) and (6.98) have been used. Finally, independent equations for the components of the heat flux vector must be given in the form

$$
\begin{cases}
Q_X = Q_X(E_{XX}, E_{YY}, E_{ZZ}, E_{XY}, E_{YZ}, E_{ZX}, \theta, \dfrac{\partial \theta}{\partial X}, \dfrac{\partial \theta}{\partial Y}, \dfrac{\partial \theta}{\partial Z}; X, Y, Z) \\[2mm]
Q_Y = Q_Y(E_{XX}, E_{YY}, E_{ZZ}, E_{XY}, E_{YZ}, E_{ZX}, \theta, \dfrac{\partial \theta}{\partial X}, \dfrac{\partial \theta}{\partial Y}, \dfrac{\partial \theta}{\partial Z}; X, Y, Z) \\[2mm]
Q_Z = Q_Z(E_{XX}, E_{YY}, E_{ZZ}, E_{XY}, E_{YZ}, E_{ZX}, \theta, \dfrac{\partial \theta}{\partial X}, \dfrac{\partial \theta}{\partial Y}, \dfrac{\partial \theta}{\partial Z}; X, Y, Z),
\end{cases}
\tag{6.183}
$$

such as Fourier's law of heat conduction stipulating a direct relation between the heat flux and the temperature gradient.

At this point, it is useful to remark that the unknown fields to be solved for are the components of the displacement field and the absolute temperature field, namely

$$\begin{cases} u = u(X,Y,Z,t) \\ v = v(X,Y,Z,t) \\ w = w(X,Y,Z,t) \\ \theta = \theta(X,Y,Z,t). \end{cases} \tag{6.184}$$

If we substitute all our constitutive equations into the equations of balance of linear momentum (6.177) and energy (6.179), we eventually obtain a system of four simultaneous partial differential equations for our four fields (6.184). Indeed, the stresses are ultimately algebraic functions of the displacement gradients and of the temperature, as prescribed by (6.181). Since the balance of linear momentum contains only first derivatives of the stresses, it turns out that the equations of balance of linear momentum are of the second order in the displacement field and at most of the first order in the temperature field. The acceleration terms involve second derivatives of the displacement field with respect to time. The highest derivatives (the second-order ones) appear linearly in the resulting equations, which are accordingly called *quasi-linear* partial differential equations. As far as the balance of energy is concerned, it contains derivatives of up to second order with respect to the body coordinates X,Y,Z and only up to first order with respect to time. It is also a quasi-linear equation. We will not write these equations explicitly.

The fact that we wind up with four (second-order) partial differential equations for the four unknown fields (6.184) is certainly physically satisfying. Moreover, it is clear that in obtaining these final equations we had to make use of both the balance equations and the constitutive laws, the latter being the trademark of phenomenological theories, where the internal structure of the material constitution is subsumed under macroscopic descriptors varying from one material to another. The fact that the number of equations matches the number of unknown fields is not a guarantee of well-posedness. Theorems of existence, uniqueness and stability need to be proven for specific material classes. As a result, restrictions that go beyond those imposed by frame indifference and thermodynamic consistency may have to be imposed on the constitutive response functions (such as convexity requirements).

We turn, finally, to the description of typical boundary and initial conditions in thermoelasticty. The boundary ∂B of the body may consist of two disjoint parts, ∂B_1 and ∂B_2, such that

$$\left. \begin{cases} u(X,Y,X,t) \\ v(X,Y,X,t) \\ w(X,Y,X,t) \end{cases} \right\} \quad \text{are prescribed on } \partial B_1, \tag{6.185}$$

and

$$\left\{ \begin{matrix} \hat{T}_x \\ \hat{T}_y \\ \hat{T}_z \end{matrix} \right\} = \begin{bmatrix} T_{xX} & T_{xY} & T_{xZ} \\ T_{yX} & T_{yY} & T_{yZ} \\ T_{zX} & T_{zY} & T_{zZ} \end{bmatrix} \left\{ \begin{matrix} N_X \\ N_Y \\ N_Z \end{matrix} \right\} \quad \text{are prescribed on } \partial B_2, \tag{6.186}$$

where N_X, N_Y, N_Z are the components of the exterior unit normal to ∂B in the reference configuration. Since the stresses are algebraic functions of the displacement gradients, the traction conditions (6.186) are of the Neumann type, as discussed above, whereas the displacement conditions (6.185) are of the Dirichlet type. Clearly, one could also specify conditions of the Robin type, such as

$$[\mathbf{A}]_1 \left\{ \begin{matrix} u \\ v \\ w \end{matrix} \right\} + [\mathbf{A}]_2 \begin{bmatrix} T_{xX} & T_{xY} & T_{xZ} \\ T_{yX} & T_{yY} & T_{yZ} \\ T_{zX} & T_{zY} & T_{zZ} \end{bmatrix} \left\{ \begin{matrix} N_X \\ N_Y \\ N_Z \end{matrix} \right\} \quad \text{prescribed on } \partial\mathcal{B}, \qquad (6.187)$$

where $[\mathbf{A}]_1$ and $[\mathbf{A}]_2$ are matrices establishing a possible coupling between forces and displacements.

Similarly, in terms of thermal energy, the boundary may be divided into two parts, on one of which the temperature is prescribed (Dirichlet-like) and on the other the heat flux is prescribed (Neumann-like). More generally (Robin-like), we have

$$B_1\theta + B_2(Q_X N_X + Q_Y N_Y + Q_Z N_Z) \quad \text{prescribed on } \partial\mathcal{B}, \qquad (6.188)$$

where B_1 and B_2 are given.

The initial conditions are

$$\left. \begin{matrix} u(X,Y,Z,t_0), v(X,Y,Z,t_0), w(X,Y,Z,t_0) \\[2mm] \left.\frac{\partial u}{\partial t}\right|_{t_0}, \left.\frac{\partial u}{\partial t}\right|_{t_0}, \left.\frac{\partial u}{\partial t}\right|_{t_0} \\[3mm] \theta(X,Y,Z,t_0) \end{matrix} \right\} \quad \text{prescribed on } \mathcal{B}, \qquad (6.189)$$

where t_0 is an initial time. The equations being essentially diffusive, the solution can be expected to exist only for $t \geq t_0$.

References

Epstein, M. (2010) *The Geometrical Language of Continuum Mechanics*. Cambridge: Cambridge University Press.

Epstein, M. and Elżanowsi, M. (2007) *Material Inhomogeneities and Their Evolution*. Berlin: Springer-Verlag.

Gasser, T.C., Ogden, R.W. and Holzapfel, G.A. (2006) Hyperelastic modelling of arterial layers with distributed collagen fibre orientations. *Journal of the Royal Society Interface* **3**, 15–35.

Humphrey, D.J. (2002) *Cardiovascular Solid Mechanics*. New York: Springer-Verlag.

Noll, W. (1965) Proof of the maximality of the orthogonal group in the unimodular group. *Archive for Rational Mechanics and Analysis* **18**, 100–102.

Ogden, R.W. (1984), *Non-linear Elastic Deformations*. Chichester: Ellis Horwood. Reprinted in 1997 by Dover Publications.

Reiner, M. (1948) Elasticity beyond the elastic limit. *American Journal of Mathematics* **70**(2), 433–446.

Treloar, L.R.G. (1958) *The Physics of Rubber Elasticity*. Oxford: Clarendon Press.

Treloar, L.R.G. (1973) The elasticity and related properties of rubber. *Reports on Progress in Physics* **36**(7), 755–826.

Truesdell, C. and Noll W. (1965) *The Non-linear Field Theories Of Mechanics*. Volume III/3 of *Handbuch der Physik* (S. Flügge, ed.). Berlin: Springer-Verlag.

Valanis, K.C. and Landel, R.F. (1967) The strain-energy function of a hyperelastic material in terms of of the extension ratios. *Journal of Applied Physics* **38**, 2997–3002.

7

Remodelling, Ageing and Growth

7.1 Introduction

Living organisms adapt continually to their environment in response to chemical and mechanical stimuli. The results of this constant process of adaptation, specific as they are to each particular instance, can be grouped under the three major headings of remodelling, ageing and growth. This subdivision is particularly suited to the purposes of mathematical modelling, but the boundary between these areas is at best fuzzy.[1]

Remodelling is said to occur at a point of an organism whenever the material in a small neighbourhood of this point remains essentially unchanged and the mass remains constant, while there is a process of re-accommodation alone. An example of remodelling is provided by trabecular bone (Figure 7.1). In a mature bone, the cells responsible for production (osteoblasts) and resorption (osteoclasts) of bone tissue have achieved a state of dynamic equilibrium, so that no net mass is created or destroyed, but the trabeculae may be slowly rearranged so as to align themselves with the directions of maximum tensile or compressive stress (Figure 7.2). This naïve view of the complex process of bone remodelling is known as Wolff's law. Formulated in 1870 by the German physician Julius Wolff (1836–1902), it is now understood in more general terms as asserting that bone architecture responds to mechanical excitations. This general principle is also applicable to other tissues. Recent work demonstrates the application of Wolff's law in plants, specifically in the observed alignment of microtubules with the largest principal stress in the arabidopsis shoot apex[2] (Figure 7.3).

As a result of remodelling and the attendant rotation and/or distortion of the material neighbourhood, internal (residual) stresses may be relieved or created, always without creation or destruction of mass and without any change in the material properties. A dramatic example of residual stresses caused by remodelling in soft tissue is shown in

---

[1] In describing biological processes involving tissue mechanics, biologists distinguish between *growth*, *remodelling* and *morphogenesis*. For a complete account, up to 1995, of the biomechanical aspects, see the comprehensive review paper by Taber (1995). Morphogenesis (describing the generation of animal form and applied usually to embryonic development) is beyond our scope. Ageing, on the other hand, is sometimes understood as a remodelling process rather than as a separate category. Parts of this chapter are the basis for a forthcoming article in *The Encyclopedia of Life Support Systems* (in press), (J Merodio and G Saccomandi, eds.), UNESCO-EOLSS.

[2] Hamant et al. (2008).

The Elements of Continuum Biomechanics, First Edition. Marcelo Epstein.
© 2012 John Wiley & Sons, Ltd. Published 2012 by John Wiley & Sons, Ltd.

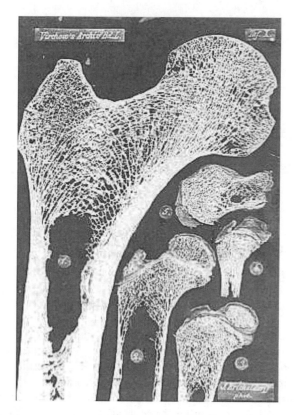

Figure 7.1 Wolff's photographs of sections of the spongy region of: (1) an adult man's femur, (2) a three-year-old girl's femur, (3) a 15-month-old boy's femur, (4) a newborn's femur and (5) a five-year-old girl's calcaneus. The trabeculae are clearly visible. From Wolff (1870), with permission from Springer

Figure 7.4. When cutting a ring-like segment of an artery radially, the residual stresses manifest themselves in a pronounced instantaneous opening angle that increases slowly with time.[3] Processes of remodelling are not limited to living organisms. Metal plasticity, although not a biological phenomenon, can be considered as the remodelling process par excellence.

Ageing occurs whenever, in contradistinction to remodelling, the material properties undergo changes with the passage of time. An example is provided by ageing itself, in the usual sense of the word. As we age, the epidermis loses its pristine elasticity and bone becomes brittle. Leaves wither away in the autumn. Strictly speaking, rejuvenation is also an ageing process, albeit in a supposedly desirable direction. Inert materials also age, as witnessed by such non-biological processes as hardening of concrete, metal corrosion, fatigue and damage. In the common bone disease of osteoporosis, bone becomes dangerously brittle due in part to its increased porosity (Figure 7.5). In this case, even if the remaining solid were of the same mineral quality as in the original young bone, the increase in porosity would lead to lower values of the average mechanical properties (such as the modulus of elasticity) of a representative volume element. In the vegetable

---

[3] Fung and Liu (1989).

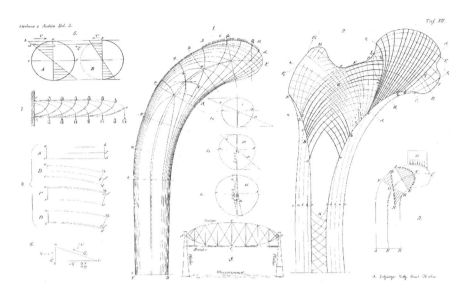

Figure 7.2 Wolff's law, suggesting an analogy between the trabecular patterns of the adult human femur (2) and the principal-stress trajectories in various structural elements. From Wolff (1870), with permission from Springer

kingdom, plant senescence is a common, often cyclical, phenomenon. Although from the biochemical point of view the processes of ageing in plants differ considerably from their counterparts in animals, the resulting changes in mechanical properties are similar. Various mineral deficiencies are the cause of yellowing and brittleness of leaves (Figure 7.6).

Growth occurs whenever the mass of an organism or of any part thereof changes in time. A negative growth is known as *resorption*. Although the treatment of growth and resorption seems to challenge one of the fundamental tenets of physics, namely mass conservation, it should be clear that biological organisms are usually open systems. To clarify this point, one may assume that a mixture of two chemically reacting substances has been placed in a perfectly rigid, hermetic and adiabatic container. As the chemical reaction proceeds, the mass of each of the two components of the mixture changes, while the total mass of the mixture is preserved. But suppose now that an observer is intent on following the evolution of a single component of the mixture, the other component being of no interest or perhaps less amenable to observation. For this observer, the body under study will clearly increase or decrease in mass. Consequently, the law of mass conservation must be reformulated as an equation of mass balance. Concomitantly, the law of momentum balance must be modified to account for the momentum contribution of the entering or exiting mass. A rather mundane example of this situation is provided by the propulsion of a rocket as it ejects fuel, thereby becoming lighter and accelerating. Similarly, the slow growth of a human bone from childhood to maturity would be very difficult to describe in terms of nutrients carried by blood flow, exposure to sunlight, and so on. To take into account all these ingredients of the mixture would be an impossible task. Their interaction with the bone itself is expressed instead in terms of external forces and supplies of free energy, entropy and mass.

Material growth of a continuous medium can be of two kinds: volumetric (or bulk) growth and surface growth. In the case of *volumetric growth*, the body is assumed to be

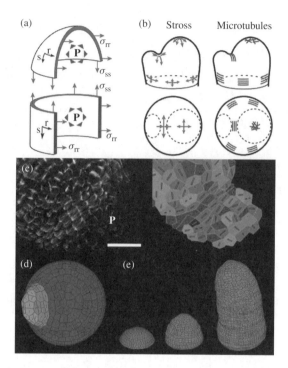

Figure 7.3 Microtubule alignment with maximum principal stress. From Hamant et al. (2008), with permission from the American Association for the Advancement of Science

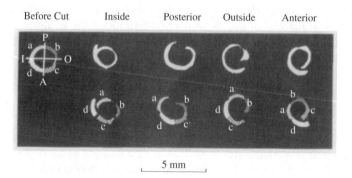

Figure 7.4 Opening of a rat aorta due to radial cuts at four different locations (I, P, O, A) along the ring. As shown in the lower row, subsequent cuts have no further effect in relieving the internal stress. From Fung and Liu (1989), with permission from Wolters Kluwer

made of a fixed continuous collection of material points which preserve their identity. Growth manifests itself by changes of the mass density with time. It is important to note that changes in density are a matter of course in deformable compressive bodies, but here we are referring to changes in mass density of the body itself as it stands in a fixed

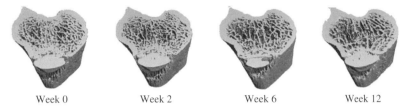

Week 0   Week 2   Week 6   Week 12

Figure 7.5 Osteoporosis: In-vivo micro-computed tomography images of bone loss in the tibia during the first 12 weeks after ovariectomy of mature female rats. From Campbell et al (2008), with permission from Springer

Figure 7.6 Plant senescence

reference configuration. *Surface growth* differs from volumetric growth in two respects. In the first place, in surface growth the addition or subtraction of mass takes place at a surface, as opposed to being distributed over the body volume. The growth surface may be a part of the body boundary, in which case one speaks of *boundary growth*. Clearly, one may also conceive of even more singular growth, which could be concentrated on lines or points. From this point of view, surface growth can be regarded as a singular case of volumetric growth, just as a concentrated force can be seen as a singular case of a distributed load. From the mathematical standpoint one would be led to a weak formulation of growth in the distributional sense.

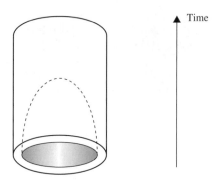

Figure 7.7   Space-time representation of the closure of a hole in an annular body

A second feature peculiar to surface growth is that the material body is no longer a fixed continuous collection of material particles, but rather a collection that varies in time as new particles are added to or taken away from the body. A possible consequence of boundary growth is, therefore, a change in the body topology. Consider, for example, a body with a hole that is being filled with new material (by a process of boundary growth) until, eventually, the hole disappears altogether. Figure 7.7 shows a space-time representation of such a body which, for simplicity of the representation, has been assumed to be two-dimensional and initially having an annular shape. The space-time picture of the process of filling the hole with new mass looks like a solid cylinder (seen from below) from which a fingertip-like cavity (indicated in dotted lines) has been extracted. At any instant of time, the body can be seen as a horizontal section of this volume. For small enough times, these sections are annular, but beyond the time represented by the apex of the fingertip-like void, the sections are solid circles. This topological catastrophe does not impair the smoothness of the space-time (higher-dimensional) representation. In a similar type of representation, the coalescence of two bodies would look like a pair of trousers, the two initial bodies being represented by the cuffs, while the final unified body is represented by the common waist.

7.2   Discrete and Semi-discrete Models

7.2.1   Challenges

The mathematical modelling of physical phenomena in general, and of those pertaining to processes of growth, remodelling and ageing in particular, raises a number of difficult philosophical as well as practical questions. The Canadian philosopher of science Mario Bunge, himself a physicist, emphasizes the problematic concept of *interaction between levels* of scientific description and the notion of scientific explanation.[4] Consider, for

---

[4] Bunge (1998, pp. 28–32) distinguishes between subsumptive and interpretive (or mechanismic) explanations: 'Subsumptive explanation, i.e. the subsumption of a singular under a generalization, smacks of *explicatio obscurum per obscurius*: it does not assuage our thirst for intellection. ... A phenomenological theory, one not representing any "mechanism", will provide subsumptive explanations; but only representational theories, i.e. theories purporting to represent the *modus operandi* of their referents, can give deeper explanations. ... The reason that interpretive explanation is deeper than subsumptive explanation, is that the former involves a *deeper analysis*, one often involving a reference to levels of reality other (usually deeper) than the level to which the explanandum belongs. Subsumptive explanation, by contrast, involves relations among items standing on the same level.'

Figure 7.8   Galileo's beam

example, a metal or wooden beam supported, say, at its ends or, more poignantly, sticking out of a mossy wall, as Galileo wanted (Figure 7.8) in one of the first attempts ever at tackling this problem in a serious manner.[5]

And what is the problem? Let us start by confessing that even this matter is not entirely clear. Is the problem one of safety? That is, do we want to make sure that the beam does not break under its service load? Or is it that we wish to predict the deformations of the beam? Or, perhaps, we want to calculate the internal forces that tend to change the average distance between molecules. Molecules, did we say? What about atoms and beyond? Is there any hope of having a theory that would work for steel, aluminium, wood and rubber? Starting with Galileo, Hooke, Euler, Bernoulli, Saint-Venant, Cauchy and ending with quantum mechanics and solid-state physics, many questions have been answered. The beam is still sticking out of the wall. Free-body diagrams are true, and so are bending-moment diagrams, aren't they? Phase-equilibrium diagrams used by material scientists are also true, and so are lattice mechanics considerations, each atom somehow sharing the burden of load transmission with its neighbours. Or, perhaps, in rubber-like materials, it is the uncoiling of long molecules that explains the development of internal stress. These different levels of explanation have all a degree of truth, the truth of one level being, as it were, confirmed when a model at the next lower level is able to justify it as an approximation. And does this inductive chain lead us eventually to the truth of the object itself?[6]

In the case of biological materials, the problem of choosing the right level of discourse is greatly exacerbated by the appearance of new intermediate levels such as cells and their

---

[5] Galilei (1638).
[6] These general considerations are part of the article by Epstein and Herzog (2003).

components as well as long and complex molecules. In bone, for example, macroscopic models of remodelling have been proposed that incorporate some detailed elements of the interactions between osteoblasts, osteoclasts and osteocytes. It is not uncommon for people working with biological materials to expect the mathematical models of living tissue to deliver results with a degree of comprehensiveness and detail not available even for the more mundane inert materials. It would be like asking Cauchy, say, after the publication in the 1820s of his seminal papers on the mechanics of deformable media, the question: 'All this is fine and well. But what would happen to an elastic steel structure if it were placed in a corrosive environment?' What we should expect from a material model is that it be based on sound scientific principles, that it be mathematically consistent and that it have predictive value. This predictive value, however, is necessarily confined to predictions at the level of departure or higher. It cannot be expected to have anything to say as far as lower levels are concerned. Thus, the bending moment at a cross section of a beam cannot provide detailed information about the inter-atomic potentials.

In general terms, it can be said that there exist two different extreme paradigms that can be and have been used in modelling biological materials, namely, discrete and continuous models, as we have already anticipated in Section 1.2. In a discrete model, typified by but not restricted to *cellular automata*, the material background is lumped into a discrete collection of point-like sites, where all events take place. These sites interact with each other according to specific rules intended to represent the physical, chemical and mechanical laws. The rules may be very simple, but the complexity of the system, reflected in terms of the large number of sites at play, can result in a very flexible overall behaviour able to capture the essential features of the macroscopic phenomena. Among the merits of this approach is its ability to incorporate directly into the model a variety of criteria which are better understood in their raw nature as discrete events. Examples of this kind are: cell subdivision and transmission of information from the genes to the environment. For purposes of illustration, we now describe two discrete-like models of growth and remodelling.

Cellular automata have been successfully used to describe the growth of cancerous tumours. Cancer develops in three successive phases: *avascular*, *vascular* and *metastatic*. Somewhat different techniques are used to model each of these phases, but the avascular phase forms the foundation. The avascular phase takes place initially when the *tumour spheroid*, as the tumour cell aggregate is called, is sustained by locally available nutrients (such as oxygen) only. The size of a tumour under these conditions is controlled by the natural dynamic balance between cell proliferation and cell death. In particular, the diffusion of the nutrients imposes a limit for tumour growth. It is generally hypothesized that, in order to continue growing, tumours recruit blood vessels from the surrounding tissues through a process known as *angiogenesis*, thus triggering the vascular phase. A comprehensive review of both discrete and continuous models of avascular tumour growth is given by Roose et al. (2007). We present first a sketch of a discrete model.

7.2.2 Cellular Automata in Tumour Growth

Recall that, in its most elementary version, a *cellular automaton*[7] consists of a finite number of fixed sites arranged in a regular fashion, each of which is in one of a finite number

---

[7] A comprehensive picture of the capabilities of cellular automata in modelling complex systems can be found in Wolfram (1994).

of possible states, whose evolution in discrete time-steps is governed by a finite set of rules of interaction of each site with its neighbours. Moreover, the state of the system at any one time is completely determined by the state of the system at the immediately preceding time. Thus, cellular automata are ideally suited for the description of the evolution of certain biological systems.

In the case of the growth of tumours, a possible automaton rule may involve a probabilistic criterion, according to which a site occupied by a cancerous cell can proliferate, become quiescent, die, or move to a neighbouring site. The probability of each of these choices depends on the local state of the system, which, in addition to the presence of a cancerous cell at a site, may include the local concentration of nutrients and other factors (mechanical, chemical, etc.) deemed to be relevant. Qi et al. (1993) construct an automaton consisting of a square lattice divided into $n \times n$ equal compartments. Each cell is affected by only four neighbours, and may be occupied by either a normal cell, a cancerous cell, a complex, a dead cancer cell, or an effector cell. The evolution of the automaton is then governed by three simple probabilistic rules, based on experimental evidence. Starting from an initial state in which there are five cancerous cells in the central compartment, and a normal cell and an effector cell in each compartment, Figure 7.9 shows the state of the tumour after about 100 steps, when it reaches its maximum size. The black compartments represent cancerous cells.

7.2.3 A Direct Model of Bone Remodelling

Another discrete (or semi-discrete) model of a different kind has been proposed by Huiskes et al. (2000) in the area of bone remodelling. Recall (from Section 4.10) that the cells involved in the remodelling processes are osteoblasts (bone producing), osteoclasts (bone removing) and osteocytes (bone maintaining). Osteocytes result from osteoblasts trapped in the bone matrix. They are considered as the cells responsible for the sensing and

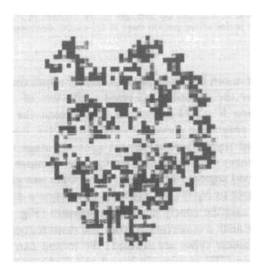

Figure 7.9   The predicted shape of a tumour. From Qi et al. (1993), with permission from The Academic Press

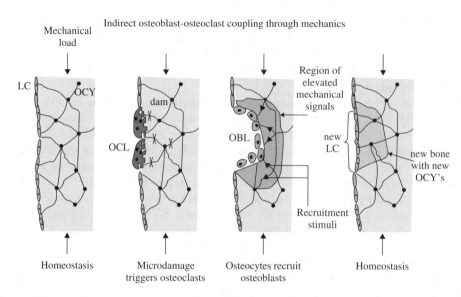

Figure 7.10 The interplay between osteoclasts (OCL), osteoblasts (OBL) and osteocytes (OCY) in response to mechanical stimuli. Some OBLs, trapped in the matrix, become OCYs, while others turn into inactive lining cells (LC). From Ruimerman et al. (2005), with permission from Elsevier

transmission of mechanical signals to the osteoblasts and osteoclasts, which control the net apposition or removal of bone tissue (Figure 7.10).

The basic assumptions of the model are as follows: (i) The mechanical variable that triggers feedback from the external forces to bone metabolism is a typical strain-energy density (SED) rate in the mineralized tissue, as produced by a recent loading history. (ii) Osteocytes react to the loading in their local environments by producing a biochemical messenger in proportion to the typical SED rate. (iii) The biochemical messenger produced by the osteocytes causes signals to be dissipated through the osteocytic network towards the bone surface, where they create an osteoblast recruitment stimulus. The strength of this signal, which is produced by all osteocytes in the environment, stimulates osteoblast recruitment and bone formation as long as it exceeds a threshold value. (iv) To portray resorption in the model, the probability of osteoclast activation per surface site at any time is considered to be regulated either by the presence of microcracks within the bone matrix (hypothesis I) or by disuse (hypothesis II). Hypothesis I is based on the idea that osteoclasts are recruited and activated where microcracks occur in the bone, possibly as an effect of signals from osteocytes. Hence, for hypothesis I the probability of resorption would be equal for all surface sites, and independent of mechanical strain; whereas for hypothesis II the local osteoclast activation probability would be high in disused areas and low in areas of high strain. These assumptions, which clearly involve a mixed bag of continuous variables (such as strain energy) and discrete quantities, need quantification, which is achieved, in part, as follows. The strength f of the signal at location x emitted by an osteocyte located at x' is assumed to decay exponentially with the distance $d(x,x')$ between the two locations according to the formula

$$f(x,x') = \exp\left(-\frac{d(x,x')}{D}\right), \tag{7.1}$$

where D is a decay constant (in units of length). This signal strength is then integrated over a region of influence around x containing a specific number of osteocytes to give a total stimulus strength $P(x,t)$. The local change in relative bone density m (1.0 for fully mineralized tissue) can then be expressed as

$$\frac{\partial m(x,t)}{\partial t} = \tau P(x,t), \tag{7.2}$$

where τ is a rate constant. This bone density rate can be further subdivided into osteoblastic and osteoclastic contributions, each subject to a separate law of evolution. Furthermore, the SED is replaced by an SED rate when cyclic loads are applied. These and other quantifying assumptions (such as the excitation threshold, the probability distributions for both hypothesis I and II, and so on) were incorporated in an iterative computer code coupled with a finite-element analysis. A typical result is shown in Figure 7.11 for a plane 2 mm square bone sample.

The first row, marked (a), in Figure 7.11 shows the result of remodelling under hypothesis I, starting from a regular grid. After 2000 iterations a homeostatic state is reached. The trabeculae are seen to align with the direction of the applied (dynamic) load. In the second row, (b) corresponds to a completely different starting architecture, which after 3000 iterations converges to the same homeostatic state. The results indicated by (c) correspond to a further rotation of the load, and shows how the final state aligns itself with the new direction. Finally, (d) shows the result under conditions of strain-controlled resorption (hypothesis II).

From the two preceding examples, it is apparent that one of the features of discrete or semi-discrete models is the presence of an enviable amount of detail at a rather low (and, hence, more fundamental) level of description. On the other hand, these models, precisely because of the degree of detail that has been incorporated in the description, tend to be somewhat tainted by the proliferation of *ad hoc* hypotheses, thus rendering the predictive ability of the model circumscribed to a very narrow window of phenomena. In other

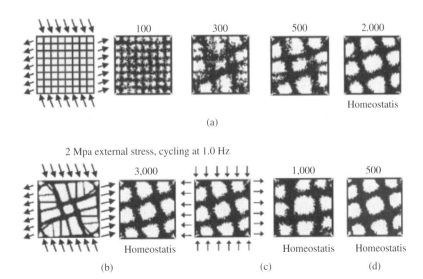

Figure 7.11 Computer simulation of bone remodelling. From Huiskes et al. (2000), with permission from the Nature Publishing Group

words, discrete models offer the advantage over continuous models (to be described below) of allowing for the transmission of signals, heterogeneity of cells and direct incorporation of relevant experimental information without the need to formulate precise physical laws of balance. On the other hand, precisely because of this lack of a foundational physicochemical underpinning, these models do not necessarily enhance the general understanding of the phenomena at hand. Each case stands on its own. Mixed models, involving a continuous substrate behaving according to the laws of continuum mechanics and a discrete model superimposed on it to represent the transmission of signals and other discrete phenomena, are very promising.

7.3 The Continuum Approach

7.3.1 Introduction

At the other paradigmatic extreme, we have continuous models. Phenomena such as the deformability of the medium and the expression of its thermomechanical properties are not easy to implement in discrete systems without a considerable loss of fidelity. It is in this domain that continuous models have the upper hand. Accordingly, modern continuum mechanics, with its sophisticated blend of mathematical generality and a centuries-old tradition of success in particular theories of solids and fluids, is the preferred setting for most present-day models of growth and remodelling.

In its standard formulation, however, continuum mechanics is not in itself sufficient to face the challenges posed by problems of growth and remodelling. Among the various reasons for this deficiency, three deserve particular mention. The first, and most obvious, reason is that one of the basic tenets of continuum mechanics is the law of conservation of mass (somewhat misleadingly called the continuity equation in fluid mechanics). This limitation is not too difficult to overcome by replacing the law of *conservation* by a law of *balance* of mass, whereby volumetric sources and surface fluxes of mass are admitted into the picture. Concomitant corrections have to be made to the other laws of balance (linear and angular momentum, energy and entropy). Such modifications are already available in the standard context in terms of the theory of chemically reactive mixtures. Indeed, focusing attention on one of the components of a mixture, it may be possible to obtain a correct version of the modified laws of balance alluded to above.

The second challenge presented to standard continuum mechanics by the analysis of growth and remodelling is that the concept of material body, as a fixed collection of particles with fixed material properties manifesting itself in space through configurations, needs to be revised. Indeed, in a process of remodelling, changes are taking place in the body even in the absence of spatial deformation or motion. In a process of ageing, for instance, the material properties are altered by the mere passage of time. The propagation of a crack may be triggered by the deformation, but results in a change in the body itself. Similar remarks apply to the propagation of phase boundaries, motion of dislocations, reorientation of the directions of anisotropy, and so on. In the last two decades, continuum mechanics has responded to this challenge by augmenting its scope to include the concept of configurational or material forces seen as the driving forces behind processes of material evolution.

Finally, a third challenge arises from the fact that (for instance, in cases of surface growth) not only the material characteristics of the body points evolve with time, but also its very topology may evolve, as already briefly described in Section 7.1. We will presently, however, concentrate on the so-called *volumetric* (or bulk) phenomena, in which, even in the presence of a process of addition or removal of mass, the body particles retain their identity. It is only the mass density and the material properties that can change with time.

7.3.2 The Balance Equations of Volumetric Growth and Remodelling

As mentioned above, the balance equations that govern processes of volumetric growth and remodelling can be regarded, at least in principle, as the result of focusing attention on a single component of a chemically reacting mixture of several substances. Mass is, therefore, not necessarily conserved, and transfers of linear momentum, angular momentum, energy and entropy appear in the equations as extra contributors to the total balance. In a process of growth, these extra terms may be, at least in part, attributed to the very addition or subtraction of mass. At best, we may have a case of 'compliant' growth, where the new mass happens to enter at the same velocity, specific energy and specific entropy as the local substratum.[8] At any rate, in most applications, the terms associated with inertia of the additional mass are negligible, since processes of growth are notoriously slow.

Balance of Mass

While at some conceivable level of analysis (such as that of chemically reacting mixtures) the appearance or disappearance of mass of one species may be accounted for by concomitant losses or gains in other species, in a volumetric growth model we accept the existence of sources and sinks of mass as part of the theory. For a fixed volume V in a reference configuration, we write

$$\frac{d}{dt} \int_V \rho_0 \, dV = \int_V \Pi \, dV + \int_{\partial V} M \, dS, \tag{7.3}$$

where ρ_0 is the possibly time-varying mass density in the reference configuration, Π is the (smooth) volumetric source of mass and M is a possible mass flux through the boundary ∂V, with exterior unit normal $\mathbf{N}$. The existence of a mass-flux vector $\mathbf{M}$ such that

$$M = \mathbf{M} \cdot \mathbf{N} \tag{7.4}$$

being guaranteed by Cauchy's theorem,[9] we can write the local form of equation (7.3) as

$$\frac{\partial \rho_0}{\partial t} = \Pi + \text{Div } \mathbf{M}. \tag{7.5}$$

In terms of the spatial counterparts of ρ_0, Π and $\mathbf{M}$, given respectively by

$$\rho = J^{-1} \rho_0, \tag{7.6}$$

---

[8] This is the terminology adopted in Epstein and Elżanowski (2007). The term 'reversible', instead of 'compliant', was used earlier in Epstein and Maugin (2000). The first instance of the inclusion of these additional terms (in the energy equation) in the context of a theory of remodelling is due to Cowin and Hegedus (1976a).

[9] Or assumed *a priori*.

$$\pi = J^{-1}\,\Pi, \tag{7.7}$$

and

$$\mathbf{m} = J^{-1}\,\mathbf{F}\,\mathbf{M}, \tag{7.8}$$

with

$$m = \mathbf{m} \cdot \mathbf{n}. \tag{7.9}$$

we obtain the following Eulerian balance law:

$$\frac{D\rho}{Dt} = \pi + \operatorname{div}\mathbf{m} - \rho\operatorname{div}\mathbf{v}, \tag{7.10}$$

where $\mathbf{v}$ is the velocity field and where we have distinguished between the spatial divergence operator (div) and its referential counterpart (Div).

In converting the equation of mass conservation into an equation of mass balance, we have followed the generic formal prescription presented in Section 6.2.1. In so doing, we have naturally assumed the existence of sources or sinks of mass (π) distributed over the volume and of mass fluxes ($\mathbf{m}$) per unit area. From the physical viewpoint, mass flux can be regarded as a process of mass self-diffusion, similar to the spread of a drop of milk in a cup of coffee, but involving one substance only. Not surprisingly, this is a controversial issue.[10] It is prudent, therefore, from this point on to make the simplifying constitutive assumption that the mass flux vanishes identically, namely

$$\mathbf{M} = \mathbf{m} = \mathbf{0}. \tag{7.11}$$

Remark 7.3.1 Notice that only under this assumption does equation (7.10) reduce to the mass balance equation of one component of a mixture (equation (4.17)). In other words, in standard mixture theory mass fluxes are not explicitly postulated, but result from diffusive processes of one species within the others.

Balance of Linear Momentum

Starting with the Eulerian formulation, the balance of linear momentum in a moving material volume v in an inertial frame states that

$$\frac{d}{dt}\int_{v} \rho\,v^{i}\,dv = \int_{v} b^{i}\,dv + \int_{\partial v} \hat{t}^{i}\,da + \int_{v}(\pi\,v^{i} + \bar{p}^{i})\,dv. \tag{7.12}$$

The meaning of the symbols b^{i} and $\hat{t}^{i}$ is the same as in equation (6.38). The term πv^{i} is the momentum contributed by the entrant mass π, entering at the background velocity v^{i},

---

[10] The issue of mass flux in theories of growth and remodelling was raised by Epstein and Maugin (2000); see also Kuhl and Steinmann (2003). A different point of view, according to which growth and remodelling are placed within the context of mixture theory involving at least two species and mass flux is used only in the context of diffusion of nutrients in the underlying tissue, is adopted in the major paper by Garikipati et al. (2004). Yet another perspective, based on a single component and the introduction of an additional kinematic variable, is presented by Epstein and Goriely (2012).

while the term $\bar{p}^i$ represents the non-compliant source of volumetric momentum. Invoking Cauchy's tetrahedron argument, we now deduce the existence of a (Cauchy) stress tensor **t** such that

$$\hat{t}^i = t^{ij} n_j. \tag{7.13}$$

Taking into consideration equation (7.13) and using the balance of mass in the form of equation (7.10), we obtain the following local Eulerian version of the balance of momentum:

$$\rho \frac{Dv^i}{Dt} = b^i + \bar{p}^i + t^{ij}_{,j}. \tag{7.14}$$

Notice that, just as in mixture theory, the compliant contribution πv^i does not appear in the final local equation.

The Lagrangian counterpart of equation (7.14) is

$$\rho_0 \frac{\partial v^i}{\partial t} = B^i + \bar{p}_0^i + T^{il}_{,I}. \tag{7.15}$$

In the last two equations we have used (Cartesian) components to avoid any ambiguity of the notation. The relation between the Lagrangian and Eulerian densities is

$$B^i = J b^i, \tag{7.16}$$

$$\bar{p}_0^i = J \bar{p}^i, \tag{7.17}$$

and

$$T^{il} = J F^{-I}_{\ j} t^{ij}, \tag{7.18}$$

where **T** is the Piola stress.

Balance of Angular Momentum

In the absence of body and surface couples, the global balance of angular momentum is expressed as

$$\frac{d}{dt} \int_v \mathbf{r} \times (\rho \mathbf{v}) \, dv = \int_v \mathbf{r} \times (\mathbf{b} + \pi \mathbf{v} + \bar{\mathbf{p}}) \, dv + \int_{\partial v} \mathbf{r} \times (\mathbf{tn}) \, ds, \tag{7.19}$$

where **r** is the spatial position vector with respect to the inertial frame. In components, this equation is expressed as

$$\frac{d}{dt} \int_v \varepsilon_{ijk} x^j (\rho v^k) \, dv = \int_v \varepsilon_{ijk} x^j (b^k + \pi v^k + \bar{p}^k) \, dv + \int_{\partial v} \varepsilon_{ijk} x^j t^{km} n_m \, ds. \tag{7.20}$$

Enforcing the balance of mass and linear momentum, the local form of this equation boils down to the symmetry of the Cauchy stress:

$$t^{ij} = t^{ji}, \tag{7.21}$$

just as in the standard formulation without growth.[11] For the Piola stress, we have

$$T^{il} F^j{}_l = T^{jl} F^i{}_l. \tag{7.22}$$

Balance of Energy

Denoting by u the internal energy per unit mass, we can write the global Eulerian form of the energy balance as

$$\frac{d}{dt} \int_v \rho \left(u + \frac{1}{2} \mathbf{v}.\mathbf{v} \right) dv$$

$$= \int_v (\mathbf{b} \cdot \mathbf{v} + \rho r)\, dv + \int_v \left[\pi \left(u + \frac{1}{2} \mathbf{v} \cdot \mathbf{v} \right) + \bar{u} + \bar{\mathbf{p}} \cdot \mathbf{v} \right] dv$$

$$+ \int_{\partial v} \left(\hat{\mathbf{t}} \cdot \mathbf{v} + h \right) dv \tag{7.23}$$

where $\bar{u}$ is the non-compliant volumetric contribution to the internal energy.[12] The term h, representing the rate of non-mechanical energy supply per unit area ('conduction') can be expressed (again, through the tetrahedron argument) in terms of a heat-flux vector $\mathbf{q}$:

$$h = -\mathbf{q} \cdot \mathbf{n}. \tag{7.24}$$

Using equations (7.9), (7.10), (7.13) and (7.14), the local form of equation (7.23) is obtained as

$$\rho \frac{Du}{Dt} = \rho r + \bar{u} + t^{ij} v_{i,j} - q^j{}_{,j}. \tag{7.25}$$

Its Lagrangian counterpart is

$$\rho_0 \frac{\partial u}{\partial t} = \rho_0 r + \bar{U} + T^{il} v_{i,I} - Q^I{}_{,I}, \tag{7.26}$$

with

$$Q^I = J F^{-I}{}_i q^i \tag{7.27}$$

and

$$\bar{U} = J \bar{u}. \tag{7.28}$$

The Clausius–Duhem Inequality

Defining the entropy content per unit mass, s, we write the Eulerian version of the global Clausius–Duhem inequality as follows:

$$\frac{d}{dt} \int_v \rho s\, dv \geq \int_v \left(\frac{\rho r + \bar{u} - \bar{h}}{\theta} + \pi s \right) dv + \int_{\partial v} \left(\frac{h}{\theta} \right) ds, \tag{7.29}$$

---

[11] The retention of the mass flux terms would have altered this result.

[12] In principle, this term could have been absorbed within the heat source r, with the understanding that it may eventually be specified constitutively, rather than just externally.

where θ is the absolute temperature. We note that we have added a further volumetric sink $\bar{h}$ of non-compliant entropy, which should be specified constitutively. The resulting local form is

$$\rho \frac{Ds}{Dt} \geq \frac{\rho r + \bar{u} - \bar{h}}{\theta} - \left(\frac{q^i}{\theta}\right)_{,i}. \tag{7.30}$$

The Lagrangian version is

$$\rho_0 \frac{\partial s}{\partial t} \geq \frac{\rho_0 r + \bar{U} - \bar{H}}{\theta} - \left(\frac{Q^I}{\theta}\right)_{,I}, \tag{7.31}$$

with an obvious notation.

Introducing the Helmholtz free energy per unit mass,

$$\psi = u - \theta s, \tag{7.32}$$

we can use the balance of energy to rewrite equation (7.31) in the form

$$\rho_0 \dot{\psi} + \rho_0 \dot{\theta} s - \bar{H} - T^{il} v_{i,I} + \frac{1}{\theta} Q^I \theta_{,I} \leq 0. \tag{7.33}$$

The *entropy sink* $\bar{H}$ may be present even if growth does not occur. It may, indeed, be responsible for processes of remodelling.

Exercise 7.3.2 Compare the balance equations of growth with their standard counterparts, as presented in Section 6.2.2, and with the balance equations of the components of a mixture in Chapter 4. Be aware of some merely notational differences.

7.4 Case Study: Tumour Growth

While more general continuous models will be discussed in the remainder of this chapter, some of the main features of these models may be gathered from the following formulation, which can be used for predicting the time evolution of tumours. As a valuable introductory example, instead of directly applying the equations of growth just proposed, we will revert to mixture theory. The theoretical continuum mechanics background should be provided by the theory of chemically reacting mixtures.[13] In applications to growth of tumours, however, it is not uncommon to retain the equations of mass balance only. By discarding the balance of forces and energy, these models must therefore implement an *ad hoc* constitutive equation expressing the diffusive velocities in terms of the concentrations alone, such as Fick's law, and add a considerable number of other simplifications.

As a starting point, recall that in mixture theory it is assumed that any spatial element of the mixture, no matter how small, is occupied by particles of all the M species of the mixture. Denoting by $\mathbf{v}_\alpha$ and ρ_α the velocity and the mass density (per unit mixture volume) of the αth species, the Eulerian equations of mass balance are given by

$$\frac{\partial \rho_\alpha}{\partial t} + \text{div}\left(\rho_\alpha \mathbf{v}_\alpha\right) = \pi_\alpha, \quad \alpha = 1, \ldots, M, \tag{7.34}$$

---

[13] See Chapter 4 and references therein.

where π_α stands for the mass production, per unit mixture volume, associated with the species α (due to chemical reactions between the various species). Adding up the individual balance laws, we obtain, as expected, the conservation law

$$\frac{\partial \rho}{\partial t} + \operatorname{div}(\rho \mathbf{v}) = 0, \tag{7.35}$$

where

$$\rho = \sum_{\alpha=1}^{M} \rho_\alpha \tag{7.36}$$

is the spatial density of the mixture and

$$\mathbf{v} = \frac{1}{\rho} \sum_{\alpha=1}^{M} \rho_\alpha \mathbf{v}_\alpha \tag{7.37}$$

is the mean (barycentric) velocity of the mixture. The total mass production vanishes, since the chemical reactions are assumed to be (stoichiometrically) balanced. Introducing the *diffusive velocities*

$$\mathbf{u}_\alpha = \mathbf{v}_\alpha - \mathbf{v}, \quad \alpha = 1, \ldots, M, \tag{7.38}$$

equation (7.34) can be rewritten as

$$\frac{\partial \rho_\alpha}{\partial t} + \operatorname{div}(\rho_\alpha \mathbf{v}) = -\operatorname{div}(\rho_\alpha \mathbf{u}_\alpha) + \pi_\alpha, \quad \alpha = 1, \ldots, M. \tag{7.39}$$

It is at this point that *Fick's law of diffusion*, which we have already encountered in Section 2.5, is introduced, namely

$$\rho_\alpha \mathbf{u}_\alpha = -K_\alpha \nabla \rho_\alpha, \quad \alpha = 1, \ldots, M, \tag{7.40}$$

where ∇ is the spatial gradient operator and K_α are positive definite scalars (or, more generally, tensors) representing the diffusivity properties of each constituent of the mixture. Substituting in equation (7.39), we obtain the final form:

$$\frac{\partial \rho_\alpha}{\partial t} + \operatorname{div}(\rho_\alpha \mathbf{v}) = \operatorname{div}(K_\alpha \nabla \rho_\alpha) + \pi_\alpha, \quad \alpha = 1, \ldots, M. \tag{7.41}$$

It is customary, in a chemical context, to express the various mass densities in equation (7.41) in terms of fixed molecular masses multiplied by the 'number of particles' of each species per unit mixture volume. If the mean velocity $\mathbf{v}$ is known, and if the chemical reaction rates are expressed in terms of the concentrations, equation (7.41) becomes a system of diffusion equations to be solved for the density of each species. In practice, however, our interest lies in solving also for the velocity field. We observe that, even with the introduction of Fick's law for each constituent, the system of equations (7.41) falls short in terms of number of equations versus number of unknown functions. To close the system, further assumptions are made, which are not always explicit or mutually consistent.

The first common assumption is that we are in the presence of a perfectly spherically symmetric problem, so that the only non-vanishing component of the mean velocity $\mathbf{v}$ is the radial one. The second assumption is that the total density ρ of the mixture is spatially and temporally constant. The third assumption is that the chemical productions are not perfectly balanced. To explain this better, let us assume that there exist just two components, which we denote by L and D, for 'live cells' and 'dead cells', respectively. Although the increase in D-cells is completely accounted for by the decrease in L-cells, the increase in L-cells is supplied externally by an unspecified source (not part of the mixture). Thus we have that the sum $\pi_L + \pi_D$ is always non-negative and seldom zero. It is also assumed that the (scalar) coefficients of diffusivity in Fick's law are independent of each other and, in particular, that K_D vanishes. Finally, chemically inspired constitutive equations (see Casciari et al., 1992) are supplied connecting the production with the velocity component. Since the density is assumed to be constant, an equation can be ultimately obtained and solved numerically for the increase in the radius of the spheroid as a function of time.

Quite apart from some of the questionable theoretical features of the model, the complexity of the underlying phenomena is such that an accurate experimental substantiation of the constitutive assumptions is virtually impossible. This is the case, in fact, for most biomechanical models. More sophisticated continuum models of tumour growth exist (e.g., Greenspan, 1976; Byrne and Chaplain, 1996) that incorporate mechanical constitutive equations, usually–but not necessarily (e.g., Roose et al., 2003)–in terms of an average pressure and an assumed strain-energy function. These models allow us to represent a tumour that grows in a non-spherically symmetric fashion and remove some of the other limitations of the purely mass-balance-driven models.

To gain an idea of the kind of problems that can be solved with this approach, we will follow Ward and King (1997), who consider a mixture of living cells (L), dead cells (D) and nutrients (N). We will assume the nutrients to occupy a negligible amount of space. The densities ρ_L and ρ_D of living and dead cells are governed, respectively, by the following mass balance equations:

$$\frac{\partial \rho_L}{\partial t} + \mathrm{div}(\rho_L \mathbf{v}) = (k_m(c) - k_d(c))\rho_L \qquad (7.42)$$

and

$$\frac{\partial \rho_D}{\partial t} + \mathrm{div}(\rho_D \mathbf{v}) = k_d(c)\rho_L. \qquad (7.43)$$

These equations are nothing but a particular case of (7.41) under the following four assumptions: (i) There is no diffusion and, therefore, the velocity of each species equals the mixture velocity $\mathbf{v}$. (ii) The mass supplies (π_L, π_D) are linear in the density of live cells with coefficients governed by two positive functions, $k_m(c)$ and $k_d(c)$, of the nutrient concentration c. These functions are supposed to represent, respectively, the processes of *mitosis* (cell division) and *death*. The supply of dead cells in equation (7.43) is the same as the death of cells in equation (7.42). (iii) The tumour cells are incompressible and there are no interstitial voids in the mixture. (iv) The tumour remains spherical during the process of growth and the velocity of all cells is radial.

Assuming that the mass of the individual cells does not change, the densities (ρ_L and ρ_D) can be regarded as constantly proportional to the number of cells (n_L and n_D) of each type per unit volume of the mixture, a policy that is usually adopted in biological applications. In terms of these quantities, we can rewrite the balance of mass equations as

$$\frac{\partial n_L}{\partial t} + \mathrm{div}(n_L \mathbf{v}) = (k_m(c) - k_d(c))n_L \qquad (7.44)$$

and

$$\frac{\partial n_D}{\partial t} + \text{div}(n_D \mathbf{v}) = k_d(c) n_L. \tag{7.45}$$

The assumed incompressibility of the cells, with individual cell volumes V_L and $V_D < V_L$, and the absence of voids, permits us to claim that

$$n_L V_L + n_D V_D = 1. \tag{7.46}$$

Taking advantage of this 'volume additivity' constraint (encountered in a different context in Section 4.5.3, while studying cartilage), we multiply equation (7.44) by V_L and equation (7.45) by V_D and add the results to obtain the single equation

$$\text{div } \mathbf{v} = \left[k_m(c) V_L - k_d(c)(V_L - V_D) \right] n_L. \tag{7.47}$$

The nutrient concentration c is assumed to be governed by an additional equation of the diffusive type, coupling c with $\mathbf{v}$ and n_L. Because of the crucial assumption that the velocity field is radial, the system of equations (7.44) and (7.47), together with the additional equation for c, constitutes a system of three partial differential equations for the three unknown fields n_L, v_r, c (where v_r is the radial speed) as functions of the radial coordinate r and time t. Equation (7.46) can be solved *a posteriori* for n_D (while equation (7.45) will be satisfied identically). Thus, under the adopted symmetry assumptions, there is no further need to couple the growth with the mechanics, either by balance or by constitutive equations.

Notice that in the equations above, the divergence of a radial vector field, such as the velocity, is given by

$$\text{div } \mathbf{v} = \frac{\partial v_r}{\partial r} + \frac{2}{r} v_r. \tag{7.48}$$

We need to specify initial and boundary conditions. This is obviously a problem with a moving boundary of radius $R = R(t)$. At the initial time $t = 0$, we assume the sphere to consist of a single live cell, whose radius is

$$R(0) = \left(\frac{V_L}{4\pi/3} \right)^{\frac{1}{3}}. \tag{7.49}$$

Since this volume is occupied by a single cell, the number of live cells per unit volume is given initially by

$$n_L(r,0) = \frac{1}{V_L}, \quad \leq r \leq R(0). \tag{7.50}$$

The nutrient concentration is assumed to have a constant value c_0 outside the tumour. At the initial time, in order to get the process started, we set

$$c(r,0) = c_0, \quad 0 \leq r \leq R(0). \tag{7.51}$$

By symmetry, the cell at the origin remains fixed at all times,

$$v_r(0,t) = 0, \tag{7.52}$$

and the outer boundary moves with the cells, namely

$$\frac{dR(t)}{dt} = v_r(R(t), t). \tag{7.53}$$

We impose the condition that at the outer boundary the concentration of nutrient is continuous, namely

$$c(R(t), t) = c_0. \tag{7.54}$$

At the origin we have

$$\frac{\partial c}{\partial r}(0, t) = 0. \tag{7.55}$$

Exercise 7.4.1 Study the article by Ward and King (1997), which supplies detailed technical and numerical information. Develop a numerical procedure to solve the initial moving boundary value problem and compare with the results reported in the paper.

7.5 Case Study: Adaptive Elasticity of Bone

Adaptive elasticity is a term coined by Cowin and Hegedus (1976a) to designate a model of bone remodelling and growth that constitutes the first formulation of a fully fledged thermomechanical theory of volumetric growth based on a complete and rigorous expression of the underlying physical and mathematical laws within the context of continuum mechanics. The main assumptions of the theory are as follows: (i) bone is considered as an elastic porous matrix (made of extracellular material) whose pores are filled with a liquid perfusant; (ii) the slow chemical reactions (mediated by the bone cells) responsible for growth and remodelling are controlled by the state of strain of the matrix; (iii) the addition or removal of solid mass resulting from the chemical reactions takes place exclusively at the expense of the porosity, thus causing no residual stresses; (iv) the porosity is included as one of the kinematic variables of the theory; (v) the balance equations are formulated on the basis of the solid phase alone (namely, the matrix), which is an open system immersed in an isothermal perfusant bath. Some of the features of the model are illustrated schematically in Figure 7.12.

Since the addition (growth) and removal (resorption) of material take place exclusively at the pores, the total volume occupied by the bone (in a stress-free configuration at constant temperature which is assumed to exist) remains invariable.

Denoting by γ the intrinsic density of the matrix material and by ϕ the porosity, the effective density ρ of the porous structure is given by

$$\rho = \gamma \zeta = \gamma(1 - \phi), \tag{7.56}$$

where $\zeta = 1 - \phi$ is the *solid volume fraction*. In Section 7.3.2 we have already dealt with the equations of balance in a general theory of growth. The equations of balance in the theory of adaptive elasticity are, therefore, a particular case. For the sake of clarity, however, and at the risk of unnecessary repetition, we will follow the original lines of Cowin and Hegedus (1976a). A crucial part of a volumetric growth theory is the incorporation of a distributed mass source π measuring the mass produced per unit time

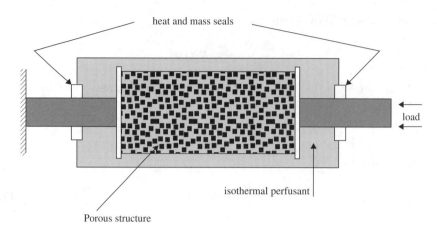

Figure 7.12 Schematic diagram of the model. From Cowin and Hegedus (1976a), with permission from Springer

and per unit spatial volume. The local Eulerian form of the balance of mass can then be written in terms of the (spatial) velocity field **v** as

$$\frac{\partial \rho}{\partial t} + \text{div}(\rho \mathbf{v}) = \pi. \tag{7.57}$$

The remaining balance equations must take into consideration two types of contributions to the linear momentum, angular momentum, internal energy and entropy, as follows: (i) the contributions arising from the entrant new mass π per unit time; (ii) the contributions arising from the interaction of the matrix with the perfusant. Consider, for example, the balance of linear momentum. In its global Eulerian form (and, for brevity, in the absence of body forces) it reads:

$$\frac{d}{dt} \int_v \rho v^i dv = \int_{\partial v} t^{ij} n_j da + \int_v \pi v^i dv + \int_v p^i dv. \tag{7.58}$$

In this equation, v denotes the current volume occupied by the body, v^i represents the spatial components of the velocity field, t^{ij} are the components of the Cauchy stress, and p^i are the components of the momentum contributed by the perfusant. At this point it is worth making a minor point which, although not important in theories of bone growth due to the small velocities at play, may be of some theoretical importance. The entrant mass at the rate π is tacitly assumed to enter at precisely the same velocity as the existing substrate. Otherwise, its contribution would have been multiplied by a different velocity. Although Cowin and Hegedus (1976a) justifiably disregard this point, we may think that the contribution p^i incorporates both the perfusant portion and the differential contribution of the entrant mass.

Using the standard technique for obtaining the local version of balance equations, and invoking the local form (7.57) of the mass balance, equation (7.58) results in

$$\rho \frac{Dv^i}{Dt} = t^{ij}{}_{,j} + p^i, \tag{7.59}$$

where a comma is used to indicate partial (or covariant) differentiation. Notice that the local form does not contain a contribution from the entrant mass (due, clearly, to the assumption that it entered at the same velocity as the substrate). Cowin and Hegedus (1976a) tacitly assume that there are no contributions from the perfusant or elsewhere to the balance of angular momentum, which can thus be understood to imply the symmetry of the Cauchy stress tensor.

When dealing with the balance of energy, Cowin and Hegedus (1976a) do explicitly discuss the possibility of an internal energy $\bar{u}$ contributed by the entrant mass by virtue of its content being different from that of the substrate, as well as from other sources (such as the interaction with the perfusant). A similar analysis to that carried out for the balance of momentum yields the local form

$$\rho \frac{Du}{Dt} = t^{ij} D_{ij} - q^i_{,i} + \bar{u}, \tag{7.60}$$

where u is the internal energy content per unit mass of the porous matrix, D_{ij} are the components of the symmetric part of the spatial velocity gradient, and q^i are the components of the heat-flux vector. We have disregarded a possible body radiation term.

Finally, the local form of the Clausius–Duhem entropy inequality is given by

$$\rho \frac{Ds}{Dt} \leq - \left(\frac{q^i}{\theta} \right)_{,i} + \frac{\bar{h}}{\theta}. \tag{7.61}$$

In this equation, s is the entropy density per unit porous matrix mass and θ denotes the absolute temperature. The term $\bar{h}$ is introduced in the same spirit as its counterpart in the energy equation to account for the fact that, for example, the entering mass may not have the same content of specific entropy as the substratum. It is interesting to note that, if one introduces the specific Helmholtz free energy defined as $\psi = u - \theta s$, the standard combination of the entropy inequality with the energy balance equation yields

$$\rho \frac{D\psi}{Dt} + \rho s \frac{D\theta}{Dt} - t^{ij} D_{ij} + \frac{q}{\theta} q^i \theta_{,i} - h \leq 0. \tag{7.62}$$

Only the difference $h = \bar{u} - \bar{h}$ makes an appearance in this equation. In open thermodynamical systems, these additional terms in the equations and in the entropy inequality represent the crucial interactions between the system under consideration and the surroundings. They may consist of entropy sinks and control mechanisms that guarantee the healthy and orderly existence of a living organism.

Having presented the laws of balance of the theory of adaptive elasticity, it is necessary to discuss its constitutive equations. As already pointed out, one of the main assumptions (and limitations) of the theory of adaptive elasticity is that the new material occupies only the space available due to the porosity of the matrix. In other words, if a configuration is free of stress, and if no forces or temperature changes are imposed, then the fact that growth is taking place will not produce any stresses and the configuration will remain unchanged, except for the fact that its porosity is changing. Adopting a local stress-free configuration as reference, we denote the intrinsic density of the bone material by γ_0 and its solid volume fraction by $\zeta_0 = 1 - \phi_0$. This quantity will undergo changes in time due to the entrant mass. Enforcing the mass balance equation (7.57), we obtain

$$\frac{D\zeta_0}{Dt} = \frac{\Pi}{\gamma_0} = \frac{J\pi}{\gamma_0}. \tag{7.63}$$

where J is the determinant of the deformation gradient $\mathbf{F}$.

Remark 7.5.1 Intrinsic behaviour of the matrix material. If we assume that the deformation gradient $\mathbf{F}$ of the porous material applies on average also to the solid itself, as one would do in a non-diffusive mixture, the free-energy density per unit mass of the porous medium is the same as the free-energy density of the solid for the same $\mathbf{F}$. Accordingly, the intrinsic density γ_0 of the undeformed matrix material will in general change to a new value $\gamma = J^{-1}\gamma_0$ in the deformed state. In reality, though, because of the presence of the pores (which means the imposition of traction-free boundary conditions over the intricate geometry of the voids), the actual local deformation gradient acting on the matrix material will be different from the effective deformation gradient applied to the averaged structure. As a result, there is no reason to believe that the densities γ_0 and γ are directly related by the determinant J of the applied deformation gradient. For example, if we assume that the matrix material is incompressible, in which case $\gamma = \gamma_0$, it is not difficult to show that the corresponding solid volume fractions are related by $\zeta_0 = J\zeta$.

The usual constitutive variables (free energy, entropy, stress, heat flux) and both the mass production π and the entropy production term h introduced above are assumed to be functions of the temperature, its gradient, the volume fraction ζ_0 and the deformation gradient. This generic assumption presupposes that the chemical reactions of growth and remodelling (in bone, at least) are triggered, from the mechanical standpoint, by the state of strain of the matrix measured with respect to a putative unstressed configuration. This configuration is assumed to be uniquely defined (modulo a rigid rotation) for each state of uniform temperature, regardless of the particular value of the porosity. From the standard thermodynamic treatment of the Clausius–Duhem inequality, it is possible to conclude that the free energy is independent of the temperature gradient, and that the free energy acts as a potential for both the entropy and the stress, which are also independent of the temperature gradient. Moreover, the constitutive functions are further restricted by the residual inequality

$$\rho\dot{\zeta}_0 \frac{\partial \psi}{\partial \zeta_0} + \frac{1}{\theta} q^i \theta_{,i} - h \leq 0. \tag{7.64}$$

7.5.1 The Isothermal Quasi-static Case

To obtain some specific results, we now specialize the equations of the theory of adaptive elasticity to purely mechanical (isothermal) processes and to very slow motions, so that the the velocities and accelerations can be neglected. Neglecting also the body forces, the equation of balance of momentum reduces to the following equilibrium equation in the Lagrangian formulation:

$$T^{il}{}_{,I} = 0. \tag{7.65}$$

The equation of mass balance is

$$\gamma_0 \frac{D\zeta_0}{Dt} = \Pi. \tag{7.66}$$

The equation of balance of energy is trivially satisfied by virtue of our simplifying assumptions.

The constitutive equations are completely controlled by two scalar functions: the free-energy density $\psi = \psi(\mathbf{F}, \zeta_0)$ and the mass production $\Pi = \Pi(\mathbf{F}, \zeta_0)$. The quantity ψ is measured per unit mass. The stress is obtained by differentiation as

$$\mathbf{T} = \gamma_0 \phi_0 \frac{\partial \psi(\mathbf{F}, \phi_0)}{\partial \mathbf{F}}. \tag{7.67}$$

The referential solid volume fraction ζ_0 acts as an *internal state variable* in this theory. We will discuss variables of this kind in more detail in Section 7.6.7. For now, let us declare that the change in internal energy density brought about by a change in the internal variable, namely, the derivative

$$A = \frac{\psi(\mathbf{F}, \zeta_0)}{\partial \zeta_0}, \tag{7.68}$$

is the *driving force* or *configurational force* behind the time evolution of the internal variable. This aspect is emphasized by the residual Clausius–Duhem inequality (7.64), which can be written as

$$\dot{\zeta}_0 A \leq 0. \tag{7.69}$$

Using equation (7.63), we obtain

$$\Pi A \leq 0. \tag{7.70}$$

One of the simplest ways to satisfy this thermodynamic restriction is to adopt a constitutive evolution equation of the form

$$\Pi = \Pi(\mathbf{F}, \phi_0) = -k A, \tag{7.71}$$

where k is a positive material constant. On the other hand, if, as suggested in Remark 7.5.1, the free-energy density is assumed to be independent of the porosity, then $A = 0$, and the Clausius–Duhem inequality is satisfied identically in a reversible manner for any assumed evolution law. In Cowin and Hegedus (1976b) the energy and evolution functions are developed in a Taylor series up to quadratic terms in the (small) strain, with coefficients depending on the porosity. The reader is referred to that article for a thorough analysis of the results obtainable under those assumptions.

Exercise 7.5.2 Let $\bar{\psi}(\mathbf{F})$ denote the intrinsic free-energy density of the solid material referred to a stress-free state. Assume that the following free-energy density function is proposed for the porous medium:

$$\psi(\mathbf{F}) = g(\zeta_0)\,\bar{\psi}(\mathbf{F}), \tag{7.72}$$

where $g(\cdot)$ is a given function. (a) Show that this constitutive equation satisfies the desired condition that for $\mathbf{F} = \mathbf{I}$ the stress vanishes regardless of the value of the solid fraction. (b) Write the explicit expression for the stress in the porous medium if the intrinsic constitutive equation of the solid is given by equation (6.157). (c) If a chunk of porous material with this constitutive equation is subjected to a uniform spherical stress state $\mathbf{T} = a(t)\mathbf{I}$, where $a(\cdot)$ is a given function of time, find an explicit expression for the deformation gradient $\mathbf{F} = f(t)\mathbf{I}$. (d) Assuming an evolution equation of the form (7.71), write the differential equation of evolution for $\zeta_0(t)$.

7.6   Anelasticity

7.6.1   Introduction

The theory of adaptive elasticity relies heavily on the fact that growth takes place at the expense of porosity and causes no residual stresses, a feature that it shares with theories of boundary growth. While this point of view is clearly justifiable in the case of bone, the growth of other tissues, such as arteries, cartilage and muscle, cannot be modelled in this way. Moreover, the idea that the process of growth is triggered and controlled by the state of strain alone is most likely contradicted in the case of arteries and the heart muscle, where stress is known to play an important role. An article by Rodriguez et al. (1994) first proposed to use a different paradigm for growth and remodelling based on the notion of anelastic material behaviour, a paradigm that had already been used successfully in the modelling of large-deformation metal plasticity.

We recall that a material is said to be elastic if its stress is completely determined by the present value of the deformation. In materials with memory, on the other hand, the history of the deformation plays a role in determining the present value of the stress, as we have seen in Section 3.8. In *anelastic* materials the influence of the whole past history is assumed to be mediated by the present value of an additional single tensor variable entering the stress constitutive equation in a peculiar way that we will discuss in great detail later. It is worth remarking at this stage that, viewed in this light, the theory of anelasticity is a particular case of the general theory of materials with *internal state variables*.[14] Physically, we may say that, as it evolves in time, this internal state variable renders the material response time-dependent. Nevertheless, an anelastic material preserves its chemical identity and all its intrinsic material parameters as time goes on. The variation of the material response is due exclusively to a re-accommodation of the (old and new) material in its surroundings.

When applied to the theory of material growth, the anelastic paradigm conveys the physical meaning that the new material is of the same kind as the substrate, but the competition for space in the body requires the surroundings to be squeezed to accommodate the newcomer in the reference configuration. From the technical point of view, we say that throughout its life the material response of an anelastic medium changes, but remains *materially isomorphic* to its initial response.

7.6.2   The Notion of Material Isomorphism

A good starting point to understand the idea of *material isomorphism* is to consider how the constitutive equation of an elastic point X depends on the reference configuration chosen to express it. For a simple (first-grade) elastic material the constitutive equation can be expressed in terms of the Cauchy stress $\mathbf{t}$ as a function of the deformation gradient $\mathbf{F}$:

$$\mathbf{t} = \mathbf{t}(\mathbf{F}), \tag{7.73}$$

where $\mathbf{F}$ is measured from a given reference configuration, as shown schematically in Figure 7.13. We have already discussed, in Section 5.5.12, the issue of change of reference configuration. For a different reference configuration, indicated in Figure 7.13 with primed

---

[14] See, for example, Coleman and Gurtin (1967); see also Maugin and Muschik (1994).

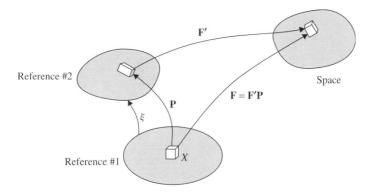

Figure 7.13 Change of reference configuration

symbols, the constitutive equation of the same point will be given by a different function $\mathbf{t}'$ as

$$\mathbf{t} = \mathbf{t}'(\mathbf{F}'). \tag{7.74}$$

How are the two constitutive equations of the same material point mutually related? The answer to this question is an elementary application of the chain rule of differentiation of functions. Let ξ denote the change of reference configuration and let $\mathbf{P} = \nabla\xi$ be its gradient evaluated at the point of interest. In accordance with equation (5.221), the relation between the constitutive functions $\mathbf{t}$ and $\mathbf{t}'$ is, therefore, given by

$$\mathbf{t}'(\mathbf{F}') = \mathbf{t}(\mathbf{F}'\mathbf{P}). \tag{7.75}$$

This equation must be satisfied identically for all possible deformation gradients $\mathbf{F}'$. Thus, the transformation between the constitutive equations of *one and the same point* with respect to two different reference configurations is given by right multiplication of the argument by a fixed non-singular linear map (or matrix) $\mathbf{P}$.

Consider now *two different points*, X_1 and X_2, with constitutive functions $\mathbf{t}_1$ and $\mathbf{t}_2$, respectively, of the form (7.73). For concreteness, let us assume that the two points belong to the same body and that we are working in a fixed global reference configuration, as shown in Figure 7.14. We ask: are the two points made of the same material? According to our previous discussion, it should be clear that the answer to this question is unequivocally that the two points are made of the same material if, and only if, there is a map[15] ξ from a neighbourhood of X_1 to a neighbourhood of X_2 such that

$$\mathbf{t}_2(\mathbf{F}) = \mathbf{t}_1(\mathbf{FP}), \tag{7.76}$$

identically for all $\mathbf{F}$, with $\mathbf{P} = \nabla\xi$. We say that the (non-singular) map $\mathbf{P}$ is a *material isomorphism*[16] between X_1 and X_2. When the mass density $\rho_0(X)$ of the reference configuration is given, it is customary to include in the definition of material isomorphism

---

[15] Technically, this map must be a *local diffeomorphism*.
[16] This definition is due to Noll (1967).

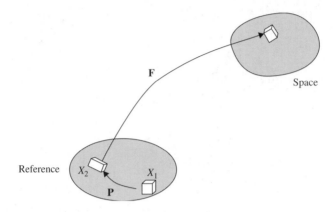

Figure 7.14   Material isomorphism or transplant

the *mass consistency condition*,

$$|J_P| = \frac{\rho_0(X_1)}{\rho_0(X_2)}, \tag{7.77}$$

where J_P denotes the determinant of **P**.

A more poignant description of material isomorphism can be gathered by regarding the two-point tensor **P** as some kind of *material transplant*, in the surgical sense of the word. Assume that a small neighbourhood of X_2 has been damaged and needs to be repaired. To achieve this goal, a small neighbourhood of X_1 is cut out and transplanted to the small neighbourhood of X_2. Since the two points would, in general, have been at different states of strain, it would be necessary to rotate and/or distort the material before grafting it at the target point. This rotation and /or distortion is precisely what the map **P** represents. If the material at the source and the target is indeed the same, and if the proper rotation and/or distortion have been effected, then clearly a perfect graft has been achieved and there is no longer a way to detect any difference by purely mechanical experiments. The mass consistency condition (1.24) guarantees that even the mass density of the grafted material has become identical to that of the material originally found at X_2.

Yet another way to visualize the notion of material isomorphism is to consider a crystalline body, such as a chunk of sodium chloride (NaCl). If we were to look at two different locations under the microscope, as suggested in Figure 7.15, we would observe in both cases the expected alternation of ions of sodium and chloride. The fact that at one location the configuration of ionic distances and angles is different from the configuration of ionic distances and angles at the other, would be of no consequence in determining that both locations are occupied by the same substance (NaCl). The map that would bring one ionic configuration into the other is precisely a material isomorphism **P**.

We remark at this point that the definition of material isomorphism is not arbitrary but based on the only possible mathematical expression that corresponds exactly to the physical meaning that the material at the two points is the same. The reasoning leading to this definition is based exclusively on the exploitation of the degree of freedom afforded by the choice of reference configuration. It simply establishes that, if the material is the same at both points, it surely should be possible to find suitable reference configurations where the responses are identical. If, on the other hand, the material is not the same, then no

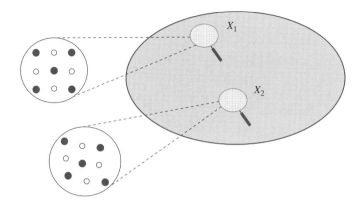

Figure 7.15 Visualization of a material isomorphism

tinkering with the reference configuration can possibly render the two responses identical. The resulting multiplicative rule in equation (7.76) is, therefore, an inevitable consequence of this definition. Later, when we compare the response at one at the same point but at two different instants of time and we claim that, in a precise sense, the material has remained 'the same', the multiplicative rule will again emerge naturally. This has absolutely nothing to do with such concepts as 'intermediate configuration', existence of a 'stress-free state determined up to a rotation', or even with elasticity. These misleading concepts, which have their origin in the theory of metal plasticity and in some attempts at its intuitive interpretation, have unfortunately permeated and sometimes undermined various theories of growth and remodelling at their very foundations.

7.6.3 Non-uniqueness of Material Isomorphisms

The maps **P** defining the material isomorphism between two points are not unique in general. To understand the physical basis for this assertion it is convenient to consider the particular case in which both points (source and target) are identified with each other. For this special case we reserve the term *material automorphism*. It is absolutely clear that the identity map is, trivially, a material automorphism. Is it unique? Consider the physical meaning of a material automorphism **G**. According to equation (7.76) we should have

$$\mathbf{t}(\mathbf{FG}) = \mathbf{t}(\mathbf{F}),\tag{7.78}$$

for all possible deformation gradients **F**. But this is precisely the condition for **G** to be a material symmetry at the point in question, as we gather from equation (6.114). In other words, a material automorphism is nothing but an alternative definition of a material symmetry![17] Consequently, the collection of material automorphisms coincides with the material symmetry group.

Let **P** now be a material isomorphism from X_1 to X_2, and let $\mathbf{G}_1$ be a material symmetry at X_1. Then, according to equations (7.76) and (7.78), we have

$$\mathbf{t}_2(\mathbf{F}) = \mathbf{t}_1(\mathbf{FP}) = \mathbf{t}_1(\mathbf{FPG}_1) = \mathbf{t}_1(\mathbf{F}(\mathbf{PG}_1)),\tag{7.79}$$

---

[17] Note that the mass consistency condition (7.77) implies that the determinant of a material automorphism is ± 1, a condition usually assumed for material symmetries.

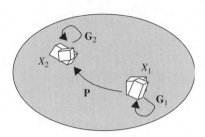

Figure 7.16   Interplay between symmetry and isomorphism

which implies that $\mathbf{P}\mathbf{G}_1$ too is a material isomorphism from X_1 to X_2. It follows that material isomorphisms are not unique and that the freedom in the choice of material isomorphisms between two points is given precisely by multiplication on the right by the material symmetry group of the source. It is not difficult to prove that the same degree of freedom is expressible as multiplication on the left by the material symmetry group of the target. Moreover, the symmetry groups of two materially isomorphic material points are conjugate to each other (see Exercise 7.6.1), the conjugation being achieved via any one material isomorphism. These assertions can be easily proved by a direct application of the definitions of the concepts involved (Figure 7.16).

Exercise 7.6.1   Symmetry, isomorphism and conjugacy. Two subgroups, $\mathcal{G}_1$ and $\mathcal{G}_2$, of a group $\mathcal{H}$ are *conjugate* if there exists an element $h \in \mathcal{H}$ such that

$$\mathcal{G}_2 = h\,\mathcal{G}_1 h^{-1}. \tag{7.80}$$

Thus, the conjugate subgroup $\mathcal{G}_2$ is generated by keeping h fixed and running through every element $g_1 \in \mathcal{G}_1$ with the operation $h g_1 h^{-1}$. Show that, as subgroups of the general linear group of non-singular matrices, the symmetry groups of two materially isomorphic points are necessarily conjugate and that that the elements h that accomplish the conjugation are precisely the material isomorphisms between the two points.

7.6.4   Uniformity and Homogeneity

A body is *materially uniform* if all its points are mutually materially isomorphic (Figure 7.17(a)). Uniformity, therefore, corresponds to the notion that the body is made of the same material at all points. Material isomorphism is an equivalence relation. Indeed, it satisfies the following three properties: (i) it is *reflexive*, in the sense that a material point is materially isomorphic to itself (trivially, via the identity map); (ii) it is *symmetric*, in the sense that if a point X_1 is materially isomorphic to another point X_2 (via a material isomorphism $\mathbf{P}$), then the point X_2 is materially isomorphic to X_1 (via the inverse map $\mathbf{P}^{-1}$); (iii) it is *transitive*, that is, if X_1 is isomorphic to X_2 (via $\mathbf{P}_{12}$), and X_2 is isomorphic to yet another point X_3 (via $\mathbf{P}_{23}$), X_1 then is isomorphic to X_3 (via the product $\mathbf{P}_{23}\mathbf{P}_{12}$). It follows, therefore, that in a materially uniform body all points are materially isomorphic to any fixed arbitrarily body point X_0, as shown schematically in Figure 7.17(b).

A third, equivalent, way to look at a materially uniform body is to place the reference point outside of the body altogether, as shown in Figure 7.17(c). In so doing, we obtain a

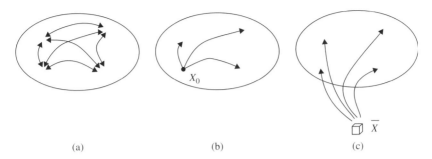

Figure 7.17 Material uniformity: (a) pairwise isomorphisms; (b) isomorphisms from a reference point; (c) isomorphisms from an archetype, or implants

sort of *material archetype*, a museum piece, as it were, that typifies the material response of all the points of the body. The material isomorphisms from the archetype to each of the body points X can be regarded as an *implant* $\mathbf{P}(X)$. If we denote with an overbar the constitutive quantities of the archetype, the constitutive law of a materially uniform body can be expressed as follows:

$$\mathbf{t}(\mathbf{F}; X) = \bar{\mathbf{t}}(\mathbf{F}\mathbf{P}(X)). \tag{7.81}$$

Remark 7.6.2   Components of the archetypal implant. We have adhered to the convention that indices in a referential coordinate system are denoted by capital letters $(I, J, K, \ldots)$ and in a spatial coordinate system by lower-case letters $(i, j, k, \ldots)$. Thus, for example, the components of the deformation gradient $\mathbf{F}$ are written as F^i_I, since this tensor acts on referential vectors to produce spatial vectors. The archetype, on the other hand, lives in its own museum. If we indicate a basis therein with Greek indices $(\mathbf{E}_\alpha \ (\alpha = 1, 2, 3))$, then the map $\mathbf{P}$ should be indicated in components by the matrix with entries P^I_α, since it maps vectors in the archetype to vectors in the reference configuration.

Remark 7.6.3   The material archetype can be in any state of deformation. In particular, there is no implication whatsoever that it might be in a stress-free state (which may not even exist). Notice also that a rotation of the material archetype would imply a change of *all* the material implants by multiplication on the right by that rotation. The same remark applies to any change of material archetype, rotationally or otherwise deformed.

   The notion of material uniformity should not be confused with that of *homogeneity*. A materially uniform body is materially homogeneous if it can be brought to a global reference configuration such that the material isomorphisms are all equal to the identity map (in the Euclidean sense). It may be possible to satisfy this condition not globally for the whole body but only piece by piece. In this case, we say that the body enjoys *local homogeneity*. Another way to express this condition is by saying that a uniform body is homogeneous if one can find implants of the archetype that 'fit well' within pieces of the given uniform body, as suggested in Figure 7.18. From the physical point of view, lack of homogeneity can be reflected in the appearance of residual stresses, which are of prime relevance to processes of growth and remodelling.

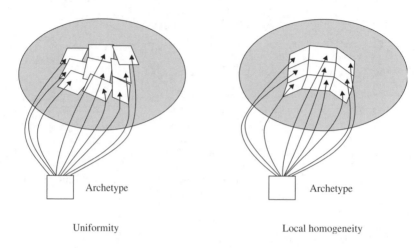

Figure 7.18   Uniformity and homogeneity

Example 7.6.4   A locally homogeneous body. Consider the case of an initially straight homogeneous stress-free elastic bar. If the bar is bent and the two ends are permanently and smoothly joined so as to form a ring, the new body will no longer be homogeneous, since (except for cutting) there is no smooth change of configuration that will render all the material isomorphisms translational only. Residual stresses, for example, cannot be removed simultaneously everywhere. On the other hand, this example can be considered as a case of *local homogeneity*, in the sense that, for each point, there exists a neighbourhood (or 'chunk') that can be brought to a homogeneous configuration by re-straightening that particular chunk.

Example 7.6.5   An inhomogeneous body. It is not difficult to produce examples of bodies that are not even locally homogeneous. This is the case of a materially uniform cylindrical shell with transversely isotropic properties such that the axis of transverse isotropy is everywhere in the radial direction. If this shell is initially stress-free (so that the material isomorphisms are pure rotations) there is no way to straighten the shell without producing stresses.

The question as to when a uniform body is homogeneous is one of integrability of a certain geometric structure. Briefly stated, a uniform body is locally homogeneous if, and only if, there exists a local (generally curvilinear) coordinate system such that its natural base vectors at different points (within the coordinate patch) are related via a material isomorphism.

7.6.5   Anelastic Response

The dictionary definition of the term 'anelasticity' states that 'the strain depends on the stress and its time rate', or that 'there is no unique relation between stress and strain'. For our purposes, however, we will understand the term in a restricted sense that goes back to the work of Eckart (1948). A material point is said to be anelastic, or to undergo *anelastic evolution*, if its constitutive equation depends on elapsed time and can be reduced to the form

$$\mathbf{t}(\mathbf{F}, t) = \mathbf{t}(\mathbf{FP}(t), t_0), \qquad (7.82)$$

with $\mathbf{P}(t_0) = \mathbf{I}$. Here, t_0 denotes a referential time.

In terms of the notion of material isomorphism, we may say that a material point undergoes anelastic evolution if and only if the responses at any two instants in time are materially isomorphic. In other words, anelasticity is the time-like counterpart of uniformity. The first rigorous combination of both notions can be traced back to a mathematical article by Wang and Bloom (1974). From the physical point of view, anelastic response corresponds exactly to what one may loosely call preservation of chemical identity of the material. Although this kind of imagery should not be pushed too far, it certainly conveys the idea that the material remains the same, except that it undergoes a process of re-accommodation, which can be due to any kind of adaptive remodelling or to growth, as more material *of the same kind* enters the body and forces itself and the (identical) surrounding material to get squeezed into place.

In equation (7.82) the material response at time t is expressed in terms of its response at a reference time t_0. Using instead the convenient concept of material archetype, we may write the equivalent expression

$$\mathbf{t}(\mathbf{F}, t) = \bar{\mathbf{t}}(\mathbf{FP}(t)), \qquad (7.83)$$

where, as before, an overbar is used to indicate the constitutive functional of the archetype. The physical essence of anelastic evolution consists, therefore, in a time-dependent implant of the archetype into (the neighbourhood of) a body point. In a somewhat misleading picture, the 'spring' representing the material response preserves its stiffness while changing its rest length with time.

If the body is not only anelastic but it also happens to be uniform, then the constitutive equation of the body acquires the form

$$\mathbf{t}(\mathbf{F}; X, t) = \bar{\mathbf{t}}(\mathbf{FP}(X, t)). \qquad (7.84)$$

7.6.6 Anelastic Evolution

Having established or assumed that a material is anelastic, an *anelastic evolution law* is a constitutive determination of the time-dependent map $\mathbf{P}(t)$ in equation (7.83). In its simplest form, an anelastic evolution law may be given by a first-order ordinary differential equation as

$$\dot{\mathbf{P}} = \mathbf{g}(\mathbf{P}, \mathbf{a}). \qquad (7.85)$$

In this equation, $\mathbf{g}$ is some constitutive function and $\mathbf{a}$ represents a generic list of additional arguments (such as stress, strain, and time)

In principle, one may imagine that the time evolution of the implant $\mathbf{P}(t)$ is controlled by an arbitrary external prescription, such as by nanoprocessors activated by a computer following some capriciously established rule (related to the weather, the state of mind of the operator, etc.). But when we say that the function $\mathbf{g}$ is a *constitutive* descriptor of the evolution, we mean that the evolution obeys some definite deterministic law as the outcome of the values of the arguments in equation (7.85). This being the case, it is not unreasonable to expect that certain formal limitations, based on physical and mathematical consistency, will apply to any possible evolution law and result in a *reduction* of the form of any proposed law. Clearly, such a reduction should arise, among other factors, by the need to respect the material symmetries of the underlying constitutive equation. Indeed, the symmetry group of an anelastically evolving material can only change by conjugation, which means that, from the physical standpoint, the material preserves its symmetry features. Thus, for example, if the material is initially transversely isotropic it remains so for the rest of its anelastic life.

Reduction to a Common Archetype

Before exploring the issue of symmetry just raised, however, there is a more general, and quite drastic, reduction of any proposed evolution law. For simplicity, let us consider the more common case in which the time variable t does not appear explicitly in the list of arguments $\mathbf{a}$ of the evolution function $\mathbf{g}$ in equation (7.85). Assume that in some reference configuration a given body point coincides, at a given instant of time t_0, with the archetype. In this case we have that, for the given point, $\mathbf{P}(t_0) = \mathbf{I}$. Assume, moreover, that the constitutive evolution function for this point is known for all possible values of the argument list $\mathbf{a}$. What we are assuming, then, is the complete knowledge of the (archetypal) function defined by

$$\bar{\mathbf{g}}(\bar{\mathbf{a}}) = \mathbf{g}(\mathbf{I}, \bar{\mathbf{a}}). \tag{7.86}$$

We have been careful to use an overbar on the list of arguments $\bar{\mathbf{a}}$ to emphasize the fact that these arguments are to be evaluated at the archetype. Thus, for example, such tensor quantities as the Piola stress or the deformation gradient, if they do appear in the list of arguments, need to be appropriately evaluated using the archetype as a local reference configuration. Under these conditions, we will argue that this knowledge of the function $\bar{\mathbf{g}}(\bar{\mathbf{a}})$ alone should be necessary and sufficient to completely determine the constitutive evolution function $\mathbf{g}$ for all values of the argument list $\mathbf{a}$ and for all values of the argument $\mathbf{P}$. Indeed, let us assume that $\mathbf{P}(t_0) = \mathbf{P}_0 \neq \mathbf{I}$ at the given point, instead of the previously assumed unit value. This is tantamount to changing the reference configuration of the body in such a way that the gradient of the change of configuration is equal to $\mathbf{P}_0$ at the point in question at time t_0. We must then require that the evolution in both cases be the same except that it is multiplied by the constant factor $\mathbf{P}_0$. In other words:

$$\mathbf{P}(t) = \mathbf{P}_0 \, \bar{\mathbf{P}}(t), \tag{7.87}$$

with an obvious notation. Consequently,

$$\dot{\mathbf{P}}(t_0) = \mathbf{P}_0 \bar{\mathbf{g}}(\bar{\mathbf{a}}) \tag{7.88}$$

or

$$\mathbf{P}_0^{-1} \, \dot{\mathbf{P}}(t_0) = \bar{\mathbf{g}}(\bar{\mathbf{a}}). \tag{7.89}$$

Since the reference time t_0 is arbitrary, this equation is actually valid for any time, namely

$$\dot{\mathbf{P}}(t) = \mathbf{P}(t) \, \bar{\mathbf{g}}(\bar{\mathbf{a}}), \tag{7.90}$$

which is the desired reduced form of the evolution equation. Defining the (infelicitously named) *inhomogeneity velocity gradient at the archetype* as

$$\bar{\mathbf{L}}_{P(t)} = \mathbf{P}^{-1}(t)\,\dot{\mathbf{P}}(t),\tag{7.91}$$

we can write the reduced form of the evolution equation in the form

$$\bar{\mathbf{L}}_P = \bar{\mathbf{g}}(\bar{\mathbf{a}}).\tag{7.92}$$

In hindsight, the fact that an evolution equation cannot entail an arbitrary dependence on $\mathbf{P}$ should not be surprising, since the implants do not have an absolute physical meaning, but rather represent a means of comparing the material at different instants of time. The situation is somewhat reminiscent of (but by no means equal to) the reduction of constitutive equations by the principle of material frame indifference, by virtue of which the dependence on the rotation cannot be arbitrary but acquires a very specific form.

Remark 7.6.6   Remodelling or growth? Consider the mass consistency condition (7.77) expressed now in the form

$$|J_{P(t)}| = \frac{\rho_0(t_0)}{\rho_0(t)}.\tag{7.93}$$

Assuming, without loss of generality, that the implants are orientation preserving (i.e., $J_P > 0$), the time derivative of (7.93) yields the result

$$\Pi = \dot{\rho}_0(t) = -\rho_0(t)\,\text{trace}(\mathbf{L}_{P(t)}).\tag{7.94}$$

As shown by this formula, in a theory of pure anelastic evolution the mass consistency condition implies a specific form for the volumetric mass supply Π in terms of the inhomogeneity velocity gradient. Moreover, the inhomogeneity velocity gradient $\mathbf{L}_{P(t)}$ is traceless if, and only if, pure remodelling without volumetric growth or resorption is taking place. In our terminology, a traceless evolution function $\bar{\mathbf{g}}$ will, accordingly, account exclusively for remodelling without growth. More generally,

$$\text{trace } \mathbf{L}_P \begin{cases} < 0 & \text{growth and remodelling} \\ = 0 & \text{pure remodelling} \\ > 0 & \text{resorption and remodelling.} \end{cases}\tag{7.95}$$

Frame Indifference

A frame (i.e., an observer) is an essentially *spatial* notion, whereas evolution is a fundamentally *material* phenomenon. Thus, neither $\mathbf{P}$ nor $\bar{\mathbf{L}}_P$ is affected in any way by a change of frame. Therefore, if we demand that the evolution function $\bar{\mathbf{g}}$ be frame-indifferent, it is only the list of extra arguments $\bar{\mathbf{a}}$ that needs special consideration. Clearly, a sufficient condition for the evolution law to be frame-indifferent is that each and every member of this list must remain unaffected by a change of frame. Purely referential tensors, such as the right Cauchy–Green tensor $\mathbf{C}$ and the Eshelby stress $\mathbf{b}$, to be introduced later, belong to this class.

Example 7.6.7   As an example of the combined effect of frame indifference and reduction to an archetype, consider the case in which the list of arguments in equation (7.93) consists of just $\mathbf{C}$. Since this is the right Cauchy–Green tensor *evaluated at the archetype*, we have

$$\bar{\mathbf{C}} = \mathbf{P}^T \mathbf{C} \, \mathbf{P}, \qquad (7.96)$$

where $\mathbf{C}$ is evaluated at the reference configuration. Thus, the evolution law (7.93) converted in terms of quantities in the reference configuration reads:

$$\dot{\mathbf{P}} = \mathbf{P}\bar{\mathbf{g}}(\mathbf{P}^T \mathbf{C} \, \mathbf{P}). \qquad (7.97)$$

Defining the *referential inhomogeneity velocity gradient* by

$$\mathbf{L}_P = \dot{\mathbf{P}}\mathbf{P}^{-1} = \mathbf{P}\bar{\mathbf{L}}_P\mathbf{P}^{-1}, \qquad (7.98)$$

equation (7.97) can also be expressed as

$$\mathbf{L}_P = \mathbf{P}\bar{\mathbf{g}}(\mathbf{P}^T \mathbf{C} \, \mathbf{P})\mathbf{P}^{-1}. \qquad (7.99)$$

Exercise 7.6.8   Indices. Write equations (7.96)–(7.99) in terms of components. Be particularly mindful of the placement of Greek and Roman indices.

The Principle of Actual Evolution

We turn now to the implications of material symmetry on the form of the evolution function. One of these, the principle of *actual evolution*, is designed to prevent the solution of the evolution equation being of the form

$$\mathbf{P}(t) = \mathbf{P}(t_0) \, \mathbf{G}(t), \qquad (7.100)$$

where $\mathbf{G}(t)$ is some time-dependent element of the symmetry group of the archetype. The reason for this prescription is the following. Since, by definition, material symmetries leave the material response unchanged, they become essentially a degree of freedom (i.e., non-uniqueness) in the very determination of the implant maps $\mathbf{P}(t)$, as we have already pointed out. Therefore, an evolution that takes advantage of this degree of freedom alone is not an evolution at all.

Consider first the case in which the material symmetry group is *discrete*, or made of a finite number of elements.[18] Since the evolution has been assumed to be continuous (and differentiable), as evidenced by the fact that it is governed by a differential equation, the only possible solution of the form (7.100) must result from using $\mathbf{G}(t) = \mathbf{I}$, seeing that there are no 'nearby' elements of the symmetry group. In this case, therefore, there is no possibility that the undesirable evolution may arise, except in the trivial case of an identically vanishing evolution function.

Consider, on the other hand, the case of a *continuous* symmetry group (technically, a *Lie group*), such as that of an isotropic or transversely isotropic solid or a crystal fluid. Taking the derivative of (7.100) with respect to time evaluated at $t = t_0$, we obtain

$$\bar{\mathbf{L}}_{P(t_0)} = \dot{\mathbf{G}}(t_0). \qquad (7.101)$$

---

[18] An example of this kind is a general orthotropic material, while a transversely isotropic material has a continuous group of symmetries, since all rotations around an axis are permitted.

Since clearly, from equation (7.100), $\mathbf{G}(t_0) = \mathbf{I}$, the right-hand side of equation (7.101) can be seen as some kind of 'velocity vector' issuing form the unit element of the group. Technically, this is called an element of the *Lie algebra* (or an *infinitesimal generator*) of the symmetry group. Recalling that the time t_0 is arbitrary, we conclude that the principle of actual evolution must stipulate that the evolution function $\bar{\mathbf{g}}$, evaluated at any value of its arguments, *should not belong to the Lie algebra* of the symmetry group of the archetype.

Example 7.6.9   The Lie algebra of the orthogonal group. An important example is provided by the isotropic solid, whose symmetry group is the orthogonal group. Physically, this is the group of all rotations (possibly followed by a reflection). The corresponding Lie algebra, therefore, must encapsulate the idea of infinitesimal rotations or, more accurately, of angular velocities. It is well known that angular velocities can be represented by skew-symmetric matrices (as we have already discovered in Section 5.5.13, for example). Mathematically, therefore, we may say that the Lie algebra of the orthogonal group is the space of skew-symmetric matrices. We conclude that, in the case of isotropic solids, the principle of actual evolution requires the function $\bar{\mathbf{g}}$ to consistently produce matrices with a non-vanishing symmetric part.

As a result of the principle of actual evolution, two different evolution laws whose results differ by an element of the material Lie algebra must be considered equivalent. In the example of the isotropic solid just presented, this means that only the symmetric part of the resulting inhomogeneity velocity gradient matters as far as the evolution is concerned. Therefore, two evolution laws (for an isotropic solid) whose results differ by a skew-symmetric matrix are equivalent. For this statement to make sense not just instantly but for finite times, we must ensure that any evolution law produces equivalent evolution processes, regardless of the particular implant chosen at the initial time. In other words, if two implants at the initial time differ by an element of the symmetry group, then the whole evolution for all subsequent times must also differ by (a possibly time-dependent) element of the symmetry group. This result will be guaranteed by a further requirement of consistency between the evolution law and the material symmetry group, a requirement to be discussed presently.

The Principle of Material Symmetry Consistency

Material evolution is a matter of comparison between the constitutive responses of a body point at two different times. The archetype *per se* is not involved in this process, but is only used as an auxiliary device to obtain a concrete expression of the form (7.92). For this reason, let us consider two different archetypes related by some linear map $\mathbf{A}$, as shown in Figure 7.19. Indicating with primes the quantities associated with the second archetype, the corresponding evolution equation is expressed as

$$\bar{\mathbf{L}}'_{P'} = \bar{\mathbf{g}}'(\bar{\mathbf{a}}').\tag{7.102}$$

To establish the connection between the functions $\bar{\mathbf{g}}$ and $\bar{\mathbf{g}}'$, we note that $\mathbf{P}' = \mathbf{PA}$ and, since $\mathbf{A}$ is independent of time, we have

$$\bar{\mathbf{L}}'_{P'} = \mathbf{A}^{-1}\bar{\mathbf{L}}_P\,\mathbf{A},\tag{7.103}$$

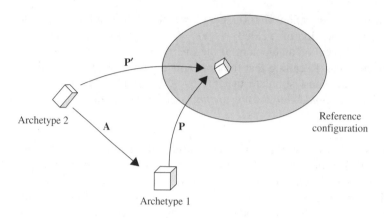

Figure 7.19   Change of archetype

whence

$$\bar{g}'(\bar{a}') = A^{-1}\bar{g}(\bar{a})\,A. \tag{7.104}$$

A change A of archetype is a *symmetry of the evolution law* if the functions $\bar{g}$ and $\bar{g}'$ are one and the same function. In other words, A is a symmetry of the evolution law if and only if

$$\bar{g}(\bar{a}) = A^{-1}\bar{g}(\bar{a})\,A, \tag{7.105}$$

for all values of $\bar{a}$. Recall that $\bar{a}$ and $\bar{a}'$ are not independent of each of other, but are related by systematically replacing P by PA in all formulas such as (7.96).

The collection of all the linear maps A satisfying equation (7.105) identically is never empty, since the identity always does. In fact, this collection is a multiplicative group, as can be verified directly. We call this group the *evolution symmetry group with respect to the given archetype*. Since an evolution equation is the reflection of an intrinsic material behaviour, we must demand that every constitutive symmetry be also a symmetry of the constitutive law. We call this requirement the *material symmetry consistency condition*. It is summarized by saying that *the material symmetry group is a subgroup of the evolution symmetry group*.

An important corollary of symmetry consistency (so important, in fact, as to almost justify the consistency requirement itself), has already been alluded to above. It stems from realizing that an evolution law is necessarily formulated in terms of the evolution of an implant map (from a given archetype). We know, however, that implant maps are not unique in general, their multiplicity being controlled by the material symmetry group of the archetype. A crucial question is, therefore, the following: if $P = P(t)$ is the solution of equation (7.92) for a given initial condition $P(t_0) = P_0$ and for any arbitrarily given $a = a(t)$, is the solution $P'(t)$ of equation (7.92) (for a small enough interval of time) with initial condition $P'(t_0) = P_0 G(t_0)$ and with the same $a = a(t)$ related to the previous solution by the formula $P'(t) = P(t)G(t)$? Here, $G(t)$ is some, possibly time-dependent, element of the material symmetry group of the archetype. The symmetry consistency condition guarantees that the answer to this question is positive. The verification of this statement is left as a rather straightforward exercise. In the course of carrying it out, we obtain the additional, somewhat surprising, information that $\mathbf{G}(t) = \mathbf{G}(t_0)$, namely, that the separation between the two solutions remains constant in time, just like synchronized swimmers.

Exercise 7.6.10 Carry out the proof of the above statements.

Remark 7.6.11 The combination of the principle of actual evolution with the principle of symmetry consistency can, somewhat unexpectedly, lead to a relaxation of the latter. Indeed, if two evolution laws are to be considered identical when they differ by an element of the Lie algebra of the material symmetry group, we may extend equation (7.105) by adding to its right-hand side an arbitrary element of this Lie algebra. As a result, more evolution laws will become acceptable than otherwise would.

Exercise 7.6.12 As an instructive exercise to summarize all the restrictions imposed so far, consider an evolution law given implicitly as

$$\phi(\mathbf{P}, \dot{\mathbf{P}}, \mathbf{t}, \mathbf{F}, \dot{\mathbf{F}}) = 0. \tag{7.106}$$

Assume the archetype to be an isotropic solid in a natural state. Reduce this evolution law as much as possible using the following criteria: reduction to the archetype, actual evolution, symmetry consistency and material frame indifference. Such evolution laws are often invoked in models of growth, whose mechanism is hypothesized to be possibly triggered by signals sent by biochemical sensors of strain or strain rate.

Further reduction of evolution laws can be claimed by invoking some form of the second law of thermodynamics. Before proceeding to discuss this important topic, we introduce the concept of configurational forces in general and of the Eshelby stress in particular.

7.6.7 The Eshelby Stress

Internal State Variables and Configurational Forces

One of the most fruitful techniques to expand the scope of application of continuum mechanics to the description of dissipative effects of varied nature (plasticity, viscoelasticity, damage, remodelling, ageing, and many other phenomena) consists of the augmentation of the space of independent constitutive variables to include additional parameters known as *internal state variables*. Governed by independently postulated *constitutive equations of evolution*, these variables are designed in each particular application to convey the intended physical meaning of 'configurational' quantities other than the histories of the deformation and the temperature. Because of their constitutive nature, the functionals defining the evolution laws (usually, first-order ordinary differential equations) must abide by the general constitutive principles adopted for all constitutive laws (causality, determinism, equipresence, frame indifference and thermodynamic consistency).

For specificity, let $\mathbf{a} = \mathbf{a}(X, t)$ represent a generic finite list of internal state variables $a_1, \ldots, a_M$, and let the (Lagrangian, say) constitutive laws of a material point be assumed to take the following (thermoelastic-like) form:

$$\mathbf{T} = \mathbf{T}(\mathbf{F}, \theta, \mathrm{Grad}\,\theta, \mathbf{a}), \tag{7.107}$$

$$\mathbf{Q} = \mathbf{Q}(\mathbf{F}, \theta, \text{Grad } \theta, \mathbf{a}), \tag{7.108}$$

$$s = s(\mathbf{F}, \theta, \text{Grad } \theta, \mathbf{a}), \tag{7.109}$$

$$\psi = \psi(\mathbf{F}, \theta, \text{Grad } \theta, \mathbf{a}), \tag{7.110}$$

and

$$\dot{\mathbf{a}} = \mathbf{f}(\mathbf{F}, \theta, \text{Grad } \theta, \mathbf{a}), \tag{7.111}$$

the last equation being the already anticipated evolution equation stipulating that the time derivative of the vector of internal variables is functionally tied to the independent constitutive variables, including the internal variables themselves.

Following the steps of Section 6.3.3, we obtain the following constitutive restrictions:

$$\frac{\partial \psi}{\partial \theta_{,J}} = 0, \tag{7.112}$$

$$T_j^J = \rho_0 \frac{\partial \psi}{\partial F_J^j}, \tag{7.113}$$

$$s = -\frac{\partial \psi}{\partial \theta}, \tag{7.114}$$

and the *residual inequality*

$$-\bar{H} + \frac{1}{\theta} Q^J \theta_{,J} + \sum_{\Lambda=1}^{M} \rho_0 \frac{\partial \psi}{\partial a_\Lambda} \dot{a}_\Lambda \le 0. \tag{7.115}$$

As expected, the free-energy density ψ, the Piola stress $\mathbf{T}$ and the entropy density s are independent of the temperature gradient. The residual inequality (7.115) contains, in addition to the heat conduction term, a term arising from the dissipation due to the internal variables.

It is useful to define the *configurational force* A^Λ associated with the internal variable a_Λ as the derivative

$$A^\Lambda = \rho_0 \frac{\partial \psi}{\partial a_\Lambda}. \tag{7.116}$$

Just as the Piola stress is a measure of the change in free energy brought about by a change in the deformation gradient, the configurational force A^Λ is a measure of the change in free energy produced by a change in the internal variable a_Λ, which can be regarded as a 'configurational' change. In terms of the configurational forces, the residual (dissipation) inequality reads:

$$-\bar{H} + \frac{1}{\theta} Q^J \theta_{,J} + \sum_{\Lambda=1}^{M} A^\Lambda \dot{a}_\Lambda \le 0. \tag{7.117}$$

The configurational forces, therefore, contribute to the dissipation via their power on the time rate of the corresponding internal variables. Using the evolution equation (7.110), the residual inequality can be also written as

$$-\bar{H} + \frac{1}{\theta} Q^J \theta_{,J} + \sum_{\Lambda=1}^{M} A^\Lambda f_\Lambda \le 0, \tag{7.118}$$

which clearly highlights the need for the evolution function $\mathbf{f}$ to satisfy constitutive restrictions.

The Eshelby and Mandel Stresses

As presented in Section 7.6.5, anelasticity is a material model with internal state variables. Indeed, the material response having been assumed of the form

$$\mathbf{t} = \bar{\mathbf{t}}(\mathbf{FP}(t)), \tag{7.119}$$

we observe that the time-dependent implant tensor $\mathbf{P}(t)$ acts as a matrix of nine internal variables P_α^I, $i, \alpha = 1, 2, 3$. But these internal variables appear in the constitutive law in a very special way, namely, as a right matrix multiplication of the deformation gradient. It is this particular form of the involvement of the internal variables that determines a particular form of the associated configurational forces known as the Eshelby stress.[19]

Remark 7.6.13 If a matrix of internal variables $\mathbf{P}$ were to appear in a general way as, say,

$$\mathbf{t} = \bar{\mathbf{t}}(\mathbf{F}, \mathbf{P}(t)), \tag{7.120}$$

the configurational force associated with a change in the matrix $\mathbf{P}$, although certainly definable by the general prescription (7.116), would not have the special form of the Eshelby stress that we are about to derive.

To isolate the most relevant terms, we assume the independent variables to be the deformation gradient $\mathbf{F}$, the temperature and the implant map $\mathbf{P}$.[20] The free energy per unit mass is

$$\psi = \bar{\psi}(\mathbf{FP}(t), \theta), \tag{7.121}$$

where $\bar{\psi}$ is the archetypal free-energy density. In particular, the Piola stress is obtained as

$$
\begin{aligned}
T_i^{\,J} &= \frac{\partial(\rho_0 \psi)}{\partial F^i{}_J} = \rho_0 \frac{\partial \psi}{\partial F^i{}_J} \\[2mm]
&= \rho_0 \frac{\partial \bar{\psi}}{\partial (F^m_M P^M_\alpha)} \frac{\partial (F^m_N P^N_\alpha)}{\partial F^i{}_J} \\[2mm]
&= \rho_0 \frac{\partial \bar{\psi}}{\partial (F^m_M P^M_\alpha)} \delta^m_i \delta^J_N P^N_\alpha \\[2mm]
&= \rho_0 \frac{\partial \bar{\psi}}{\partial (F^i_M P^M_\alpha)} P^J_\alpha
\end{aligned}
\tag{7.122}
$$

---

[19] The Eshelby stress was first proposed in Eshleby (1951).

[20] More general constitutive schemes can be considered, but it is imperative to realize that, were we to consider non-simple materials (e.g., materials whose kinematic constitutive variables include second gradients of the deformation), even within a generalized anelastic response the nature and involvement of the implants would be of a more complicated nature and, again, the Eshelby stress would be generalized accordingly. For a more general treatment, see Epstein and Elżanowski (2007).

or, more compactly, as

$$\mathbf{T} = \rho_0 \left(\frac{\partial \bar{\psi}}{\partial (\mathbf{FP})} \right) \mathbf{P}^T. \tag{7.123}$$

According to the general treatment of internal variables, the configurational forces associated with the implant $\mathbf{P}$ are given by

$$\hat{m}_I^{\ \alpha} = \rho_0 \frac{\partial \psi}{\partial P^I_{\ \alpha}}, \tag{7.124}$$

or, more explicitly,

$$
\begin{aligned}
\hat{m}_I^{\ \alpha} &= \rho_0 \frac{\partial \psi}{\partial P^I_{\ \alpha}} \\
&= \rho_0 \frac{\partial \bar{\psi}}{\partial (F^m_{\ M} P^M_{\ \beta})} \frac{\partial (F^m_{\ N} P^N_{\ \beta})}{\partial P^I_{\ \alpha}} \\
&= \rho_0 \frac{\partial \bar{\psi}}{\partial (F^m_{\ M} P^M_{\ \beta})} F^m_{\ N} \delta^N_I \delta^\alpha_\beta \\
&= \rho_0 \frac{\partial \bar{\psi}}{\partial (F^m_{\ M} P^M_{\ \alpha})} F^m_{\ I}.
\end{aligned} \tag{7.125}
$$

In block notation, we have

$$\hat{\mathbf{m}} = \rho_0 \mathbf{F}^T \left(\frac{\partial \bar{\psi}}{\partial (\mathbf{FP})} \right). \tag{7.126}$$

Since $\mathbf{P}$ is invertible, we can read off from equation (7.123):

$$\rho_0 \left(\frac{\partial \bar{\psi}}{\partial (\mathbf{FP})} \right) = \mathbf{T} \, \mathbf{P}^{-T}. \tag{7.127}$$

Introducing this result in equation (7.126), we obtain the following tensorial expression for the configurational force $\bar{\mathbf{m}}$ associated with the implant $\mathbf{P}$:

$$\hat{\mathbf{m}} = \left[\mathbf{F}^T \, \mathbf{T} \right] \mathbf{P}^{-T}. \tag{7.128}$$

This is a mixed (two-point) tensor from the archetype to the reference configuration, which is essentially the so-called Mandel stress. It is customary, however, to define the purely referential *Mandel stress* $\mathbf{m}$ as the bracketed quantity in (7.128), namely

$$\mathbf{m} = \hat{\mathbf{m}} \, \mathbf{P}^T = \mathbf{F}^T \, \mathbf{T}. \tag{7.129}$$

According to the preceding derivation, the Mandel stress is the natural configurational force that drives the implants in the anelastic material model. In other theories of material evolution, the configurational forces driving the evolution of the internal variables will, in general, be different.

By the mass consistency condition (7.93), the density at a point in the reference configuration is related to the density of the archetype, $\bar{\rho}$, by the formula

$$\rho_0(t) = J_{P(t)}^{-1} \, \bar{\rho}, \tag{7.130}$$

and, therefore, the free energy *per unit volume* in the reference configuration is

$$\rho_0 \psi = J_{P(t)}^{-1} \, \bar{\rho} \, \bar{\psi}(\mathbf{FP}(t), \theta). \tag{7.131}$$

If, by analogy with the Piola stress, we evaluate the change of this internal energy (per unit referential volume) with respect to a change in implants, namely

$$\hat{b}_I{}^\alpha = \frac{\partial (\rho_0 \psi)}{\partial P^I{}_\alpha} = \frac{\partial (J_{P(t)}^{-1} \, \bar{\rho} \, \bar{\psi}(\mathbf{FP}(t), \theta))}{\partial P^I{}_\alpha}, \tag{7.132}$$

and if we recall that the derivative of the determinant is given by

$$\frac{\partial J_P}{\partial P^I{}_\alpha} = J_P \, P^{-\alpha}{}_I, \tag{7.133}$$

reproducing similar calculations as for the Mandel stress we obtain the result

$$\hat{\mathbf{b}} = -[\rho_0 \, \psi \, \mathbf{I} - \mathbf{m}] \, \mathbf{P}^{-T}. \tag{7.134}$$

This is two-point (archetype-referential) tensor. The purely referential bracketed expression, namely

$$\mathbf{b} = \rho_0 \, \psi \, \mathbf{I} - \mathbf{m}, \tag{7.135}$$

is the *Eshelby stress*.

Exercise 7.6.14   A case of ageing. In some models of bone remodelling, including but not confined to the theory of adaptive elasticity presented in Section 7.5, the strength of the bone material is related to the solid volume fraction ζ_0. As a result of this observation, the strain energy of the bone can be expressed as

$$\psi = g[\zeta_0(t)] \, \bar{\psi}(\mathbf{F}, \theta). \tag{7.136}$$

In this equation, $g(\cdot)$ is a constitutive function of the single internal variable ζ_0, and $\bar{\psi}$ is the constitutive equation of the bone matrix material without the pores. Determine the configurational force associated with the internal variable ζ_0. Verify that this is not the Mandel stress. In our terminology, since the material properties do not evolve by material isomorphism, a process of *ageing* is taking place, although the term remodelling is commonly used to describe this particular situation. Any measure of quantities such as 'moduli of elasticity' is not preserved with the passage of time.

Constitutive Restrictions

It remains for us to investigate the restrictions imposed by the principle of thermodynamic consistency (in other words, the second law of thermodynamics) on the anelastic constitutive function $\mathbf{g}$. Considering the case where the list of arguments $\mathbf{a}$ in equation (7.94) consists of the deformation gradient and the temperature, namely

$$\dot{\mathbf{P}} = \mathbf{g}(\mathbf{P}, \mathbf{F}, \theta), \tag{7.137}$$

the residual inequality (7.118) applied to the anelastic case reads:

$$-\bar{H} + \text{trace}\left(\mathbf{m} \; \mathbf{L}_P^T\right) \leq 0, \tag{7.138}$$

where $\mathbf{L}_P$ is the inhomogeneity velocity gradient defined by equation (7.98). Equation (7.138), when expressed in terms of (7.137), represents the desired restriction on the evolution function $\mathbf{g}$.

The entropy sink $\bar{H}$ may serve as an externally imposed mechanism of control of a living process that would otherwise cease to exist or lose its equilibrium stability, just like the voluntary mechanics of balancing an upside-down broom on one's finger. If we set $\bar{H} = 0$, the constitutive restriction becomes

$$\text{trace}\left(\mathbf{m} \quad \mathbf{L}_P^T\right) \leq 0. \tag{7.139}$$

To understand the implications of this restriction, imagine that an isotropic and homogeneous sphere of a material susceptible to growth and resorption has been uniformly compressed into a smaller sphere so as to fit it into a rigid spherical container. Under these conditions, since both the Cauchy stress $\mathbf{t}$ and the left stretch $\mathbf{V}$ are proportional to the spatial identity, the Mandel stress $\mathbf{m} = \mathbf{F}^T\mathbf{T}$ is spherical or, more precisely, a negative number times the identity tensor in the reference configuration. The satisfaction of (7.139) requires, therefore, that the trace of $\mathbf{L}_P$ be positive, which implies that resorption will develop (see Remark 7.6.6). In other words, the material will tend to partially evaporate so as to relax the compressive state of stress. This conclusion satisfies our intuition, although it remains to be checked against experimental results in biological tissue.

7.7 Case Study: Exercise and Growth

7.7.1 Introduction

At the outset, it should be evident that, in and of itself, a theoretical analysis cannot possibly have anything to say about the positive or negative effects of exercise on bone health. What we mean by this case study, therefore, is just the discussion of the consequences of a particular mathematical model of material response. If, on the other hand, this model were to be validated independently by means of experimental evidence or by resorting to deeper levels of scientific explanation (cellular, say), then the mathematical derivations would have predictive value beyond the confines of the experiments or the deeper analysis that served to establish the material model in the first place.

With this caveat, our first step is the proposal of an evolution function $\bar{\mathbf{g}}$ that would satisfy all the formal constitutive restrictions discussed in Section 7.6.6. One of the simplest choices is

$$\bar{\mathbf{g}}(\bar{\mathbf{m}}) = -k\,\bar{\mathbf{m}}, \tag{7.140}$$

where k is a positive constant (or, more generally, a positive function of the temperature). This formula is in the best tradition of laws in physics that postulate a linear relation between causes and effects (Ohm's law, Fourier's law, Fick's law, Darcy's law, Hooke's law, Newton's viscosity law, and many others).

7.7.2 Checking the Proposed Evolution Law

The adoption of the Mandel stress as the only independent variable implies that the evolution function depends ultimately on the same constitutive arguments as the stress, so that the principle of *equipresence* is certainly not violated.

In accordance with the principle of *reduction to the archetype*, as expressed in equation (7.92), we must evaluate the Mandel stress at the archetype, which we have indicated with an overbar.

Exercise 7.7.1   The Mandel stress at the archetype. Show that the Mandel stress at the archetype is related to the Mandel stress at the reference configuration by the formula

$$\bar{\mathbf{m}} = J_P \, \mathbf{P}^T \, \mathbf{m} \, \mathbf{P}^{-T}. \tag{7.141}$$

If we were to change the archetype, the form of the evolution equation would be altered, as shown in equation (7.104).

Exercise 7.7.2   Show that under a change of archetype given by a matrix $\mathbf{A}$, the proposed evolution function preserves the form (7.140) if, and only if, $\mathbf{A}$ is orthogonal.

From this exercise, we conclude that the symmetry group $\mathcal{A}$ of the proposed evolution law is the orthogonal group. If the material is a solid and if we adopt an archetype in a natural state, we conclude that the *material symmetry consistency* condition is satisfied. Moreover, because of the choice of the Mandel stress, which is a purely referential or archetypal tensor, as the only independent variable of the evolution law, the principle of material *frame indifference* is automatically satisfied.

To verify the satisfaction of the principle of *actual evolution*, assume that the symmetry group of the material is continuous. Starting at the unit $\mathbf{I}$, we can, therefore, move continuously within the group along some curve $\mathbf{G}(\tau)$, where τ is a parameter and $\mathbf{G}(0) = \mathbf{I}$. By definition of material symmetry, for all values of the parameter τ within some interval containing 0, the free-energy density must satisfy

$$\psi(\mathbf{F}\mathbf{G}(\tau), \theta) = \psi(\mathbf{F}, \theta). \tag{7.142}$$

Taking the derivative with respect to τ (intuitively, moving smoothly within the group, so that nothing happens to the free energy), we obtain

$$\frac{\partial \psi(\mathbf{F}\mathbf{G}(\tau), \theta)}{\partial \tau} = 0. \tag{7.143}$$

Using the chain rule yields

$$\frac{\partial \psi(F^i{}_K G^K{}_J, \theta)}{\partial F^j{}_M G^M{}_N} \frac{\partial F^j{}_R G^R{}_N}{\partial G^P{}_Q} \frac{\partial G^P{}_Q(\tau)}{\partial \tau} = \frac{\partial \psi(F^i{}_K G^K{}_J, \theta)}{\partial F^j{}_M G^M{}_N} F^j{}_P \frac{\partial G^P{}_N(\tau)}{\partial \tau} = 0. \tag{7.144}$$

We now set $\tau = 0$ and obtain

$$\frac{\partial \psi(F^i{}_K, \theta)}{\partial F^j{}_N} F^j{}_P \frac{\partial G^P{}_N(\tau)}{\partial \tau} = \frac{1}{\rho_0} T_j{}^N F^j{}_P \frac{\partial G^P{}_N(\tau)}{\partial \tau} = 0. \tag{7.145}$$

This result can written as

$$\text{trace} \left(\mathbf{m} \, \frac{\partial \mathbf{G}}{\partial \tau} \right) = 0. \tag{7.146}$$

In other words, the Mandel stress is perpendicular to (the transpose of) every element of the Lie algebra of the symmetry group. In particular, unless it vanishes, it cannot belong to this algebra. This is a general result. But, having assumed an evolution equation by which the inhomogeneity velocity gradient is proportional to the Mandel stress, we conclude that the former cannot belong to the Lie algebra (of infinitesimal generators) of the symmetry group. This is precisely the statement of the principle of actual evolution!

The only remaining restriction to be satisfied by the proposed evolution law is that of the *dissipation inequality* (7.139), which in our case is satisfied by construction.

7.7.3 A Numerical Example

For definiteness, we will adopt the modified neo-Hookean constitutive law for the free energy that we have already used in Section 6.5.5, namely

$$\bar{\rho}\bar{\psi} = \frac{G}{2}\left(\text{trace }\bar{\mathbf{B}} - \ln(\det \bar{\mathbf{B}})\right). \tag{7.147}$$

This is the constitutive law for the archetype. In a global reference configuration of a uniform evolving material body, the constitutive equation is obtained via the implant field $\mathbf{P} = \mathbf{P}(X,t)$ as

$$\rho_0\psi = \frac{G}{2 J_P}\left(\text{trace }\mathbf{P}^T \mathbf{C} \mathbf{P} - \ln(\det \mathbf{P}^T \mathbf{C} \mathbf{P})\right). \tag{7.148}$$

The corresponding Piola stress is

$$\mathbf{T} = \frac{G}{J_P}\left(\mathbf{F} \mathbf{P} \mathbf{P}^T - \mathbf{F}^{-T}\right). \tag{7.149}$$

Similarly, the evolution equation (7.140), when converted to quantities in the reference configuration, reads:

$$\dot{\mathbf{P}} = -k J_P \mathbf{P} \mathbf{P}^T \mathbf{m} \mathbf{P}^{-T}. \tag{7.150}$$

To emphasize the potential applications of the theory, we will show that an evolution model as simple as that of equation (7.140), in combination with an archetypal constitutive law such as (7.147), leads to net growth under harmonic loading around a zero-stress state. More generally, the effect of an oscillatory load is more favourable to growth and less favourable to resorption than the effect of the steady application of the average load.

Consider a spherical chunk of isotropic material subjected at each point of its boundary to a normal traction of magnitude $a = a(t)$. We are neglecting the effects of inertia. The Piola stress throughout the body is spatially constant and is given by

$$\mathbf{T} = a(t) \mathbf{I}. \tag{7.151}$$

The deformation gradient and the implant field are given, respectively, by

$$\mathbf{F} = f(t) \mathbf{I} \tag{7.152}$$

and

$$\mathbf{P} = q(t) \mathbf{I}. \tag{7.153}$$

We want to solve for the time-dependent quantities $f(t)$ and $q(t)$. The Mandel stress is given by

$$\mathbf{m} = f(t)a(t)\,\mathbf{I}. \tag{7.154}$$

The evolution law (7.150) yields

$$\dot{q} = -kq^4 fa. \tag{7.155}$$

The constitutive law (7.149) can be solved explicitly for f as

$$f = \frac{\frac{aq^3}{G} + \sqrt{\left(\frac{aq^3}{G}\right)^2 + 4q^2}}{2q^2}. \tag{7.156}$$

As a result, we obtain the following first-order ordinary differential equation for the time evolution of the implant variable $q(t)$:

$$\dot{q} = -\frac{ka}{2}q^2\left(\frac{aq^3}{G} + \sqrt{\left(\frac{aq^3}{G}\right)^2 + 4q^2}\right). \tag{7.157}$$

Using the change of variables

$$\tau = kGt, \quad \alpha = \frac{a}{G}, \tag{7.158}$$

we obtain

$$\frac{dq}{d\tau} = -\frac{\alpha}{2}q^2\left(\alpha q^3 + \sqrt{(\alpha q^3)^2 + 4q^2}\right). \tag{7.159}$$

In particular, we will consider an oscillatory time dependence of the form

$$a(t) = a_0\,\sin(\omega t), \tag{7.160}$$

the amplitude a_0 and the angular frequency ω being given.

We recall that a net growth is reflected in a decrease of the value of q from its initial value of $q(0) = 1$. Physically, the smaller the q the more compressed the implanted archetype is, resulting in more material being squeezed in. In our example, the evolution of the referential density is given by

$$\rho_0(t) = \frac{\rho_o(0)}{q(t)}, \tag{7.161}$$

so that the mass increases with a decrease of q. For very small values of the ratio $\alpha_0 = a_0/G$ the response $q(t)$ is approximately periodic, without any significant gain or loss of mass. But for somewhat larger values of this ratio a net mass gain (i.e., a net decrease in $q(t)$) can be observed from numerical evaluations. It is important to notice that a net gain is obtained regardless of the direction of the initial cycle, as can be verified by changing the sign of α_0. Figure 7.20 shows q plotted against τ for $\alpha_0 = 0.1$ and $\omega/(kG) = 20$. Calculations and graphs were produced using the Mathematica® package. The general qualitative behaviour obtained is largely independent of the particular archetypal constitutive law used.

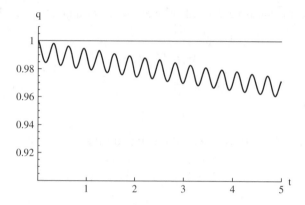

Figure 7.20   Growth enhanced by oscillatory load

Exercise 7.7.3   Tendon growth. The example just shown may possibly apply to a piece of bone in the vicinity of a synovial joint, where the state of stress is closer to hydrostatic. Consider instead the case of a straight organ such as a tendon or ligament, where the state of stress is nearly uniaxial. Reproduce the steps of the previous example for the case of uniaxial loading using the same constitutive laws. Plot the graph of referential mass density as a function of time.

Exercise 7.7.4   Back to adaptive elasticity. Building upon the results of Exercise 7.5.2, adopt some specific form (perhaps quadratic) for the constitutive function $g(\cdot)$ and specific values for the constants involved. With the loading given by equation (7.160), solve numerically for the solid fraction ζ_0 as a function of time.

7.8   Case Study: Bone Remodelling and Wolff's Law

Always in the spirit of testing the ability of continuum mechanics to respond to the challenges posed by the modelling of living organisms, we propose to explore a classical problem, to which we have already alluded in Section 7.1. One of the widely held tenets in the theory of bone remodelling and growth is the so-called Wolff's law. Formulated at the end of the nineteenth century, it is a qualitative statement to the effect that the internal architecture of an organ is, at least in part, determined by the stresses acting upon it. In particular, one extreme interpretation of this law is that in trabecular bone the main trabeculae tend to align themselves with the prevalent direction of the maximum principal stress. Few specialists in biomechanics nowadays believe this statement to be completely accurate, but rather hold the view that there must exist some connection between bone remodelling and mechanical stresses and other functional factors. Be that as it may, we propose to consider a highly idealized situation on a purely hypothetical basis and then compare the results thus obtained with the general statement of Wolff's law.

In order to represent the inherent anisotropy of bone, we will model it as a transversely isotropic medium, the strong direction (represented by the unit vector **n**) being aligned with the local orientation of the main trabeculae. We have already encountered, in Section 6.5.5, a modified (transversely isotropic) version of the archetypal isotropic constitutive law (7.147) in the form

$$\bar{\psi} = \frac{G}{2}\left(\text{trace}(\mathbf{M}(\bar{\mathbf{C}} - \mathbf{I})\mathbf{M} + \mathbf{I}) - 2\ln\left(\sqrt{\det(\mathbf{M}(\bar{\mathbf{C}} - \mathbf{I})\mathbf{M} + \mathbf{I})}\right)\right), \qquad (7.162)$$

where $\mathbf{M}$ is a bias tensor. Adopting a basis in the archetype such that the first base vector is aligned with the strong material direction, the tensor $\mathbf{M}$ will have the following component form:

$$[M] = \begin{bmatrix} \alpha & 0 & 0 \\ 0 & 1 & 0 \\ 0 & 0 & 1 \end{bmatrix}, \qquad (7.163)$$

where $\alpha > 1$ is a material constant representing the ratio between the strength in the trabecular direction and its transverse counterpart.

As far as the evolution law is concerned, we will consider a purely rotational remodelling, without distortion and without net growth. To achieve this end, we will assume the inhomogeneity velocity gradient to be proportional to the skew-symmetric part of the Mandel stress, both pulled back to the archetype. In other words, denoting by $\bar{\mathbf{m}}_W$ the skew-symmetric part of the Mandel stress in the archetype, the evolution law reads:

$$\mathbf{P}^{-1}\dot{\mathbf{P}} = -k\bar{\mathbf{m}}_W, \qquad (7.164)$$

where k is a material constant. As far as the sign of this constant is concerned, if we assume that the non-compliant entropy sink $\bar{H}$ vanishes, it follows from the entropy inequality that k must be non-negative. This is a crucial remark for the correct interpretation of our final result.

Exercise 7.8.1 Prove that the general solution of equation (7.164) consists of a (time-dependent) orthogonal matrix multiplied on the left by a (time-independent) symmetric and positive definite matrix. [Hint: use the left polar decomposition theorem.] Conclude that, if the initial condition is a (proper) orthogonal matrix, the solution remains (proper) orthogonal as time goes on.

We assume the initial implant to be of the form

$$[P] = \begin{bmatrix} \cos\theta & \sin\theta & 0 \\ -\sin\theta & \cos\theta & 0 \\ 0 & 0 & 1 \end{bmatrix}, \qquad (7.165)$$

where θ represents the angle of rotation of the implant, assumed to take place about the third coordinate axis. Notice that a positive θ represents a *clockwise* rotation (of the trabecular axis $\mathbf{n}$ with respect to the longitudinal axis of the bone, i.e., the first coordinate axis).

Assuming that a cylindrical specimen is subjected to a constant axial tensile force that preserves its magnitude and direction in space (a direction that we assume aligned with the first coordinate axis), the components of the Piola stress are given by

$$[T] = \begin{bmatrix} a & 0 & 0 \\ 0 & 0 & 0 \\ 0 & 0 & 0 \end{bmatrix}, \qquad (7.166)$$

where a is a constant and where we have assumed the spatial and referential coordinate systems to coincide.

Under these circumstances, we seek a solution of the problem when the deformation gradient has the form

$$[F] = \begin{bmatrix} b(t) & c(t) & 0 \\ 0 & d(t) & 0 \\ 0 & 0 & e(t) \end{bmatrix}. \tag{7.167}$$

This form of the solution is chosen so that a material die originally aligned with the coordinate axes is forced to preserve the directions of the first and the third axes and to keep the deformed second axis perpendicular to the third. We remark that this form of the solution is consistent with the necessary symmetry of the Cauchy stress.

The numerical solution of this problem yields a somewhat startling result. Regardless of the initial value of θ (different from zero), the solution tends asymptotically to a final value $\theta = \pi/2$. In other words, the putative main trabeculae end up aligning themselves with the *smaller* principal stress! Moreover, the zero solution (obtained by setting the initial angle to zero) is unstable.[21] We recall, however, that the entropy sink $\bar{H}$ has been assumed to vanish. If it does not (by virtue of some externally controlled mechanism) we can possibly change the sign of the constant k, thereby reversing the result and recovering Wolff's law. Alternatively, even assuming the vanishing of $\bar{H}$, one may argue that the remodelling process is initiated only after the Eshelby stress exceeds a certain threshold, in which case spots of stress concentration would play a determining role in the process. In particular, if the voids within the bone are regarded as very elongated in the axial direction, the location of the maximal principal stress may well be the apices of these voids. More importantly, the direction of these principal stresses is perpendicular to the axis of the bone, a result that would be consistent with Wolff's law, albeit in a rather indirect and unexpected way.

To dispel the impression that the numerical results obtained may be qualitatively dependent on the particular transversely isotropic constitutive law adopted, we note that the evolution equation (7.164) reduces in our case to the following differential equation:

$$\frac{d\theta}{dt} = \frac{k}{2} ac(t), \tag{7.168}$$

as can be obtained by direct substitution of the assumed forms of **P**, **T** and **F**. In other words, the time evolution of the implants is dictated entirely by the 'shear' component $c(t)$ of the deformation gradient. Although the numerical value of this component certainly depends on the particular constitutive law of the bone material, it is clear that its sign does not. If the main (strong) axis of anisotropy is rotated counter-clockwise with respect to the bone axis (as shown in Figure 6.7), the cross section of the specimen rotates in the same direction. In other words, the sign of c is the same as the sign of $\tan\theta$. By symmetry, c vanishes only when $\theta = 0, \pi$ or $\theta = \pm\pi/2$. In both of these cases we obtain a constant solution, but this solution is unstable for the former case ($\theta = 0, \pi$), and stable for the latter.

Exercise 7.8.2   Wolff's law? Write some computer code (e.g., for Mathematica®) to solve numerically the system of algebraic and ordinary differential equations of this case

---

[21] A similar result was found by DiCarlo et al. (2005).

study for the functions $\theta(t), b(t), c(t), d(t), e(t)$. Example 6.5.15 may be useful. Plot $\theta(t)$ and $c(t)$ using various initial conditions and verify that θ tends to $\pi/2$ regardless of the initial conditions.

References

Bunge, M.A. (1998) *Philosophy of Science*, Vol. II. New Brunswick, NJ: Transaction Publishers.

Byrne, H.M. and Chaplain, M.A.J. (1996) Modelling the role of cell-cell adhesion in the growth and development of carcinomas. *Mathematical and Computer Modelling* **24**, 1–17.

Campbell, G.M., Buie, H.R. and Boyd, S.K. (2008) Signs of irreversible architectural changes occur early in the development of experimental osteoporosis as assessed by in-vivo micro-CT. *Osteoporosis International* **19**, 1409–1419.

Casciari, J.J., Sotirchos, S.V. and Sutherland, R.M. (1990) Nutrient concentration and pH gradients in multi-cellular tumor spheroids and their effects on growth and metabolism. In *Proceedings of the AIChE Annual Meeting, Chicago*, Paper 192c.

Coleman, B.D. and Gurtin, M.E. (1967) Thermodynamics with internal state variables. *Journal of Chemical Physics* **47**, 597–613.

Cowin, S.C. and Hegedus, D.H. (1976a) Bone remodelling I: Theory of adaptive elasticity. *Journal of Elasticity* **6**(3), 313–326.

Cowin, S.C. and Hegedus, D.H. (1976b) Bone remodelling II: Small strain adaptive elasticity, *Journal of Elasticity* **6**(4), 337–352.

DiCarlo, A., Naili, S., Quiligotti, S. and Teresi, L. (2005) Modeling bone remodeling. In J.M. Petit and J. Daluz, (eds), *Proceedings of the COMSOL Multiphysics Conference*, Paris, pp. 31–36.

Eckart, C. (1948) The thermodynamics of irreversible processes, IV: The theory of elasticity and anelasticity. *Physical Review* **73**, 373–382.

Epstein, M. and Elżanowski, M. (2007) *Material Inhomogeneities and their Evolution*. Berlin: Springer-Verlag.

Epstein, M. and Goriely, A. (2012) Self-diffusion in remodelling and growth. *Zeitschrift für Angewandte Mathematik und Physik* **63**, 339–355.

Epstein, M. and Herzog, W. (2003) Aspects of skeletal muscle modelling. *Philosophical Transactions of the Royal Society of London* **B358**, 1445–1452.

Epstein, M. and Maugin, G.A. (2000) Thermomechanics of volumetric growth in uniform bodies. *International Journal of Plasticity* **16**, 951–978.

Eshelby, J.D. (1951) The force on an elastic singularity. *Philosophical Transactions of the Royal Society of London* **A244**, 87–112.

Fung, Y.C. and Liu, S.Q. (1989) Change of residual strains in arteries due to hypertrophy caused by aortic constriction. *Circulation Research* **65**, 1340–1349.

Galilei, G. (1638) *Discorsi e Dimostrazioni Matematiche intorno a Due Nuove Scienze*. Leiden: Elsevirii.

Garikipati, K., Arruda, E.M., Grosh, K., Narayanan, H. and Calve, S. (2004) A continuum treatment of growth in biological tissue: The coupling of mass transport and mechanics. *Journal of the Mechanics and Physics of Solids* **52**, 1595–1625.

Greenspan, H.P. (1976) On the growth and stability of cell cultures and solid tumors. *Journal of Theoretical Biology* **56**, 229–242.

Hamant, O., Heisler, M., Jonsson, H., Krupinski, P., Uyttewaal, M., Bokov, P., Corson, F., Sahlin, P., Boudaoud, A., Meyerowitz, E., Couder, Y. and Traas, J. (2008) Developmental patterning by mechanical signals in arabidopsis. *Science* **322**, 1650–1655.

Huiskes, R., Ruimerman, R., van Lenthe, G.H. and Janssen, J.D. (2000) Effects of mechanical forces on maintenance and adaptation of form in trabecular bone. *Nature* **405**, 704–706.

Kuhl, E. and Steinmann, P. (2003) Mass- and volume-specific views on thermodynamics for open systems. *Proceedings of the Royal Society of London* **A459**, 2547–2568

Maugin, G.A. and Muschik, W. (1994) Thermodynamics with internal variables. *Journal of Non-equilibrium Thermodynamics* **19**, 217–289.

Noll, W. (1967) Materially uniform bodies with inhomogeneities. *Archive for Rational Mechanics and Analysis* **27**, 1–32.

Qi, A.-S., Zheng, X., Du, C.Y. and An, B.S. (1993) A cellular automaton model of cancerous growth, *Journal of Theoretical Biology* **161**, 1–12.

Rodriguez, E.K., Hoger, A. and McCulloch, A.D. (1994) Stress-dependent finite growth in soft elastic tissues. *Journal of Biomechanics* **27**, 455–467.

Roose, T., Netti, P.A., Munn, L.L., Boucher, Y. and Jain, R.K. (2003) Solid stress generated by spheroid growth estimated using a linear poroelasticity model. *Microvascular Research* **66**, 204–212.

Roose, T., Chapman, S.J. and Maini, P.K. (2007) Mathematical models of avascular tumor growth. *SIAM Review* **49**, 179–208.

Ruimerman, R., Hilbers, P., van Rietbergen, B. and Huiskes, R. (2005) A theoretical framework for strain-related trabecular bone maintenance and adaptation. *Journal of Biomechanics* **38**, 931–941.

Taber, L.A. (1995) Biomechanics of growth, remodeling and morphogenesis. *Applied Mechanics Reviews* **48**(8), 487–545.

Wang, C.-C. and Bloom, F. (1974) Material uniformity and homogeneity in anelastic bodies. *Archive for Rational Mechanics and Analysis* **53**, 246–276.

Ward, J.P. and King, J.R. (1997) Mathematical modelling of avascular-tumour growth. *IMA Journal of Mathematics Applied in Medicine and Biology* **14**, 39–69.

Wolff, J. (1870) Ueber die innere Architectur der Knochen und ihre Bedeutung für die Frage vom Knochenwachsthum. *Virchows Archiv* **50**(3), 389–450.

Wolfram, S. (1994) *Cellular Automata and Complexity*. Reading, MA: Addison-Wesley.

8

Principles of the Finite-Element Method

8.1 Introductory Remarks

You do not need to know thermodynamics to drive a car. Or, as Oliver Heaviside (1850–1925) famously said when commenting on the supposed lack of rigour of his operational calculus: 'Shall I refuse my dinner because I do not fully understand the process of digestion?' Similarly, you do not need to read this chapter to be able to use one of the many available commercial finite-element packages. A user-friendly interface allows you to input the geometry of a material body and the loads acting on it and, as a result, the program produces beautiful stress and temperature plots.

The objective of this chapter, on the other hand, is to make you aware of some of the physical and mathematical ideas that lie behind the finite-element method (FEM). And what really is the FEM? A short answer is the following: it is one of many methods available to the scientist and engineer to obtain approximate solutions of the systems of equations that arise in the modelling of natural phenomena. These equations are usually integral or differential equations for the fields of interest (such as displacements, velocities, temperature, electromagnetic field, and so on). Typically, they are partial differential equations (PDEs), whose solutions are seldom obtainable analytically.

What makes the FEM particularly attractive from the engineering standpoint is that it can be approached from so many different angles that it virtually becomes an arena in its own right where engineering and mathematics engage in a friendly and mutually respectful competition. The winner is undoubtedly the spectator, who stands to gain a deeper understanding of some ideas that, without the context of the FEM, may seem too abstract to grasp (such as the notion of weak solutions, or the calculus of variations) or too concrete to be of general interest (such as the principle of virtual work or the stiffness method for structural analysis). That the FEM blends these ideas into a harmonious whole is partially a reflection of its very peculiar origins and historical development. It was created by engineers for engineers, mainly with structural aeronautics applications in mind. Originally, it was viewed as a kind of generalization of the stiffness method for the analysis of linear framed structures, a method which had been well established by the end of the nineteenth century and which, with the advent of electronic computers in the early 1950s, had attained wide popularity. It did not take too long, however, for the engineering community at large, and for the academic community in particular, to realize that the FEM could also be regarded as a clever extension of certain numerical techniques known as *direct methods of the calculus of variations*. These techniques are not limited to structural analysis, and thus the door was opened for the extension of the FEM to other areas of engineering and physics.

The FEM is, therefore, the result of the blending of time-tested engineering ideas, sound mathematical techniques and miraculously fast electronic computing devices. It is one of the great success stories of the human mind and, not surprisingly, one of the great financial success stories for many talented engineers who saw the opportunity to satisfy a market hungry for a versatile and almost universal technique for the numerical analysis of engineering problems. Many commercial FEM packages compete in this market, each one offering various degrees of generality, pre- and post-processing features, user friendliness, prices, and so on. At the very core of these codes lie the foundations of the method that we will discuss in this chapter.

8.2 Discretization Procedures

Most of the traditional methods, including the FEM, for the solution of field problems are essentially *discretization procedures*. By means of these procedures, an approximate solution of the field equations governing some phenomenon defined on a continuous domain (e.g., a deformable solid or fluid) is sought with the following two features:

1. The approximate solution can be described by means of a finite (although perhaps very large) number of parameters.
2. The approximate solution can be shown to converge to the exact solution of the field equations as the number of parameters is increased in a systematic way.

The term 'discretization' alludes to the fact that continuous fields are replaced by discrete arrays of parameters. As part and parcel of any discretization procedure, therefore, one must provide a way to reconstruct the continuous fields approximately by some definite technique, such as interpolation. In Figure 8.1 we have summarized the general scheme common to these methods. Examples of discretization procedures are: power series expansion, Fourier series, finite differences and finite elements.

8.2.1 Brief Review of the Method of Finite Differences

In the *finite-difference method* the domain of interest is 'meshed' by means of a (usually regular) rectangular grid, whose nodes are used as keypoints on which (and only on which) the unknown functions will be evaluated approximately to impose the field equations. As the grid becomes finer and finer, the solution is expected to converge to the exact solution, a fact that needs to be either rigorously proved or, at least, tested by more or less convincing numerical experiments.

As an example, consider the evaluation of first and second derivatives in a two-dimensional problem with some unknown function $\psi = \psi(x,y)$ governed by a second-order PDE. The domain of interest is meshed by an orthogonal grid, as partially shown in Figure 8.2. At a typical point not belonging to the boundary, the various partial derivatives

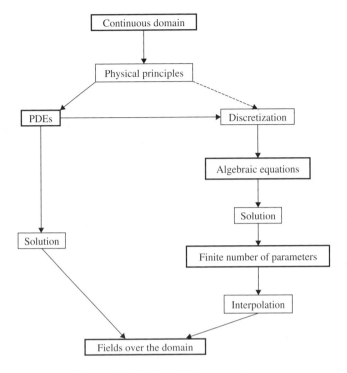

Figure 8.1 Discretization procedures

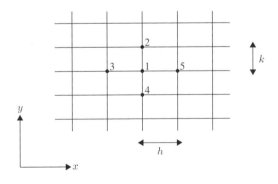

Figure 8.2 A finite-difference mesh

can be approximated, according to the very definition of derivative as the limit of a difference quotient, as follows:

$$\left[\frac{\partial \psi}{\partial x}\right]_1 \approx \frac{\psi_5 - \psi_3}{2h}, \tag{8.1}$$

$$\left[\frac{\partial \psi}{\partial y}\right]_1 \approx \frac{\psi_2 - \psi_4}{2k}, \tag{8.2}$$

$$\left[\frac{\partial^2 \psi}{\partial x^2}\right]_1 \approx \frac{\psi_5 - 2\psi_1 + \psi_3}{h^2}, \tag{8.3}$$

$$\left[\frac{\partial^2 \psi}{\partial y^2}\right]_1 \approx \frac{\psi_2 - 2\psi_1 + \psi_4}{k^2}. \tag{8.4}$$

In this way, the differential equation is enforced approximately at the mesh-points, while at the boundary points one implements the boundary conditions. The result is a system of *algebraic* equations for the values of the function at the mesh points.

Exercise 8.2.1 Deflections of a membrane. An elastic membrane is subjected to a uniform tension N in all directions and then attached to a rigid frame, such as in the case of a drum of an orchestra. The tension N is measured in terms of force per unit length of the perimeter. If the membrane is acted upon by a static normal force p, measured in terms of force per unit area, the resulting transverse (small) displacement field $w = w(x, y)$, expressed as a function of Cartesian coordinates x, y placed on the plane of the membrane, abides by the following PDE, known as *Poisson's equation*:

$$\frac{\partial^2 w}{\partial x^2} + \frac{\partial^2 w}{\partial y^2} = \frac{p}{N}. \tag{8.5}$$

Assuming the frame to be rectangular, solve the membrane deflection problem by the finite-difference method. You may want to take advantage of symmetry to reduce the number of equations. Use two different grids to check convergence. Use numerical data of your choice.

The finite-difference formulas (8.1)–(8.4) are known as *central differences*, since they provide approximations to the derivatives at a point in terms of symmetrically arranged neighbouring points of the grid. In many applications, particularly those involving the time variable, it proves convenient to consider *forward differences* or other asymmetric schemes. Thus, for example, equation (8.2) in a forward-difference scheme would be replaced by

$$\left[\frac{\partial \psi}{\partial y}\right]_1 \approx \frac{\psi_2 - \psi_1}{k}. \tag{8.6}$$

In fact, we have already suggested the use of this scheme in the approximation of the time derivative of the one-dimensional diffusion equation in Section 2.5.2.

Remark 8.2.2 Diffusive processes and cellular automata. The use of forward differences for the time variable in diffusive processes leads to an intriguing connection with cellular automata. Indeed, according to this scheme, the state of the system at the next time-step is completely determined by the present state of the system, which is the basic tenet of a cellular automaton.

In spite of its great generality and its solid mathematical foundation, one of the inconveniences of the finite-difference method is its relative lack of flexibility to handle arbitrary geometries. If we need to model, say, an airplane wing or a piece of cartilage at the knee joint, practical difficulties arise at the boundaries. Similarly, if we need to join different types of structural elements, or study the interaction of a solid and a fluid, special consideration has to be given to the interfaces between phenomena governed by different field equations. To overcome these limitations, structural engineers in the 1950s started modelling complex structures by a more or less systematic substitution of simpler ones. For instance, a plate could be approximately modelled as an assemblage of beams, whose stiffness properties could be adjusted so as to represent the original structure faithfully. Out of these first attempts, the tremendous impetus provided by the space programme and the advent of faster and larger electronic computers in the 1960s, the finite-element method began to take shape.

8.2.2 Non-Traditional Methods

One final word needs to be said about some of the non-traditional numerical techniques. If you look at Figure 8.1, you will notice a broken-stem arrow pointing from the physical principles directly to the discretization procedures. For example, the technique of replacing a plate with a system of beams falls into this category, since no field equations are formulated to begin with. More modern techniques, such as the method of cellular automata, which we have already encountered in Section 7.2.2, suggest doing away with the physical laws altogether and modelling the phenomena directly as the interaction of particles following certain simple rules of behaviour while letting the complexity of the system take care of itself. Spectacular results have been obtained by this and other methods, such as genetic algorithms and neural networks, particularly in modelling turbulence in fluids, the mechanics of sand piles and the growth of plants. Some believe that these ideas herald the advent of a new physics without universal laws in the traditional sense of the term. It is indeed very frustrating to realize that, even though we may know all the laws of nature, we still cannot model the details of the beautiful swaying of the leaves of a tree with the wind. But whether or not this frustration justifies the abandonment of one of the greatest intellectual hopes of the human race is debatable.

8.3   The Calculus of Variations

8.3.1   Introduction

Although originally seen as a clever and powerful structural technique, it was soon realized that the foundations of the FEM had already been established in a branch of mathematical analysis known as the calculus of variations and its several attendant numerical techniques, already available by the late nineteenth and early twentieth centuries, such as the Rayleigh–Ritz method, the Galerkin procedure and the method of weighted residuals. The awareness of this momentous pedigree allowed the method of finite elements to be applied to a much wider class of problems than originally intended. We will, therefore, undertake a brief tour through the calculus of variations, a very important topic in its own right with remarkable practical applications in many areas of science and engineering, and discuss the numerical techniques just mentioned, before embarking upon the study of the FEM proper.

From its very beginnings, infinitesimal calculus was conceived, at least in part, as a method for finding maximum and minimum values of smooth functions. So much so, that the first paper on calculus ever to appear in print was entitled 'Nova methodus pro maximis et minimis ...'. The author was Gottfried Wilhelm Leibniz (1646–1716) and the article appeared in the journal *Acta Eruditorum* in October 1684. Today, more than

300 years later, we are all confident that we know how to find relative maxima and minima of differentiable functions by precisely the method devised by Newton (1643–1727) and Leibniz. All we need to do is calculate the derivative of the function and equate it to zero. But even at the time of Newton and Leibniz some people, including the Bernoulli brothers Jakob (1654–1705) and Johann (1667–1748), had come to realize that there were cases of perfectly smooth problems where the usual method was not suitable for the purpose of finding maxima and minima.

The first such problem to be explicitly stated, known as the problem of the *brachystochrone*, appeared in the form of a challenge to mathematicians posed by Johann Bernoulli in the June 1696 issue of the *Acta Eruditorum*. The problem reads as follows: 'Given two points A and B in a vertical plane, find the path AMB that a moving point (mass) M must follow such that, starting from A, it reaches the other point B in the shortest time descending under its own gravity.' This clearly sounds like a standard problem of finding a minimum of a function, but it is not. Among the several correct solutions given at the time, it is interesting to record that Johann Bernoulli's own solution established a clear connection between this problem and a problem in optics that had been proposed and solved earlier by Pierre de Fermat (1601–1665). This is the problem of finding the path of a ray of light travelling through a medium of varying refractive index. Fermat had suggested that, following the ideas of Heron of Alexandria (first century), one should find the trajectory that minimizes the time of travel between two given points.

In the eighteenth century, Leonhard Euler (1707–1783), the great Swiss mathematician, developed the foundations of a general theory to handle problems of this kind, a theory that was later to be known as the *calculus of variations*. The eighteenth century, the century of Bach, Haydn, Mozart, Rousseau and Voltaire, was to bring not only mathematical progress in the calculus of variations, but also a related philosophical attitude towards the laws of nature that can be regarded as revolutionary. Pierre Louis Moreau de Maupertuis (1698–1759), for example, maintained that the Creator of the universe is not only wise, but also efficient, and that Nature must accordingly function in the most economical way possible. In other words, every law of nature must be such that it can be formulated in terms of the minimization of a certain quantity, which he called *action*.

This bold optimality criterion found confirmation in many physical theories. The works of Joseph-Louis Lagrange (1736–1813) and Sir William Rowan Hamilton (1805–1865) established that the laws of Newton can be replaced by a principle stating that a particle moving in a conservative force field follows a trajectory that minimizes the difference between its kinetic and potential energies. We have already mentioned that in geometrical optics, the rays of light satisfy a criterion of extremality of the time of travel. In the theory of elasticity, it can be shown that a structure deforms in such a way as to render the total potential energy minimum. In Albert Einstein's (1879–1955) theory of general relativity, it was found by David Hilbert (1862–1943) that Einstein's equations can be derived from a single principle, whereby the scalar curvature of the universe is minimized, taking account of the presence of matter and other fields. Similar variational principles are also used today in more mundane applications, such as optimal control theory and biomechanics.[1]

Since many physical theories can be formulated in terms of optimality criteria, it is reasonable to expect that there exist approximate numerical methods of solution that are

---

[1] It is sometimes postulated, for instance, that many biological processes attempt to minimize the consumption of metabolic energy.

directly based on the approximate enforcement of such criteria. This is indeed the case, and these so-called *direct methods of the calculus of variations* are in many ways the precursor of the finite-element method. But even without this close mathematical and physical connection with the FEM, the calculus of variations is worthy of study for its own sake, as it sheds light on many areas of science and engineering.

8.3.2   The Simplest Problem of the Calculus of Variations

To understand why we cannot just invoke the usual methods of infinitesimal calculus to solve the kinds of problems mentioned above, let us consider the case of an ant starting at a point A on a leaf and attempting to reach a grain of sugar at another point, B, situated on the same leaf, following the shortest possible path (Figure 8.3). What we mean by this is that the actual path γ should be shorter than any other path, always starting at A and ending at B, in the vicinity of γ. If we know the equation of the shape of the leaf, a surface in space, it is clearly possible to write an expression for the length of a curve lying on this surface and joining the points A and B. And how will this expression look? It will certainly be some complicated integral. For example, if the leaf happens to be plane and if we introduce in it some x, y Cartesian coordinates (a luxury we cannot afford in a real curved leaf!), and if we write the equation of the curve γ as $y = y(x)$, then, by a direct use of Pythagoras' theorem in its infinitesimal version, we obtain the length of the curve as

$$L(\gamma) = \int_{x_A}^{x_B} \sqrt{1 + y'^2}\, dx, \tag{8.7}$$

where the prime denotes the derivative with respect to x.

Equation (8.7) reveals that the quantity to be minimized (L) is indeed a function of the variable curve (γ). But what kind of a function is this? Its independent variable is not just one or several real numbers, but a whole function, namely $y = y(x)$, which can be roughly seen as an infinite number of data: one y for each x in the interval $[x_A, x_B]$. Moreover, the *derivative* of this argument function appears in the expression for the function L. So, if we would like to follow the lead of classical calculus, we should take the derivative of the dependent variable (L), with respect to … what? And what kind of result do we expect to obtain that will tell us what is the equation $y = y(x)$ of the shortest path? The answer to these questions is the subject of the calculus of variations.

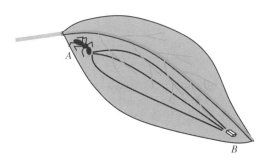

Figure 8.3   The shortest path

And among the numerical techniques that can be used to solve questions of this type approximately, the finite-element method figures prominently.

To distinguish between ordinary real-valued functions of a finite number of real variables and the real-valued functions which, like in equation (8.7), have as arguments one or more functions and their derivatives, we will reserve the term *functional* for the latter.[2] The simplest non-trivial problem of the calculus of variations is the following: to find a differentiable function, $y = y(x)$, that renders a functional of the form

$$J[y(x)] = \int_{x_0}^{x_1} F(x, y, y') \, dx \tag{8.8}$$

maximum or minimum (more generally, *stationary*), while satisfying the end conditions

$$y(x_0) = y_0, \quad y(x_1) = y_1, \tag{8.9}$$

where (x_0, y_0) and (x_1, y_1) are the coordinates of two given points. In other words, we are looking for a curve passing through these two points and rendering the functional stationary with respect to all neighbouring paths passing through those points.

To avoid any possible confusion, we need to clarify at this point what the function $F(x, y, y')$ means. This function, assumed to be differentiable, is an ordinary function of three independent variables. It simply tells us what we have to do with these three variables to obtain a specific expression under the integral. For instance, in the example of equation (8.7), F happens not to depend on x or y, but only on the third variable y'. It is important to stress that the fact that y' is the derivative of $y(x)$ is completely irrelevant as far as the function F is concerned, since F is there just to tell us how to combine these three arguments (x, y, y') in a mathematical expression. As a differentiable function of three variables, F has (partial) derivatives with respect to each of them. Again, these variables being independent, as far as the function F is concerned, these partial derivatives are calculated independently of each other. For instance, for equation (8.7) we would have

$$\frac{\partial F}{\partial x} = 0, \quad \frac{\partial F}{\partial y} = 0, \quad \frac{\partial F}{\partial y'} = \frac{y'}{\sqrt{1 + y'^2}}. \tag{8.10}$$

The first thing we need to do to check whether or not a candidate function $y = y(x)$, as shown in Figure 8.4, produces, say, a minimum value of J, is to compare the values that J attains for $y = y(x)$ and for other functions in its vicinity. To create functions in the vicinity of $y = y(x)$, we proceed as follows. We define an arbitrary differentiable function $g(x)$ that vanishes at the end points. If we now multiply this function by an arbitrary real number ε and form the sum

$$y + \delta y = y(x) + \varepsilon g(x), \tag{8.11}$$

we see that by keeping the function $g(x)$ fixed and varying the coefficient ε, we visit an infinite number of functions in a neighbourhood of $y(x)$, all of which satisfy the end conditions (8.9). We have suggestively denoted

$$\delta y = \varepsilon g(x), \tag{8.12}$$

---

[2] We have already used this terminology in Section 3.8.1.

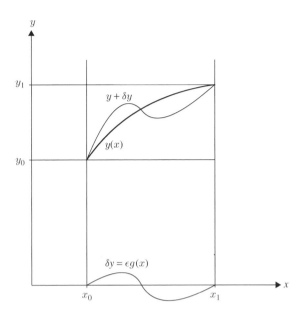

Figure 8.4   The neighbourhood of a function

to elicit the idea of a 'differential' of the 'independent variable' of a functional.[3] We call $\delta y(x)$ a *variation* of the function $y(x)$.

Let us keep in mind, however, that so far we have not visited the whole neighbourhood of our candidate function, but only those functions which differ from it by a function of the chosen shape of $g(x)$. We will come back to this point later. Be that as it may, if our functional J is to attain a minimum value, say, when compared to the value it attains for *all* the functions in the neighbourhood, then its value must certainly be smaller than the value attained for every function in the restricted family that we are discussing. And what is the value of J when $y(x)$ is varied to $y + \delta y$? If ε is small (after all, we need to explore just a small neighbourhood) the integrand in equation (8.8) will change by a small amount, due to the assumed continuity of all the quantities at play.

Since F is a differentiable function of its three variables, we can estimate its increment by calculating its differential. Observe that the independent variable x has not been touched at all. What has been affected is y, which has been changed, at each point x, by an increment equal to $\delta y(x) = \varepsilon g(x)$. At the same time, the derivative y' has also been changed, as we can appreciate from Figure 8.4, by an increment equal to $\delta y' = \varepsilon g'(x)$. It follows that, to a first-order approximation, the value of the integrand has changed by

$$ dF = \frac{\partial F}{\partial y}\, \varepsilon g(x) + \frac{\delta F}{\delta y'}\, \varepsilon g'(x). \qquad (8.13) $$

---

[3] Contrary to what some may think, a good notation is essential in mathematics. An impressive historical example is the great success of Leibniz's approach to calculus, which is due, in part at least, to his superior notation as compared to Newton's. Another example is provided by the extraordinary conciseness of modern differential geometry, in which the symbols almost suggest the correct results. We have hinted at an example of this kind in Section 5.3.6 in relation to Stokes' theorem. In real estate parlance, we may say that there are three important things in mathematics, and they are: 'notation, notation, notation'.

This means that the integral J has been changed by the amount

$$\delta J = \int_{x_0}^{x_1} \left(\frac{\partial F}{\partial y} \varepsilon g(x) + \frac{\partial F}{\partial y'} \varepsilon g'(x) \right) dx, \tag{8.14}$$

where the derivatives are evaluated at the values of y and y' corresponding to the chosen candidate function. Again we are using the suggestive notation δJ to denote the (first) *variation of the functional J*. It is the analogue of the differential of the dependent variable in calculus.

When we look at equation (8.14), we immediately realize that, as long as we keep the function $g(x)$ unchanged, the value of the integral depends only on the choice of the coefficient ε. Therefore, under the stated condition, the variation is an ordinary differential of a function of a single variable. More explicitly, this function is

$$J_g(\varepsilon) = \int_{x_0}^{x_1} F(x, y + \varepsilon g, y' + \varepsilon g') dx. \tag{8.15}$$

But, for a function of a single variable to attain a (relative) minimum value, it is necessary that its derivative (i.e., the ratio $\frac{dJ_g}{d\varepsilon}$ at $\varepsilon = 0$) vanish. We obtain, therefore, as a necessary condition of minimum the following expression:

$$\int_{x_0}^{x_1} \left(\frac{\partial F}{\partial y} g(x) + \frac{\partial F}{\partial y'} g'(x) \right) dx = 0. \tag{8.16}$$

This expression can be somewhat simplified by using the formula for integration by parts. It is the second term that we want to alter by exchanging derivatives. We have

$$\int_{x_0}^{x_1} \frac{\partial F}{\partial y'} g'(x) dx = \frac{\partial F}{\partial y'} g(x_1) - \frac{\partial F}{\partial y'} g(x_0) - \int_{x_0}^{x_1} \frac{d}{dx} \left(\frac{\partial F}{\partial y'} \right) g(x) dx. \tag{8.17}$$

Recalling that the function $g(x)$, by construction, vanishes at the end points, we can finally write

$$\int_{x_0}^{x_1} \frac{\partial F}{\partial y'} g'(x) dx = - \int_{x_0}^{x_1} \frac{d}{dx} \left(\frac{\partial F}{\partial y'} \right) g(x) dx. \tag{8.18}$$

Introducing this result in equation (8.16), we obtain the following necessary condition for a relative minimum of J:

$$\int_{x_0}^{x_1} \left(\frac{\partial F}{\partial y} - \frac{d}{dx} \left(\frac{\partial F}{\partial y'} \right) \right) g(x) dx = 0. \tag{8.19}$$

It is true that we have obtained this condition for a particular shape of the increment, $g(x)$. But now we realize that this shape is completely arbitrary and that we obtain exactly the same condition (8.19) regardless of the chosen function $g(x)$. This means that condition (8.19) has effectively become an *identity* to be necessarily satisfied *for all functions* $g(x)$. We will exploit this identity to obtain a condition much stronger than (8.19).

We claim that for (8.19) to hold identically for all shapes $g(x)$, the integrand must vanish. The proof of this fact (sometimes called the *fundamental lemma of the calculus of variations*) is quite straightforward. Imagine that at a certain interior point x_2 (Figure 8.5) the integrand did not vanish. We will show that this would lead to a contradiction. Indeed, since we can choose $g(x)$ of any shape we please (except for the fact that it must vanish at the end points), we can choose it as vanishing everywhere except in a very small interval containing the point x_2. But if the integrand of equation (8.19) is not zero at this point it must be either positive or negative. Let us say that it is positive. On the other hand, since this integrand is a continuous function, there must exist an interval around x_2 where the positive sign is preserved. Choosing $g(x)$ to be everywhere non-negative and to vanish identically outside this interval,[4] as shown in Figure 8.5, we conclude that the integral gives a positive result, a fact that contradicts the assumption that it vanishes identically for all $g(x)$. We have obtained, therefore, the following necessary condition for the candidate function to lead to a minimum of the functional J:

$$\frac{\partial F}{\partial y} - \frac{d}{dx}\left(\frac{\partial F}{\partial y'}\right) = 0. \tag{8.20}$$

Several remarks are in order regarding equation (8.20). The first is that this is a *differential equation*. In the classical case of finding the minimum of an ordinary function, we end up with *algebraic equations*. The second point to be made is that this is an *ordinary* differential equation (ODE) and not a *partial* differential equation. The partial derivatives appearing in equation (8.20) have only to do with manipulations that we are supposed to perform with the function F. Recall that the function F is not an unknown of the problem. In fact, it is actually given as the very definition of the problem. To arrive at the differential equation (8.20) for a particular F we need to carry out the operations indicated, and these operations happen to involve the evaluation of some partial derivatives. A third remark has to do with the total derivative with respect to x appearing in (8.20). This derivative appeared as a consequence of having used integration by parts. This means that, after performing the indicated operations with F, we now have to take derivatives with respect to x of the expression $\frac{\partial F}{\partial y'}$. In doing this, we can no longer ignore

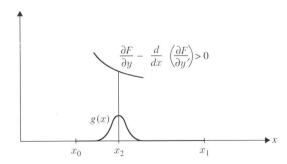

Figure 8.5   The fundamental lemma

---

[4] This can always be achieved with a function $g(x)$ as smooth as desired.

that y' is the derivative of y with respect to x. Finally, this differential equation is of the second order and, therefore, needs to be supplemented with two boundary conditions. But we already have those boundary conditions, since we have assumed that the values of the unknown function $y(x)$ are given at the end points.

The differential equation (8.20) is known as the *Euler–Lagrange equation* associated with the original variational problem. Going back to our ant seeking the shortest path to get to the grain of sugar, we see that, once it gets started in some direction, which may or may not be the right one, 'all' it has to do is to solve an ODE as it goes. This will guarantee that it is advancing along a so-called *geodesic line*, or line of shortest length. Whether or not this is the right geodesic is fated by the ant's choice of the initial direction of advance. Nevertheless, it is quite remarkable that a geodesic line, which is an essentially *global* statement, can be boiled down to a *local*, differential, statement.

Let us illustrate the solution of the Euler–Lagrange equation for the case of equation (8.7), which so far is the only explicit example we have. For the sake of sanity, we surely hope to obtain that the shortest path between two points in the plane is a straight line! In the case of equation (8.7), we have already evaluated the partial derivatives of F, shown in equation (8.10). The Euler–Lagrange equation becomes

$$\frac{\partial F}{\partial y} - \frac{d}{dx}\left(\frac{\partial F}{\partial y'}\right) = 0 - \frac{d}{dx}\left(\frac{y'}{\sqrt{1+y'^2}}\right) = -\frac{y''}{\left(1+y'^2\right)^{3/2}} = 0. \qquad (8.21)$$

The denominator of the final expression never vanishes, so the numerator must vanish, namely $y'' = 0$. But this is precisely the equation of a straight line.[5]

We have not discussed whether or not the necessary condition is also sufficient, nor have we distinguished between maximum, minimum and, perhaps, other stationary cases. These issues are beyond the scope of this elementary presentation. To avoid ambiguity, we will say that a functional J is stationary at $y = y(x)$ if its first variation δJ vanishes identically for all choices of the variation δy.

Exercise 8.3.1   A gauge transformation. Show that if the integrand in (8.8) is changed by adding a term of the form $ayy' + b$, where a and b are arbitrary constants, the Euler–Lagrange equation remains unchanged. More generally, show that the Euler–Lagrange equation of a given variational problem of the form (8.8) is indifferent to the addition to the integrand of a term of the form

$$\frac{\partial\phi}{\partial y}\,y' + \frac{\partial\phi}{\partial x}, \qquad (8.22)$$

where $\phi = \phi(x, y)$ is an arbitrary twice-differentiable function. Can you think how this idea might have been arrived at? This kind of degree of freedom afforded by a Lagrangian is called a *gauge transformation*.

Exercise 8.3.2   The brachystochrone. Show that the Euler–Lagrange equation associated with the brachystochrone problem, described in Section 8.3.1, can be written as

$$1 + y'^2 + 2yy'' = 0, \qquad (8.23)$$

---

[5] Notice, incidentally, that the final expression, denominator and all, is precisely the *curvature* of the candidate curve. The Euler–Lagrange equation tells us that, for this problem, the curve being sought has a vanishing curvature. This is an illustration of how the variational formulation of a problem is closely connected with its underlying physical meaning.

where $y = y(x)$ is measured downward from the point A of departure. Verify that the *cycloid* satisfies this differential equation. [Hint: by conservation of energy, the speed is proportional to $\sqrt{y}$.]

8.3.3 The Case of Several Unknown Functions

So far we have discussed the case of a functional that depends on one function $y = y(x)$ alone. A similar treatment shows that, when the functional depends on two or more functions of one variable, we obtain a *system* of simultaneous ODEs. For example, the functional

$$J[y(x), z(x)] = \int_{x_0}^{x_1} F(x, y, y', z, z') \, dx \tag{8.24}$$

gives rise to the following system of Euler–Lagrange equations:

$$\frac{\partial F}{\partial y} - \frac{d}{dx}\left(\frac{\partial F}{\partial y'}\right) = 0, \tag{8.25}$$

$$\frac{\partial F}{\partial z} - \frac{d}{dx}\left(\frac{\partial F}{\partial z'}\right) = 0. \tag{8.26}$$

An important example of the application of these equations can be found in the formulation of analytical mechanics. According to *Hamilton's principle* of Lagrangian mechanics, a mechanical system with a finite number of degrees of freedom, under the action of conservative forces, will follow a trajectory that renders the integral of the difference between the kinetic and potential energies stationary with respect to all neighbouring trajectories starting and ending at the same initial and final positions and times.[6] To apply this principle, we first need to choose a complete set of independent degrees of freedom for our system, and then express the kinetic and potential energies in terms of them. The rest is just a matter of applying the Euler–Lagrange equations. The advantage of this method as compared with Newton's (free-body diagram) method is that we do not need to deal with the forces of constraint (tensions in links, reactions at supports, etc.).

Example 8.3.3   A dynamical system. A cart of mass m moves along a horizontal frictionless rail while carrying the point of suspension of a pendulum of mass M and length a moving in the vertical plane x, y, as shown in Figure 8.6. As degrees of freedom of this system we choose the horizontal coordinate (x) along the rail and an angular coordinate (θ) for the pendulum. Our objective is to obtain the differential equations of motion of this system without resorting to free-body diagrams and without involving the forces of constraint or the support reactions. Denoting with a dot derivatives with respect to the independent time variable t, the kinetic energy T of this system can be calculated directly from the figure as

$$T = \frac{1}{2}m\dot{x}^2 + \frac{1}{2}M(\dot{x}^2 + a^2\dot{\theta}^2 + 2a\dot{x}\dot{\theta}\cos\theta), \tag{8.27}$$

---

[6] It is remarkable that such a *teleological* statement, namely a statement about the final purpose of a system, is in fact equivalent to a *causal* statement, namely that forces cause accelerations. This equivalence is precisely the content of the Euler–Lagrange equations, which show how the minimization of a functional (a teleological statement) is equivalent to the satisfaction of a differential equation (a causal statement).

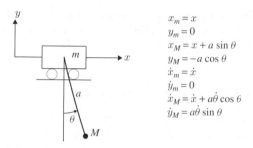

Figure 8.6 Example: a mechanical system

while the gravitational potential energy V with respect to the horizontal x-axis is clearly

$$V = -Mga \cos \theta, \tag{8.28}$$

where g denotes the acceleration due to gravity. According to Hamilton's principle, we need to render stationary (minimize, say) the integral of their difference, called the *Lagrangian* functional L, namely

$$L = \int_{t_0}^{t_1} \left(\frac{1}{2}m\dot{x}^2 + \frac{1}{2}M \left(\dot{x}^2 + a^2\dot{\theta}^2 + 2a\dot{x}\dot{\theta} \cos \theta \right) + Mga \cos \theta \right) dt \tag{8.29}$$

The fact that that neither the initial and final times nor the values attained thereat by the unknown functions are available is irrelevant to us, since all we are after is the derivation of the differential equations of motion of the system.[7] To this end, equation (8.29) has already provided us with our functional. All we have to do now is apply to it the Euler–Lagrange prescription of equations (8.25) and (8.26).[8] The result is the following system of coupled non-linear ODEs:

$$0 - \frac{d}{dt}\left((m + M)\dot{x} + a\dot{\theta} \cos \theta \right) = 0, \tag{8.30}$$

$$-Ma(\dot{\theta}\dot{x} + g) \sin \theta - \frac{d}{dt}\left(Ma^2\dot{\theta} + Ma\dot{x} \cos \theta \right) = 0. \tag{8.31}$$

We could further open the brackets to obtain two equations of the second order, but it is often convenient to leave the equations in this form. For example, in this way we can see that the first equation can be immediately integrated once, to give a constant of the motion (also called a *first integral*), namely

$$(m + M)\dot{x} + a\dot{\theta} \cos \theta = \text{constant}. \tag{8.32}$$

The physical meaning of this constant of the motion is the conservation of linear momentum in the x direction, since no external forces are applied to the system in this direction. The connection between conserved quantities and symmetries of the system (in this

---

[7] In fact, in many problems the variational method can be invoked to derive the correct form of the missing boundary or initial conditions. We will discuss this topic in the next section.
[8] Notice that the independent variable is now t, while the two dependent variables are x and θ.

case, the absence of an explicit dependence on the x variable) is of paramount importance in physics.

Equations (8.30) and (8.31) are non-linear and their analytical solution is not available, so one has to resort to numerical integration techniques, usually some variant of the finite-difference method. But the relative ease with which the equations were obtained is undoubtedly one of the main advantages of the variational formulation of mechanics.

Exercise 8.3.4   Write Mathematica® code, or equivalent, to solve numerically the system of ODEs (8.30) and (8.31), for any numerical values of the masses and the length of the pendulum and for any given initial conditions $x(0)$ and $\theta(0)$. Check that, *a posteriori*, the linear momentum (8.32) is indeed preserved. Can you find another physical constant of the motion?

8.3.4   Essential and Natural Boundary Conditions

In the formulation of the simplest problem of the calculus of variations, presented in Section 8.3.2, we systematically assumed that the functions under consideration as candidates for finding a stationary value of the functional must attain specified values at the end points, namely

$$y(x_0) = y_0, \quad y(x_1) = y_1. \tag{8.33}$$

These are also the boundary conditions to be imposed on the solution of the corresponding Euler–Lagrange equation. They are called *essential boundary conditions*.
    In the process of deriving the Euler–Lagrange equation, we have used the fact that, as a consequence of these conditions, the variations of the candidate functions must necessarily vanish at the end points, that is,

$$\delta y(x_0) = \delta y(x_1) = 0. \tag{8.34}$$

This fact played a crucial role in the derivation. We now ask what would happen if we were not to require the candidate functions to attain specified values at the end points of the interval of interest $[x_0, x_1]$. Specifically, we are asking for a function that renders the functional

$$J[y(x)] = \int_{x_0}^{x_1} F(x, y, y') \, dx \tag{8.35}$$

stationary with respect to *all* differentiable functions in a neighbourhood, regardless of what values they attain at the end points.
    To clarify the statement of the problem, let us consider an intermediate situation, where we specify the value y_0 that the candidate functions must attain at $x = x_0$ but leave the value at $x = x_1$ unspecified. Figure 8.7 is a schematic depiction of a few candidate functions. A particular example of this situation could be to find the shortest path in the plane that, starting at point $A = (x_0, y_0)$, terminates at some initially indeterminate point on the vertical line $x = x_1$. The intuitive answer to this question is clear: the horizontal line through A.

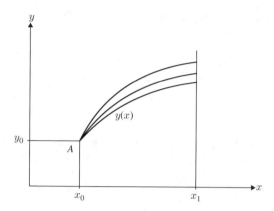

Figure 8.7   Mixed fixed-free problem

Remark 8.3.5   If we leave *both* end ordinates unspecified, then the problem of finding the shortest path has an infinite number of solutions, namely, every horizontal line.

If we apply the variation operator δ to equation (8.35), as we did before in greater detail, we obtain as a necessary condition for a stationary value the following equation:

$$\delta J = \int_{x_0}^{x_1} \left(\frac{\partial F}{\partial y} \delta y + \frac{\partial F}{\partial y'} \delta y' \right) dx = 0. \tag{8.36}$$

Integrating by parts, we obtain

$$\delta J = \int_{x_0}^{x_1} \left(\frac{\partial F}{\partial y} - \frac{d}{dx} \left(\frac{\partial F}{\partial y'} \right) \right) \delta y \, dx + \left[\frac{\partial F}{\partial y'} \delta y \right]_{x_0}^{x_1} = 0. \tag{8.37}$$

This equation must be satisfied identically *for all* variations $\delta y(x)$, regardless of whether they vanish at the ends of the interval or not. Since this is the case, we may start by considering all variations $\delta y(x)$ which happen to vanish at the end points. As a result, the bracketed term involving the end values vanishes and we are left with the integral term only. By the fundamental lemma, we conclude that

$$\frac{\partial F}{\partial y} - \frac{d}{dx} \left(\frac{\partial F}{\partial y'} \right) = 0. \tag{8.38}$$

We have obtained the somewhat unexpected result that the Euler–Lagrange equation still applies for the more general case of free boundaries. Having arrived at this conclusion, we return to equation (8.37), and deduce that

$$\left[\frac{\partial F}{\partial y'} \delta y \right]_{x_0}^{x_1} = 0. \tag{8.39}$$

But, since the boundaries are now free, we still have at our disposal all the variations $\delta y(x)$ that do not vanish at the end points. In particular, we can consider a variation such that

$$\delta y(x_0) \neq 0, \quad \delta y(x_1) = 0. \tag{8.40}$$

We conclude that for equation (8.39) not to be violated, we must have

$$\left[\frac{\partial F}{\partial y'} \right]_{x=x_0} = 0. \tag{8.41}$$

Analogously, by considering variations that vanish at x_0 but not at x_1, we obtain

$$\left[\frac{\partial F}{\partial y'} \right]_{x=x_1} = 0. \tag{8.42}$$

These last two equations provide extra conditions at the boundary of the domain. They are called *natural boundary conditions*. This name derives from the fact that, as distinct from the essential boundary conditions, the natural boundary conditions arise naturally (or automatically) from the variational formulation itself. If a function rendering the functional stationary is found, it will automatically satisfy the natural boundary condition at a boundary point at which the essential boundary condition has not been required. If, on the other hand, the desired function is to be found by means of solutions of the Euler–Lagrange equation, then the natural boundary conditions must be imposed at those points where the essential ones have not been prescribed.

In conclusion, when solving a variational problem by means of the Euler–Lagrange equation, the following alternative holds at each end:

$$\text{either } y \text{ is given or } \frac{\partial F}{\partial y'} = 0. \tag{8.43}$$

Exercise 8.3.6   The shortest path from a point to a line. What is the natural boundary condition at the free end for the problem (described at the beginning of this section) of finding the shortest path starting at a point A with coordinates (x_0, y_0) and ending on the vertical line $x = x_1$?

8.3.5   *The Case of Higher Derivatives*

In Section 8.3.3 we showed how Hamilton's principle can be used in conjunction with the calculus of variations to yield the equations of motion of dynamical systems with a finite number of degrees of freedom. For problems involving deformable elastic bodies, however, we need to deal with functionals that involve derivatives higher than the first. Consider, therefore, a functional of the form

$$J[y(x)] = \int_{x_0}^{x_1} F(x, y, y', y'') dx, \tag{8.44}$$

and apply the same ideas as we did for equation (8.8) or (8.35). Using the same scheme of notation, we can write

$$\delta J = \int_{x_0}^{x_1} \left(\frac{\partial F}{\partial y} \delta y + \frac{\partial F}{\partial y'} \delta y' + \frac{\partial F}{\partial y''} \delta y'' \right) dx = 0. \tag{8.45}$$

Integrating by parts (once for the second term and twice for the third), we eventually obtain

$$\int_{x_0}^{x_1} \left(\frac{\partial F}{\partial y} - \frac{d}{dx} \frac{\partial F}{\partial y'} + \frac{d^2}{dx^2} \frac{\partial F}{\partial y''} \right) \delta y dx + \left[\left(\frac{\partial F}{\partial y'} - \frac{d}{dx} \frac{\partial F}{\partial y''} \right) \delta y \right]_{x_0}^{x_1} + \left[\frac{\partial F}{\partial y''} \delta y' \right]_{x_0}^{x_1} = 0. \tag{8.46}$$

We now reason as follows. Since we are free to choose any variation δy, we will start by choosing variations such that they and their derivatives vanish at the ends, whether or not these conditions are imposed on the sought-after solution. By doing this, we clearly convince ourselves that the integral in equation (8.46) must vanish, since the other two expressions, evaluated as they are at the boundary points only, vanish automatically for the type of variation we are considering. But recall that equation (8.46) is not just an equation, but an identity. So, invoking the vanishing of the integral *for all* variations of this special type (which are quite arbitrary at the internal points of the interval), we can write

$$\frac{\partial F}{\partial y} - \frac{d}{dx} \frac{\partial F}{\partial y'} + \frac{d^2}{dx^2} \frac{\partial F}{\partial y''} = 0. \tag{8.47}$$

This is the generalized Euler–Lagrange equation for the case of functionals that involve second derivatives. If we exclude the latter, we recover the previous version.

Let us now go back to the remaining terms in (8.46), namely the expression

$$\left[\left(\frac{\partial F}{\partial y'} - \frac{d}{dx} \frac{\partial F}{\partial y''} \right) \delta y \right]_{x_0}^{x_1} + \left[\frac{\partial F}{\partial y''} \delta y' \right]_{x_0}^{x_1} = 0. \tag{8.48}$$

When fully expanded, this expression yields four terms, two at each end point. Assume, for example, that the value of $y(x)$ at $x = x_0$ has been specified as a known boundary condition. In this case, since the boundary condition cannot be violated, the variations must necessarily vanish at that point: $\delta y(x_0) = 0$. Similarly, if the slope has been specified as a boundary condition at that point, we must have that $\delta y'(x_0) = 0$. It follows that, whenever boundary conditions for the function or for its derivative are specified, the corresponding term in equation (8.48) vanishes automatically by construction. For this reason, these boundary conditions are called *essential boundary conditions* for the case of functionals involving first and second derivatives of the unknown function.

Let us assume, on the other hand, that one of these boundary conditions has not been specified. Consider, for example, a case in which the derivative $y'(x_1)$ has not been given as a boundary condition. We now impose a variation $\delta y(x)$ which vanishes at $x = x_0$ and at $x = x_1$ and whose derivative vanishes at $x = x_0$, but is different from zero at $x = x_1$. Recall that we are free to do so since (8.48) must be satisfied identically for all variations that satisfy the essential boundary conditions. We immediately conclude that for the identity not to be violated, the term $\frac{\partial F}{\partial y''}$ must vanish at $x = x_1$. The conclusion of this exercise is that at the end points we have the following four alternatives:

$$\text{at } x = x_0, \text{ either } y \text{ is given or } \frac{\partial F}{\partial y'} - \frac{d}{dx}\frac{\partial F}{\partial y''} = 0;$$

$$\text{at } x = x_0, \text{ either } y' \text{ is given or } \frac{\partial F}{\partial y''} = 0;$$

$$\text{at } x = x_1, \text{ either } y \text{ is given or } \frac{\partial F}{\partial y'} - \frac{d}{dx}\frac{\partial F}{\partial y''} = 0;$$ \hfill (8.49)

$$\text{at } x = x_1, \text{ either } y' \text{ is given or } \frac{\partial F}{\partial y''} = 0$$

The first choice in each alternative, as we have already mentioned, is called an essential boundary condition, while the second choice is a *natural boundary condition* for the variational problem involving up to second derivatives. In the case of a problem involving solely first derivatives of the unknown function, the second and fourth alternatives are not applicable and the first and third are simplified accordingly. In conclusion, the variational formulation is so complete that it even contains implicitly the physically meaningful boundary conditions that one should specify for any problem governed by a variational principle.

Example 8.3.7   Elastic beams. For the static deformations of an elastic structure, we can invoke the *principle of minimum potential energy* in elasticity, which we will discuss later in greater generality. To express it in a concrete example, we apply it to the case of the in-plane bending of a beam of a symmetric cross section in a regime of small deflections–the classical example covered in elementary courses on mechanics of materials. If we call the modulus of elasticity E, the moment of inertia I and the vertical deflection w, the elastic bending energy stored per unit length of the beam is given by $EIw''^2/2$, where priming denotes the derivative with respect to x, the axial coordinate of the beam. The potential energy of the external distributed load $q = q(x)$ per unit length is given by qw (assuming the load positive downwards and the deflection positive upwards). The functional of total potential energy U over the whole length L of the beam is, therefore, given by the expression

$$U = \int_0^L \left(\frac{1}{2} EIw''^2 + qw \right) dx. \hfill (8.50)$$

The principle of minimum (or, more precisely, stationary) potential energy states that the equilibrium configuration under the given load corresponds to a deflection $w = w(x)$ for which the first variation (δU) of this functional vanishes.[9] Applying our newly found Euler–Lagrange equation (8.47) to the energy functional (8.50), we obtain the following (fourth-order) differential equation:

$$EIw'''' + q = 0, \hfill (8.51)$$

where we have assumed a constant cross section and a homogeneous material. Equation (8.51), obtained from the principle of stationary potential energy by means of the calculus

---

[9] If the functional attains a relative *minimum*, the configuration is one of stable equilibrium.

of variations, is identical to the differential equation derived in courses on mechanics of materials by considerations of local equilibrium and geometry. As far as the boundary conditions are concerned, equation (8.49) tells us that at each end we either specify the deflection or the vanishing of the shear force, and we either specify the rotation or the vanishing of the bending moment.[10]

Remark 8.3.8 Concentrated loads. In deriving the Euler–Lagrange equation for the beam problem, we have assumed that the applied load is a distributed force, $q = q(x)$. In the case of concentrated forces or couples acting on the beam, whether at an interior point or on the boundary, the potential energy is easily modified by adding the product of the force (or the couple) by the displacement (or the rotation) at the point of application. If the concentrated load acts on the boundary, the natural boundary condition will reflect this fact by an expression which, instead of vanishing, equals the applied load. On the other hand, if the concentrated load is applied at an interior point, the degree of smoothness required for the application of the formula of integration by parts is lost at that point. The Euler–Lagrange equation is still valid at all other points of the domain, but it must be supplemented with appropriate 'jump conditions'. It is an important and attractive feature of the variational formulation that it covers all these cases under a single umbrella. This is reflected in the relative ease with which the numerical procedures based on the calculus of variations can handle singularities.

The following two exercises are taken from fields as diverse as optics and geophysics.

Exercise 8.3.9 Rays of light in a dense medium. The speed v of light in a dense medium is given by $v = c/n$, where c is the speed of light *in vacuo* and $n > 1$ is the index of refraction of the medium. Invoke Fermat's principle of least time to derive differential equations for a ray of light propagating in a medium with a smoothly varying index of refraction $n = n(x, y, z)$. You may use a parametric representation of the ray $x = x(s)$, $y = y(s)$, $z = z(s)$, where s measures length along the ray.

Exercise 8.3.10 Elastic ray trajectories in geophysics. In many geophysical applications, the speed of propagation of elastic waves is assumed to abide by the optical model. Assume that the medium consists of a sedimentary rock with horizontal layering, so that the index of refraction can be considered to be a function of the depth alone, namely $n = n(z)$, where z is a vertical axis (pointing downwards, say). For a ray completely contained in the vertical plane $y = 0$, we can express its equation in terms of a function $x = f(z)$. Write a differential equation for this function. Show that when the speed of propagation varies linearly with the depth z (i.e., when $v = a + bz$, where a and b are positive material constants) the rays are circular arcs. Find the height of the centre of the arc in terms of a and b.

8.3.6 *Variational Problems with More than One Independent Variable*

We turn our attention to variational principles involving unknown functions of more than one independent variable, which is the most common case in continuum mechanics.

---

[10] Recall that the second derivative of the displacement is proportional to the bending moment, and the third derivative is proportional to the shear force.

Consider the problem of finding a stationary value of the following functional:

$$J[w(x^1,\ldots,x^n)] = \int_C F\left(x^1,\ldots,x^n,w,w_{,1},\ldots,w_{,n}\right) dC, \qquad (8.52)$$

where C is a bounded domain in $\mathbb{R}^n$. For conciseness, we are using the notation

$$w_{,i} = \frac{\partial w}{\partial x^i}. \qquad (8.53)$$

Equating the first variation to zero, we obtain

$$\delta J = \int_C \left(\frac{\partial F}{\partial w}\delta w + \sum_{i=1}^n \frac{\partial F}{\partial w_{,i}}\delta w_{,i}\right) dC = 0. \qquad (8.54)$$

At this stage, in the case of a single independent variable, we had recourse to integration by parts. Instead, we now proceed to use Theorem 5.3.7 (the divergence theorem). To this effect, we note that

$$\sum_{i=1}^n \frac{\partial F}{\partial w_{,i}}\delta w_{,i} = \sum_{i=1}^n \left(\frac{\partial}{\partial x^i}\left(\frac{\partial F}{\partial w_{,i}}\delta w\right) - \frac{\partial}{\partial x^i}\left(\frac{\partial F}{\partial w_{,i}}\right)\delta w\right). \qquad (8.55)$$

Of the two expressions on the right-hand side of this equation, the first is clearly a divergence. By virtue of the divergence theorem, therefore, we can write

$$\int_C \sum_{i=1}^n \frac{\partial}{\partial x^i}\left(\frac{\partial F}{\partial w_{,i}}\delta w\right) dC = \int_{\partial C} \sum_{i=1}^n \frac{\partial F}{\partial w_{,i}}\delta w \, n_i \, da, \qquad (8.56)$$

where n_i are the components of the exterior unit normal to an element of area da on the boundary ∂C of the domain C.

Returning to the fundamental condition (8.54) and using (8.55) and (8.56), we arrive at the following identity:

$$\int_C \left(\frac{\partial F}{\partial w} - \sum_{i=1}^n \frac{\partial}{\partial x^i}\left(\frac{\partial F}{\partial w_{,i}}\right)\right) \delta w \, dC + \int_{\partial C} \sum_{i=1}^n \frac{\partial F}{\partial w_{,i}}\delta w \, n_i \, da = 0, \qquad (8.57)$$

to be satisfied identically for all differentiable functions $\delta w(x^1,\ldots,x^n)$. Following the by now familiar reasoning, this identity delivers the Euler–Lagrange equation

$$\frac{\partial F}{\partial w} - \sum_{i=1}^n \frac{\partial}{\partial x^i}\left(\frac{\partial F}{\partial w_{,i}}\right) = 0, \qquad (8.58)$$

and the alternative boundary conditions

$$\text{either } w \text{ is prescribed or } \sum_{i=1}^n \frac{\partial F}{\partial w_{,i}}n_i = 0 \text{ at each point of } \partial C. \qquad (8.59)$$

Notice that now the Euler–Lagrange equation is a PDE and the boundary condition alternative corresponds to the Dirichlet/Neumann choice.

Remark 8.3.11 Just as in the case of a single independent variable, the derivatives with respect to x^i in the second term of the Euler–Lagrange equation, coming as they do from the use of the divergence theorem, must take into consideration not only the explicit dependence of F on x^i, but also the fact that w and $w_{,j}$ depend on x^i. Explicitly, by the chain rule, we have

$$\frac{\partial}{\partial x^i}\left(\frac{\partial F}{\partial w_{,i}}\right) = \frac{\partial^2 F}{\partial x^i \partial w_{,i}} + \frac{\partial^2 F}{\partial w \partial w_{,i}} w_{,i} + \sum_{j=1}^{n} \frac{\partial^2 F}{\partial w_{,j} \partial w_{,i}} w_{,ji}. \tag{8.60}$$

Exercise 8.3.12 The membrane revisited. Verify that the membrane equation (8.5) is the Euler–Lagrange equation corresponding to a stationary value of the functional

$$U = \iint_D \left(\frac{N}{2}(w_x^2 + w_y^2) + pw\right) dx\, dy. \tag{8.61}$$

Verify, moreover, that the choice of the natural boundary condition in (8.59) corresponds in this case to the Neumann boundary condition for the membrane equation, as described in Section 6.6.

In most continuum mechanics applications, the functional of interest involves *several* functions (e.g., displacement field components) of several independent variables. As a result, the variational formulation leads to a *system* of simultaneous Euler–Lagrange PDEs. We now turn to the description of numerical methods, collectively known as *direct methods of the calculus of variations*, which allow us to tackle the approximate solution of these problems by searching directly for a function rendering the functional stationary, rather than by attempting to solve the associated system of PDEs (as the method of finite differences would do).

8.4 Rayleigh, Ritz and Galerkin

8.4.1 Introduction

Lord Rayleigh (1842–1919) is credited with the first use of a direct numerical method of the calculus of variations. To understand what we mean by 'direct method', imagine that you have been given a physical problem governed by a variational principle (e.g., the deflection of an elastic beam, governed by the principle of minimum potential energy). One clearly has the option of converting this problem into an equivalent one formulated in terms of one or more differential equations with suitable boundary conditions. Indeed, this is precisely what we have done so far, thus arriving at the so-called Euler–Lagrange equations associated with the problem.

One way to go about solving the original problem approximately is, therefore, to make use of a numerical method for the approximate solution of differential equations, such as the method of finite differences that we reviewed in Section 8.2.1. But there is a different way to approach the problem. One can forget all about the Euler–Lagrange equations and try to construct a numerical method by approximating (discretizing) *the original functional directly* and then using ordinary calculus to find a minimum of the discretized problem. What are the possible advantages of doing so?

The first advantage stems from the fact that the original functional always contains derivatives of order lower than the corresponding Euler–Lagrange equations. This fact is due to our having had to use integration by parts in the passage from the integral to the differential equations. Each time we integrate by parts, we increase the order of the derivatives of the unknown functions by one. Generally, then, the order of each of the Euler–Lagrange differential equations is double the order of the highest derivative appearing in the integrand of the functional. For example, if you consider the equation of the deflection of the elastic beam, the order is 4, while in the energy functional the highest derivative is of order 2. Clearly, lower derivatives are easier to approximate numerically than higher ones.

The second advantage of the direct methods stems from the fact that the variational principle may apply to cases in which derivatives do not even exist. We have already remarked on this point in Section 8.3.5 when dealing with the example of the beam and commenting on the possibility of having a concentrated force applied at a point. If one opts for the differential equation approach, obviously there is some mathematical trouble. For example, if you try to obtain the differential equations from the equilibrium of an infinitesimal slice, as is done in mechanics of materials courses, you have to carefully state that the equation is valid only if the slice does not carry a concentrated load. If it does, additional so-called 'jump conditions' must be enforced. These additional conditions demand special treatment. On the other hand, let us look at the variational principle itself. It says that the total potential energy is a minimum. We already have calculated the elastic potential energy stored in the body and the potential energy of the distributed load. The potential energy contributed by the concentrated force P is just Pw_c, where w_c denotes the deflection of point C. That is all we need to add, as long as we do not try to carry out the passage to the Euler–Lagrange equations. But that passage is not needed if we use a direct discretization of the integral problem. This is a truly remarkable advantage of the direct methods of the calculus of variations and of other associated methods that we will discuss. The finite-element method, of course, inherits this advantage.

Finally, among other advantages of the direct methods over the traditional approximation of differential equations, we must certainly count the fact that in the variational formulation the natural boundary conditions do not need to be explicitly stated, while only the essential ones, which are the 'easy' ones, do. Consider, for example, the case of a cantilever beam carrying a concentrated load at its free end. If we use a method based on the differential equation of equilibrium, then at the free end we must specify that the bending moment (second derivative of the deflection) vanishes and that the shear force (third derivative) is equal in magnitude to the applied concentrated load. Again, if we think of the finite-difference method, we need to do something special and quite awkward at that boundary. In a direct method of the calculus of variations, we need to do nothing at all, except for incorporating the potential energy of the tip load into the functional.

A final introductory comment is in order. You may say: yes, all this is very nice, but what if my particular problem cannot be formulated as a variational one? What happens, for instance, if the material of my beam is not elastic but viscoelastic or elasto-plastic? In that case there is energy dissipation and the principle of minimum potential energy no longer holds. This is quite true. But, as we shall see, there exist ways to mimic many of the features of the variational approach, by producing a so-called *weak formulation* of the original problem. An example of this idea is the *principle of virtual work* in structural mechanics. Instead of minimizing the potential energy, we replace this statement by the identity between the internal and the external virtual work, a weaker (but ultimately equally powerful) statement.

Lord Rayleigh (John William Strutt), was a British physicist who was awarded the Nobel Prize for his discovery of argon. But his more lasting contribution, as far as engineering is concerned, is contained in his famous, and surprisingly readable and modern, treatise on *The Theory of Sound*. In this work he conceived the idea of calculating the fundamental frequency of vibration of an elastic oscillating system directly from an energy consideration, without solving any differential equations. As pointed out by Timoshenko (1983), 'perhaps no other single mathematical tool has led to as much research in the strength of materials and the theory of elasticity'. And note that Timoshenko (1878–1972) was writing this in 1953, before he could have predicted the emergence of the finite-element method!

The basic idea of Lord Rayleigh was later elaborated by Walter Ritz (1878–1909), a brilliant Swiss physicist who died of tuberculosis at the tender age of 31. He was a classmate of Albert Einstein in Zurich, with whom he was involved in a lively, sometimes bitter, scientific dispute on the origin of the second law of thermodynamics and on the general question of reversibility of the equations of electrodynamics. A couple of weeks before Ritz's premature death, a joint article with Einstein appeared in print, in which each of them explains his position on those issues. Today, Ritz's name is mostly remembered because of his contribution to the generalization of Rayleigh's method.

8.4.2 The Rayleigh–Ritz Method

Briefly stated, the Rayleigh–Ritz method applied to a variational problem involves assuming that the solution of the problem can be sufficiently well approximated by a number of *shape functions*, combined additively with arbitrary coefficients. To fix ideas, let us apply this concept to the variational problem associated with a functional of the form (8.8), that is,

$$J[y(x)] = \int_{x_0}^{x_1} F(x, y, y')dx, \tag{8.62}$$

with the end conditions

$$y(x_0) = y_0, \quad y(x_1) = y_1. \tag{8.63}$$

To find an approximate solution of the problem $\delta J = 0$, we choose (using, perhaps, our physical intuition) a finite[11] number $(n + 1)$ of mutually independent shape functions,

$$\phi_0(x), \phi_1(x), \phi_2(x), \ldots, \phi_n(x), \tag{8.64}$$

which can be completely arbitrary, except that (i) $\phi_0(x)$ must satisfy the essential boundary conditions (8.63), and (ii) each of the remaining functions must vanish at the end points, that is,

$$\begin{cases} \phi_0(x_0) = y_0 & \phi_0(x_1) = y_1 \\ \phi_i(x_0) = 0 & \phi_i(x_1) = 0 \quad (i = 1, \ldots, n) \end{cases} \tag{8.65}$$

---

[11] In his original work, Lord Rayleigh considered just a single shape function, while Ritz's improvement consisted mainly of extending this idea to an arbitrary finite number of shape functions. The importance of Ritz's modification is that it provides for a means of refinement of the approximate solution and opens the door to the analysis of convergence considerations.

Under these conditions, we assume that the solution to the stationary value of the functional (8.62) can be sufficiently well approximated as a linear combination of the shape functions in the form:

$$y(x) \approx \phi_0(x) + A_1\phi_1(x) + A_2\phi_2(x) + \ldots + A_n\phi_n(x). \tag{8.66}$$

It is clear that, for any values of the unknown coefficients $A_1, \ldots, A_n$, the suggested approximation will satisfy the essential boundary conditions. If these conditions happen to stipulate the vanishing of $y(x)$ at the end points, we can eliminate the function ϕ_0 altogether.

Introducing this approximation in the functional, we obtain

$$J[y(x)] \approx \int_{x_0}^{x_1} F(x, \ \phi_0 + A_1\phi_1 + \ldots + A_n\phi_n, \ \phi_0' + A_1\phi_1' + \ldots + A_n\phi_n') \, dx. \tag{8.67}$$

The functional has thus become an *ordinary function* $\hat{J}$ of n variables $(A_1, \ldots, A_n)$, since everything else (namely, the shape functions) is held fixed. Consequently, we have approximated a problem for minimizing a functional by means of a problem for minimizing a function of a finite number of variables. We can also state this in the following way: instead of visiting the entire neighbourhood of a function, we limit our visit to a restricted neighbourhood, whose extent is controlled by n parameters. Instead of having an infinite number of control knobs, as required to capture the infinite flexibility of a variable function, we content ourselves with a finite number of controls, hoping that we have captured enough functions so as never to be too far from the sought-after solution. This is a simple idea, but a very ingenious and fruitful one. We effectively have obtained

$$J[y(x)] \approx \hat{J}(A_1, \ldots, A_n). \tag{8.68}$$

And how do we look for a stationary value of a function $\hat{J}$ of several variables? By equating to zero each and every partial derivative, namely, by the following system of equations:

$$\frac{\partial \hat{J}}{\partial A_1} = 0, \quad \frac{\partial \hat{J}}{\partial A_2} = 0, \quad \ldots, \quad \frac{\partial \hat{J}}{\partial A_n} = 0, \tag{8.69}$$

A solution of this system of *algebraic* equations provides us with the best we can do within the restricted family of shape functions. And this 'best' is usually quite good, provided our choice of shape functions has been cleverly made, or, failing that, provided we have chosen a sufficiently large number of shape functions.

If an essential boundary condition had been specified at one end only, then the shape functions would not have been subjected to any requirements at the other end. Put differently, the natural boundary conditions do not need any special treatment, since they are implicit in the functional itself. The method can be extended analogously to functionals involving higher derivatives as shown in the following exercise.

Exercise 8.4.1   Elastic beams. (a) Solve approximately for the deflections of the two linearly elastic beams of Figure 8.8 under the loadings shown, using a one-term approximation of the principle of minimum potential energy. Use equation (8.50) supplemented, if necessary, with the contribution of the concentrated load. The first beam is simply supported (no deflection) at both ends, while the second beam is clamped (no deflection and

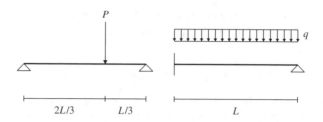

Figure 8.8   Data for Exercise 8.4.1

no rotation) at the left end and simply supported at the right end. Assume EI = constant.
(b) Improve on the solution by using a two-term approximation. Choose your own shape
functions. Compare your approximate results with the exact solutions, if available.

Exercise 8.4.2   The membrane once again. Construct a polynomial shape function
satisfying the boundary condition $w = 0$ over the boundary of a rectangular domain D
and apply the Rayleigh–Ritz method to obtain an approximate solution of the membrane
problem as formulated in Exercise 8.3.12. Compare with the results of Exercise 8.2.1.

8.4.3   The Methods of Weighted Residuals

It should be clear by now that the Rayleigh–Ritz direct method of the calculus of variations
offers definite advantages over traditional methods, such as finite differences, that operate
at the differential equation level. The question is now: can these advantages be imported
to problems governed by differential equations that are not necessarily obtainable as the
Euler–Lagrange equations of a variational problem? The partially positive answer to this
question is provided by the methods of *weighted residuals*, among which the *Galerkin
method* is the most widely used. Born in Polozk (situated in present-day Belarus), Boris
Grigorievich Galerkin (1871–1945) was a prominent Russian mathematician and engineer.
His seminal ideas have become crucial in the development of finite-element techniques.
From the mathematical point of view, the methods of weighted residuals operate on the
so-called *weak formulation* of a differential equation.

   Before attempting the more difficult problem of solving differential equations approximately by means of weighted-residuals methods, it will be useful to look at the simpler
problem of approximating a given function by these methods. In the weighted-residuals
methods the basic philosophy of the Rayleigh–Ritz method is preserved. And what is
this philosophy? It is contained in equation (8.66). It states that the function of interest,
$y = y(x)$, will be approximated, in some interval of interest $[x_0, x_1]$, by a *linear combination* of a finite number of shape functions, $\phi_0, \phi_1, \ldots, \phi_n$, with coefficients $A_1, A_2, \ldots, A_n$
to be determined by some criterion that guarantees a good fit. It is in the way these
coefficients are determined that the various methods of weighted residuals differ from
each other. The linear combinations of shape functions are often called *trial functions*,
or *candidate functions*, of which the shape functions constitute a *basis*. Recall that the
first shape function, $\phi_0(x)$, must satisfy the essential boundary conditions, that is, it must
attain the same values as the given function, $y(x)$, at the end points x_0 and x_1, while the
remaining shape functions must vanish at these points. This is done so that the approximation will automatically satisfy the essential boundary conditions, regardless of the values
of the coefficients.

The error or *residual* of a given approximation is given, point by point, by the function

$$r(x) = y(x) - \left(\phi_0(x) + A_1\phi_1(x) + \ldots + A_n\phi_n(x)\right). \tag{8.70}$$

Naturally, we want to render this residual as small as possible overall. To this end, we will demand that a number n of integrals over $[x_0, x_1]$ of the residual function $r(x)$, weighted in some appropriate manner, vanish. Otherwise stated, we will demand that

$$\int_{x_0}^{x_1} r(x)\, W_i(x)\, dx = 0, \quad i = 1, \ldots, n, \tag{8.71}$$

where $W_1(x), W_2(x), \ldots, W_n(x)$ are independent *weight functions*. Notice that if we were to demand that equation (8.71) be satisfied for *all* possible weight functions, we would conclude, by the fundamental lemma of the calculus of variations, that the residual necessarily vanishes. From this point of view, we can appreciate how, by demanding that the integral (8.71) vanish for only a finite number of weight functions, we are defining a legitimate discretized approximation procedure.

Assuming that these weight functions have been chosen, we proceed to determine the values of the coefficients so as to satisfy equation (8.71) for each weight function. Introducing (8.70) into (8.71) and using the notation

$$f_i = \int_{x_0}^{x_1} \left(y(x) - \phi_0(x)\right) W_i(x)\, dx \tag{8.72}$$

and

$$K_{ij} = \int_{x_0}^{x_1} \phi_j(x)\, W_i(x)\, dx, \tag{8.73}$$

we obtain

$$\sum_{j=1}^{n} K_{ij}\, A_j = f_i, \quad i = 1, \ldots, n. \tag{8.74}$$

This is a system of n linear equations for the determination of the values of the n coefficients $A_1, \ldots, A_n$. It can also be conveniently written in matrix form as

$$[\mathbf{K}]\{\mathbf{A}\} = \{\mathbf{f}\}, \tag{8.75}$$

with an obvious notation.

There are many meaningful possibilities for choosing the weight functions. For example, one may want to favour the data in a certain part of the domain by assigning a higher weight to that part than to the rest. The method of Galerkin is characterized by choosing as weight functions the *same functions* adopted as shape functions for the approximation, namely

$$W_i(x) = \phi_i(x), \quad i = 1, \ldots, n. \tag{8.76}$$

There are several advantages in making this choice. One of them is that the matrix $[\mathbf{K}]$ becomes *symmetric*, thus requiring fewer calculations. Another advantage, as we shall soon see when dealing with differential equations, is that if the differential equation under consideration happens to be the Euler–Lagrange equation of a variational problem, the Galerkin method delivers exactly the same result as the Rayleigh–Ritz method!

Exercise 8.4.3   Minimizing the total quadratic error. In the notation of equation (8.70), we define the total quadratic error ε of an approximation as the value of the integral

$$\varepsilon = \int_{x_0}^{x_1} r^2 \, dx. \tag{8.77}$$

Show that the Galerkin coefficients are precisely those that minimize the total quadratic error.

Exercise 8.4.4   Approximate the function $y(x) = \sin\frac{\pi x}{2}$ in the interval $[0, 1]$ by means of the shape functions $\phi_0 = x$ and $\phi_i = x(x^i - 1)$, for $i = 1, 2, 3, \ldots, n$, using Galerkin's method. Check first that these functions satisfy the required conditions. Use at least $n = 3$. You may want to use numerical integration to calculate the entries in the matrices $[\mathbf{K}]$ and $\{\mathbf{f}\}$. Compare your result (which is, after all, a power series) with the Taylor expansion of the sine function around the origin. Are the coefficients of equal powers the same in both methods?

Exercise 8.4.5   Prove that if the original function $y = y(x)$ to be approximated happens to be of the form $y(x) = \phi_0(x) + B_1\phi_1(x) + \ldots + B_n\phi_n(x)$ for some constants $B_1, \ldots, B_n$, then the method of weighted residuals delivers, via equation (8.74), the coefficients $A_i = B_i, i = 1, \ldots, n$. Although quite straightforward, this result is conceptually very important.

8.4.4   Approximating Differential Equations by Galerkin's Method

The basic idea of the Galerkin method for finding approximate solutions of differential equations is the same as in the case of approximating functions that we have just discussed. Nevertheless, some extra care must be exercised so as not to fall into mathematical traps. Consider, for example, a single ordinary differential equation of the form

$$y'' - G(x, y, y') = 0, \tag{8.78}$$

for a function $y = y(x)$ in the interval $[x_0, x_1]$ with boundary conditions

$$y(x_0) = y_0 \tag{8.79}$$

and

$$y'(x_1) = g(y(x_1)). \tag{8.80}$$

The function G in the differential equation is, at this stage, arbitrary, so that the differential equation may be non-linear, but we have assumed that the differential equation can be

brought into the form (8.78), so that it is linear in the highest derivative. As far as
the boundary conditions are concerned, we have purposely mixed an essential boundary
condition at one end with a non-essential one at the other. In a second-order differential
equation we call a boundary condition 'essential' if it stipulates the value of the unknown
function. In this sense, the boundary condition (8.79) is of the essential type. On the other
hand, the boundary condition (8.80) is not essential, since it combines the value of the
first derivative with the value of the function itself by means of a formula symbolized
by some function g. For example, in the case of a bar under axial loading, an essential
boundary condition could be the full support of one end, while a non-essential one could
be the support of an end by means of an axial spring of given stiffness.

The formulation of a physical problem by means of differential equations and boundary
conditions, such as the system represented by equations (8.78)–(8.80), is known mathematically as a *strong formulation*. But, as we will see, this is one of those cases in which
strength is not necessarily a good thing. The reason for this assertion is that, for a solution
of this problem to make sense, we must assume that it is quite smooth (at least twice
differentiable). If you happen to find a solution that violates this condition even at a single
point, you will have to reject it, since you cannot verify that the differential equation is
satisfied at that point. Since many problems, both because of their physical meaning and
because of the need for nice numerical procedures to approximate their solution, may
benefit from a relaxation of such strong conditions of smoothness, we look now for a
way to weaken the formulation.

To this end, we do something very simple: we just multiply the differential equation
(8.78) by an arbitrary function, which we will suggestively call $\delta y(x)$. For the time being,
let us assume that this function is as smooth as desired. We will only demand that it vanish
at a boundary where an essential boundary condition has been imposed. In our example,
therefore, δy must vanish at $x = x_0$. Assume now that we have found a solution $y = y(x)$
of the strongly formulated problem (8.78)–(8.80). This means that when we plug this
solution into the left-hand side of equation (8.78), we obtain zero for every value of x in
the interval $[x_0, x_1]$. When multiplied by $\delta y(x)$, we still get zero. And we still get zero if
we integrate this result over the whole interval. So, we have

$$0 = \int_{x_0}^{x_1} \left(y'' - G(x, y, y') \right) \delta y \, dx$$

$$\tag{8.81}$$

$$= \int_{x_0}^{x_1} \left(-y' \, \delta y' - G(x, y, y') \, \delta y \right) dx + g(y(x_1)) \, \delta y(x_1),$$

where we have performed an integration by parts and, exploiting the fact that $y = y(x)$
has been assumed to be a solution of the problem described by equations (8.78)–(8.80),
we have used the boundary conditions (8.79) and (8.80). What we have found, in fact, is
that every solution of the original ('strong') problem, satisfies the equation

$$\int_{x_0}^{x_1} \left(-y' \, \delta y' - G(x, y, y') \, \delta y \right) dx + g(y(x_1)) \, \delta y(x_1) = 0 \tag{8.82}$$

identically for all 'variations' $\delta y(x)$ that respect the essential boundary conditions.

We will now show the converse, namely, that every *smooth enough* solution of the identity (8.82) is also a solution of the strong formulation (8.78)–(8.80). All we have to do is retrace our steps and use the fundamental lemma with which we are already familiar. Let us do so. We start from equation (8.82) and integrate by parts:

$$
\begin{aligned}
0 &= \int_{x_0}^{x_1} \left(-y'\,\delta y' - G(x,y,y')\,\delta y\right)\,dx + g(y(x_1))\,\delta y(x_1) \\[2mm]
&= \int_{x_0}^{x_1} \left(-(y'\delta y)' + y''\delta y - G(x,y,y')\delta y\right)dx + g(y(x_1))\delta y(x_1) \\[2mm]
&= \int_{x_0}^{x_1} \left(y'' - G(x,y,y')\right)\delta y\,dx + g(y(x_1))\delta y(x_1) - \left[y'\delta y\right]_{x_0}^{x_1} \\[2mm]
&= \int_{x_0}^{x_1} \left(y'' - G(x,y,y')\right)\delta y\,dx + \left[g(y(x_1)) - y'(x_1)\right]\delta y(x_1).
\end{aligned}
$$

(8.83)

But since this identity has been assumed to be true for all variations that respect the essential boundary condition, we can first take arbitrary variations that vanish at *both* ends and thereby conclude, by a direct application of the fundamental lemma (see Figure 8.5), that the integrand in the last expression of equation (8.83) must vanish. This integrand is nothing but the differential equation (8.78). Moreover, having ensured that the integral vanishes, we can now consider variations that do not vanish at the right end, and thereby conclude that the expression within square brackets must vanish as well. This expression is precisely the non-essential boundary condition (8.80). This completes the proof.

So, what have we gained? We have obtained an alternative equivalent formulation of our original problem with the following features: (i) it encompasses in a single expression both the differential equation and the boundary conditions; (ii) the highest order of differentiation in (8.82) is 1 (whereas in the original strong formulation it was 2), a feature that can prove to be very useful in numerical approximations; (iii) most importantly, the alternative formulation (8.82) makes sense even if the candidate solutions are not very smooth. This is true not only because the highest order of differentiation required is lower, but also because all we need to do with the functions involved is carry out integration, an operation for which certain types of discontinuities are permitted. In short, although the two formulations are equivalent in the realm of smooth functions, the second formulation is valid also for weaker requirements on the smoothness of the solution. It is for this reason that the formulation given by equation (8.82) is called the *weak formulation* of the original problem.

We have only presented the case of a single ODE of the second order. The concept of weak formulation can be extended to equations of higher order, and to systems of ODEs and PDEs. In some instances, such as in the case of the principle of virtual work, or D'Alembert's principle in dynamics, the statement of the physical problem is already in the weak form.

To use Galerkin's method for the solution of a differential equation, or system of equations, we must first convert the problem into a weak form. Once this is done, we consider the collection of candidate functions and introduce into it a discretization by means of equation (8.66). What this means is that we are using a finite number of shape functions as a basis for the candidate solutions of the problem. Next, we use a similar idea for the collection of variations (also called *test functions*), namely, we discretize this collection by assuming that every variation is a linear combination of the weight functions. In the Galerkin method, as we have seen, the weight functions are identified with the shape functions. We have, therefore,

$$\delta y(x) = \delta A_1\, \phi_1(x) + \delta A_2\, \phi_2(x) + \ldots + \delta A_n\, \phi_n(x) \tag{8.84}$$

and

$$\delta y'(x) = \delta A_1\, \phi_1'(x) + \delta A_2\, \phi_2'(x) + \ldots + \delta A_n\, \phi_n'(x). \tag{8.85}$$

Let us plug these two equations into the weak statement, equation (8.82). We obtain

$$0 = \int_{x_0}^{x_1} \left(-y'\, \delta y' - G(x,y,y')\, \delta y \right) dx + g(y(x_1))\, \delta y(x_1)$$

$$= \delta A_1 \int_{x_0}^{x_1} \left(-y'\, \phi_1' - G\, \phi_1 \right) dx + \ldots + \delta A_n \int_{x_0}^{x_1} \left(-y'\, \phi_n' - G\, \phi_n \right) dx$$

$$+ g(y(x_1))[\delta A_1\, \phi_1(x_1) + \ldots + \delta A_n\, \phi_n(x_1)] \tag{8.86}$$

But, since the coefficients δA_i, $i = 1, \ldots, n$, are mutually independent,[12] we can collect all the terms having the same coefficient δA_i into one expression and conclude that each of these expressions must vanish. We obtain, therefore, the following n equations:

$$\int_{x_0}^{x_1} \left(-y'\, \phi_i' - G(x,y,y')\, \phi_i \right) dx + g(y(x_1))\, \phi_i(x_1) = 0, \quad i = 1, \ldots, n. \tag{8.87}$$

Now we can use the discretization of the candidate functions themselves, as per equation (8.66), and conclude that (8.87) can be seen as a system of algebraic equations for the unknown coefficients A_i, $i = 1, \ldots, n$.

Example 8.4.6 Find an approximate solution $y(x)$ of the following differential equation and boundary conditions:

$$y'' - \sin^2 y + 1 = 0, \quad y'(0) = 1, \quad y(2.5) = 0, \tag{8.88}$$

in the interval $[0, 2.5]$ by Galerkin's method.

Solution. This is a non-linear differential equation, so it is quite unlikely that we will find a closed-form solution. Note that the argument of the sine function is not the independent variable but the unknown function itself. To apply Galerkin's method, we start by producing a weak formulation of the problem. Following the steps of the general

---

[12] The coefficients being independent of each other, we can set one of them to the value 1 and the rest to the value 0, for example.

derivation, we integrate by parts and then substitute the non-essential boundary condition. The result is

$$\int_0^{2.5} (-y'\delta y' - (\sin^2 y - 1)\delta y)dx - \delta y(0) = 0. \tag{8.89}$$

Next, we introduce a set of weight functions $W_i(x)$ (which, for Galerkin's method, will also be the shape functions $\phi_i(x)$) with the property that each of them vanishes at $x = 2.5$. A simple choice is

$$\phi_k(x) = (x - 2.5)^k. \tag{8.90}$$

Our system of weighted equations (8.87) then becomes

$$\int_{x_0}^{x_1} (-k\, y'(x - 2.5)^{k-1} - (\sin^2 y - 1)(x - 2.5)^k)dx - (-2.5)^k = 0, \quad k = 1, \dots, n. \tag{8.91}$$

To convert this into an algebraic system of equations, we need to: (i) truncate the basis function set (8.90); (ii) express the unknown function as a linear combination of the truncated basis. Note that in the general case we would also need to supply a function $\phi_0(x)$ to satisfy the essential boundary condition, but in this particular case this function can be chosen as zero, because so is the value of the essential boundary condition. For the purpose of this illustration, we will truncate the basis at $n = 2$, that is, we assume that

$$y(x) \approx A_1(x - 2.5) + A_2(x - 2.5)^2. \tag{8.92}$$

This is, admittedly, a very crude parabolic approximation, since it assigns a constant value to y''. The Galerkin procedure will, nevertheless, tell us what is the 'best' parabolic fit to the solution. We obtain the following system of two equations:

$$\int_0^{2.5} \{(-A_1 - 2A_2(x - 2.5))$$

$$- [\sin^2(A_1(x - 2.5) + A_2(x - 2.5)^2) - 1](x - 2.5)\, \} \, dx + 2.5 = 0, \tag{8.93}$$

$$\int_0^{2.5} \{(-A_1 - 2A_2(x - 2.5))2(x - 2.5)$$

$$- [\sin^2(A_1(x - 2.5) + A_2(x - 2.5)^2) - 1](x - 2.5)^2\, \} \, dx - 6.25 = 0.$$

This is a system of non-linear algebraic equations. To solve it numerically, we resort to numerical integration and to the Newton–Raphson method. The result obtained is

$$A_1 = -0.969\,164, \quad A_2 = -0.381\,942. \tag{8.94}$$

The approximate solution delivered by the Galerkin method is, therefore,

$$y(x) \approx -0.969\,164(x - 2.5) - 0.381\,942(x - 2.5)^2. \tag{8.95}$$

To get an intuitive idea of how good this approximate solution might be, we check the error in the non-essential boundary condition, as follows:

$$y'(0) - 1 \approx A_1 + 2A_2(0 - 2.5) - 1 = 0.940\,546 - 1 = -0.059\,454. \tag{8.96}$$

Since we do not have an exact solution of the original differential equation, it is not possible to appreciate the relative magnitude of this error. Notice that it does not make much sense to plug our approximate solution in the original differential equation and to check the residual. We are not expecting second derivatives to be well approximated by this procedure, but we certainly expect that the function itself is well approximated. Solving the original equation by the method of finite differences (with a very small increment), you should be able to check that the error of the Galerkin method is indeed very small, even with the two-term approximation that we have used. (The maximum absolute value of the 'exact' solution is about 0.6, and the maximum absolute value of the error is about 0.01.) This has been an unusually complicated exercise because of the high non-linearity of the original equation.

Exercise 8.4.7   Verify that all the computations carried out in the proposed solution of the previous example are correct. You may do the calculations by hand, writing your own computer code or resorting to mathematical packages (to carry out numerical integration and solutions of systems of non-linear algebraic equations). If possible, provide the next order approximation by truncating the set (8.85) at $n = 3$ or higher. By how much does the approximate solution change? Carry out the solution of the original differential equation by the finite-difference method and compare with the approximate solutions obtained by Galerkin's method.

8.5   The Finite-Element Idea

8.5.1   Introduction

The approximations provided by the methods of Rayleigh and Ritz and, more generally, by the weighted-residuals methods (such as Galerkin's) are amenable to a systematic refinement, as all well-formulated discretization methods should be. The way this refinement can be achieved, according to what we have studied so far, is by systematically increasing the number of basis (or shape) functions in the approximation. On the other hand, any of the examples we have solved is enough to convince us that very high-order approximations of this kind are computationally impractical. Moreover, higher-order polynomials, or trigonometric functions of higher and higher frequency, have an innate tendency to oscillate somewhat uncontrollably around the average values, a feature that may turn out to be a high price to pay for a more accurate solution. The finite-element method provides a viable alternative that combines all of the advantages of variational and/or weak formulations with an ingenious refinement method that does not require the adoption of, for example, higher-order polynomials. This objective is achieved by taking advantage of the weak form of the problem so as to relax the smoothness requirements of the approximation.

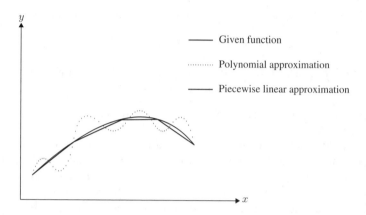

Figure 8.9 Two approximation techniques

To illustrate this concept, imagine that we have to achieve a 'decent' approximation of a function. We may think, for example, of the measured elongations of a tendon under the action of a force. One way to accomplish this approximation is by means of a number of functions affected by coefficients that are determined so as to fit a curve to the data. For example, one of the simplest techniques involves using a linear regression, whereby a straight line is made to fit the data so as to minimize the total quadratic error at a selected number of points. This procedure of least squares can be generalized for any number of basis functions. Clearly, to refine the approximation we need more and more functions, each of which extends over the whole domain of interest.

A different technique to achieve the same goal may be the following. We subdivide the domain of interest into small subdomains ('finite elements') and within each of them we use a simple approximation (e.g., a linear one). If we increase the number of subdivisions, without changing the simple nature of the approximation *within each subdivision*, we clearly improve the approximation. What have we lost? The answer is simple: 'smoothness'. We can recover some of the loss by increasing the order of the approximation within each subdomain (e.g., to a quadratic one). But we may not be interested in gaining smoothness if all we are looking for is a good approximation of the field alone (e.g., the elongation of the tendon as a function of the force). Figure 8.9 shows a schematic representation of the two techniques just described.

Of these two techniques, the first is clearly the analogue of the Rayleigh–Ritz method in that globally smooth functions (such as polynomials) are used to approximate the given functions. The admittedly not straightforward application of the second technique (involving only piecewise smooth functions, such as broken lines) to the solution of differential equations did not occur to anyone in the engineering community until the 1960s. One of the reasons for this apparent oversight is that, in the absence of fast and massive calculating devices, the solution of simultaneous algebraic equations was a daunting task. Not very long ago, many a consulting engineering office offered the visitor the frightening and depressing sight of an array of desks, at each of which there was an engineer slowly and laboriously solving systems of linear equations (arising, for example, from structural engineering or from circuit analysis and other such important applications) by a variety of approximate techniques.[13] All calculations were carried out by hand or by mechanical or even electronic (but, alas, non-programmable) devices.

The advent of the electronic computer and its rapid development into a commercial product changed all that. During the 1960s the aeronautical and aerospace industries in particular demanded more sophisticated techniques of analysis and design. Engineers responded to the challenge first with crude *ad hoc* techniques but soon with an astounding and ever more rigorous effort in which universities and industry collaborated. Thus the finite-element method was born[14] out of a generalization of techniques that were familiar to structural engineers since the turn of the twentieth century, techniques that were mainly applicable to framed structures (such as those used for skyscrapers). In particular, the so-called *stiffness-matrix method* showed the most promise of being amenable to elevation to this new level of consciousness. Slowly, researchers began to realize that they had stumbled upon something much larger than structural engineering, namely, a general technique for the solution of a vast number of problems that arise in all kinds of applications in engineering and physics. Mathematicians too began to be attracted by the exciting mathematical problems posed by the new technique and asked questions such as: Does the approximate solution converge to the exact one? If so, in what sense? Can the error be estimated *a priori*? Are there optimal subdivisions of a given domain into finite elements? What are the types of differential equations that are amenable to this treatment? They soon discovered that the method had already been used by one of the great mathematicians of the century, Richard Courant (1888–1972) as far back as 1943.[15]

8.5.2 *A Piecewise Linear Basis*

The FEM idea emerges, as we have just implied, when using either the Rayleigh–Ritz method or any method of weighted residuals, such as the Galerkin method, but allowing test functions that are not necessarily globally smooth. More specifically, the FEM takes advantage of this freedom by subdividing the domain of interest into small pieces, called finite elements, and defining a basis of test functions each of which vanishes in all but a small number of elements. This basic idea can be brought to a higher degree of precision and generality. In this section we will consider the case of a one-dimensional domain and try to exhibit most of the features of the general procedure.

---

[13] In structural engineering, one such technique for the solution of framed structures was a so-called relaxation technique known by the name of Hardy and Cross. To get started, one would assume that none of the nodes of the structure moved except one, and then one would begin to release the other nodes one by one and hope that the procedure would converge, which it usually did.

[14] The name 'finite-element method' was coined by Ray Clough in 1960. Ray Clough was a professor at the University of California (Berkeley). His work on finite elements was done in collaboration with Boeing Corporation. Earlier work along structural engineering lines had been carried out by John Argyris (1913–2004), a Greek engineer who later became professor at the University of Stuttgart, Germany.

[15] Born in Germany (in Lubliniec, today in Poland), Courant emigrated to the United States in 1934, with the rise of Nazism in Germany. In New York, he established the renowned Courant Institute. In a book published in 1922 (co-authored with Hurwitz) he had applied piecewise polynomials to some theoretical consideration on the calculus of variations. In 1943 he applied these ideas to solve a problem of torsion of a cylinder by dividing the cross section into triangles, within each of which a polynomial approximation was used. This work was completely forgotten until the 1960s, when it was rediscovered by the pioneers of the FEM.

Assume that the highest derivative appearing in the weak formulation of a problem is of the first order. If we look at our treatment in Section 8.4.4, we can appreciate that the original strong form of the problem may have been of the second order. So, we have already gained in simplicity, since our shape and test functions need only be once (rather than twice) differentiable. But there is more. In the weak formulation, we may have a finite number of discontinuities in the first derivatives, as long as the result is integrable. Even if you think of integration naïvely as calculation of the area under the graph of a function, it is clear that this graph may be a broken one, and still the area under it makes sense. The shape functions themselves, on the other hand, have to be continuous, since we want our approximation to be so. This is clearly displayed in Figure 8.9. This means that the simplest type of shape function that we will be willing to deal with for the approximation of a first-order weak problem is one made up of straight pieces, zigzag-like.

To generate these shape functions in a systematic way, we divide the domain of interest into subdomains, or finite elements. In the case of a one-dimensional problem, the interval of interest is partitioned into subintervals. These subintervals need not be equal in length. In two-dimensional situations, we can (approximately) cover the domain by means of triangles, and similarly for higher dimensions. But let us return to our one-dimensional example, in which the finite elements are just contiguous segments. The ends of these segments, shared at most by two contiguous elements, are called *nodes*. We can imagine that we have stretched a rubber band over the domain and that we have attached it by nailing it at the nodes only. If we now remove just one of the nails and erect a pole of, say, a unit length perpendicularly to the domain, we obtain a tent-like shape (Figure 8.10). This is the shape function associated with that node.

If we have an essential boundary condition, then we must discard the corresponding shape function. Thus, for example, if the boundary condition (8.79) is in force, we must discard the function ϕ_1, since shape functions must vanish at points where essential boundary conditions have been specified, as we already know.

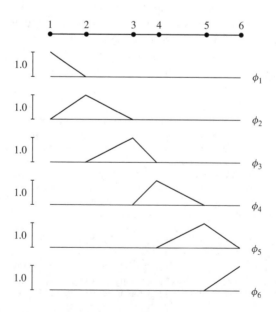

Figure 8.10   A piecewise linear basis

Notice that the basic shape functions that we have proposed enjoy the following property: each of these functions vanishes at all nodes, except one. More precisely, we have numbered the shape functions according to the numbering of the nodes. Accordingly, the shape function ϕ_i vanishes at all nodes, except at node i, where it attains the value 1.0. This feature allows us to give a nice interpretation to the coefficient A_i. Indeed, it follows directly from equation (8.66) and the property just outlined, that A_i is precisely the value that the approximated field minus the function ϕ_0 attains at the node number i. We will see later how to get rid of the function ϕ_0 altogether so that A_i will actually be the value of the field itself at node i.

When we look at equation (8.82), we immediately realize what a great simplification we have achieved by means of piecewise linear shape functions. For one thing, the derivatives of the shape functions are piecewise constant. For another, the integral for each weight function vanishes outside the corresponding tent. This feature will later be taken into consideration to create a procedure called *assembly*, whereby integrations are carried out in a standardized way first over each element, and later added together. All these integrations need not be evaluated exactly, since clearly it does not make sense to demand more precision of the integration than we have demanded of the shape functions themselves. A simple trapezoidal rule suffices. The accuracy of the FEM, as we are presenting it here, relies not so much on the quality of the shape functions or of the integration as on the number of subdivisions.

A subdivision into elements is known as a *mesh*. A mesh is *finer* than another if it is made up of more elements of smaller size. The process of increasing the fineness of a mesh is known as *mesh refinement*. From the mathematical point of view it is important to prove that, for a given class of problems, the systematic refinement of the mesh leads to convergence to the solution of the original problem. We will not deal with this difficult mathematical issue in this book.

Example 8.5.1   To illustrate the use of the piecewise linear shape functions, we will find an approximate solution of a very simple differential equation, whose exact solution is known. Consider the differential equation

$$y'' + y = 0, \tag{8.97}$$

with the boundary conditions

$$y(0) = 1, \quad y\left(\frac{\pi}{2}\right) = 1, \tag{8.98}$$

for a function $y = y(x)$ in the interval $[0, 1]$. The exact solution of this problem is

$$y(x) = \cos x + \sin x, \tag{8.99}$$

information that we will keep in mind for the purpose of checking the accuracy of the Galerkin approximation.

Solution. The weak form of this problem is obtained by the general method (see equation (8.82)) as

$$\int_0^{\pi/2} (-y'\delta y' + y\delta y)\,dx = 0. \tag{8.100}$$

Let us subdivide the interval of interest $[0, \pi/2]$ into five equal elements and construct the six corresponding shape functions, as shown in Figure 8.10. We discard the first and last shape functions, because of the essential boundary conditions given by equation (8.98). We choose our function $\phi_0(x)$ (see equation (8.66)) as

$$\phi_0(x) = 1, \tag{8.101}$$

so as to satisfy the two essential boundary conditions, and seek an approximate solution of the form

$$y(x) \approx \phi_0(x) + A_2\,\phi_2(x) + \ldots + A_5\,\phi_5(x). \tag{8.102}$$

The explicit form of the functions $\phi_2(x), \ldots, \phi_5(x)$ and of their derivatives is an easy exercise, since the functions are linear by parts. The test functions in Galerkin's approach are of the form

$$\delta y(x) \approx \delta A_2\,\phi_2(x) + \ldots + \delta A_5\,\phi_5(x), \tag{8.103}$$

where $\delta A_2, \ldots, \delta A_5$ are arbitrary independent coefficients. The resulting equations for the determination of the approximate solution (8.102) are given by equation (8.87) as

$$\int_0^{\pi/2} (-y'\phi_i' + y\phi_i)\,dx = 0, \quad i = 2, \ldots, 5. \tag{8.104}$$

As we have pointed out, however, for each value of i the integration extends over at most two contiguous elements. Moreover, the shape functions and their derivatives attain very simple expressions within that range. These equations, which you should be able to reproduce, are provided explicitly for the sake of clarity:

$$\int_0^{\pi/10} (-A_2\phi_2'\phi_2' + (1 + A_2\phi_2)\phi_2)\,dx +$$

$$\int_{\pi/10}^{\pi/5} (-(A_2\phi_2' + A_3\phi_3')\phi_2' + (1 + A_2\phi_2 + A_3\phi_3)\phi_2)\,dx = 0, \tag{8.105}$$

$$\int_{\pi/10}^{\pi/5} (-(A_2\phi_2' + A_3\phi_3')\phi_3' + (1 + A_2\phi_2 + A_3\phi_3)\phi_3)\,dx +$$

$$\int_{\pi/5}^{3\pi/10} (-(A_3\phi_3' + A_4\phi_4')\phi_3' + (1 + A_3\phi_3 + A_4\phi_4)\phi_3)\,dx = 0, \tag{8.106}$$

$$\int_{\pi/5}^{3\pi/10} (-(A_3\phi_3' + A_4\phi_4')\phi_4' + (1 + A_3\phi_3 + A_4\phi_4)\phi_4)\,dx +$$

$$\int_{3\pi/10}^{2\pi/5} (-(A_4\phi_4' + A_5\phi_5')\phi_4' + (1 + A_4\phi_4 + A_5\phi_5)\phi_4)\,dx = 0, \tag{8.107}$$

$$\int\limits_{3\pi/10}^{2\pi/5} (-(A_4\phi_2' + A_5\phi_5')\phi_5' + (1 + A_4\phi_4 + A_4\phi_5)\phi_5)\, dx +$$

$$\int\limits_{2\pi/5}^{\pi/2} (-A_5\phi_5'\phi_5' + (1 + A_5\phi_5)\phi_5)\, dx = 0. \tag{8.108}$$

We can clearly see that all the integrations that need to be carried out are contained in formulas of the following three types:

$$H_i = \int\limits_e \phi_i\, dx, \quad I_{ij} = \int\limits_e \phi_i\phi_j\, dx, \quad J_{ij} = \int\limits_e \phi_i'\phi_j'\, dx, \tag{8.109}$$

where e denotes a generic element. Numbering from the left, the element number i is contained between the nodes with numbers i and $i + 1$. Because in our example all elements are equal in length, the results are the same for all elements. The non-vanishing results of these integrations for element number i are:

$$H_i = \frac{\pi}{20}, \quad I_{ii} = \frac{\pi}{30}, \quad I_{i,\,i+1} = \frac{\pi}{60}, \quad J_{ii} = \frac{10}{\pi}, \quad J_{i,\,i+1} = -\frac{10}{\pi}. \tag{8.110}$$

Using these values, we obtain the following system of linear equations:

$$\begin{bmatrix} -6.156\,76 & 3.235\,46 & 0 & 0 \\ 3.235\,46 & -6.156\,76 & 3.235\,46 & 0 \\ 0 & 3.235\,46 & -6.156\,76 & 3.235\,46 \\ 0 & 0 & 3.235\,46 & -6.156\,76 \end{bmatrix} \begin{Bmatrix} A_2 \\ A_3 \\ A_4 \\ A_5 \end{Bmatrix} = \begin{Bmatrix} -0.314\,159 \\ -0.314\,159 \\ -0.314\,159 \\ -0.314\,159 \end{Bmatrix}, \tag{8.111}$$

whose solution is

$$\begin{Bmatrix} A_2 \\ A_3 \\ A_4 \\ A_5 \end{Bmatrix} = \begin{Bmatrix} 0.257\,292 \\ 0.392\,502 \\ 0.392\,502 \\ 0.257\,292 \end{Bmatrix}. \tag{8.112}$$

The values of the approximate solution at the $x = \pi/10$ and $x = \pi/5$ are, therefore, 1.257 292 and 1.392 502, respectively. The exact values at those points are 1.260 074 and 1.396 802, respectively. The approximate solution consists of four straight lines joining the ordinates at the nodes, which we have obtained with remarkable accuracy. The objective of this crude example has been just to illustrate the use of piecewise linear shape functions as a viable technique for the approximate solution of differential equations.

Exercise 8.5.2 Show that the differential equation (8.97) can be obtained as the Euler–Lagrange equation of the functional

$$J = \int\limits_0^{\pi/2} \left(\frac{1}{2}y'^2 - y \right) dx. \tag{8.113}$$

Use this fact to formulate the approximate solution of the problem by means of the Rayleigh–Ritz (rather than the Galerkin) method using the same mesh and the same shape functions as in the example.

Exercise 8.5.3   Solve the previous example, but changing the boundary condition at the right end to

$$y'\left(\frac{\pi}{2}\right) = -1. \tag{8.114}$$

To get a reasonable approximation, you will need to use a mesh finer than the one we used in the example.

8.5.3   Automating the Procedure

We have already mentioned that the FEM drew much of its conceptual and formal approach from the analysis of elastic framed structures. By the time the FEM was in its initial phases of development, there were two competing methods for the analysis of framed structures. These methods are dual to each other and go by the names of the *force* or *flexibility* method and the *displacement* or *stiffness method*. One of the manifestations of the duality alluded to is that the number of equations needed for the solution of a framed structure by one method behaves in the opposite way to the number required in the other method.[16] With the advent and rapid development of fast computing devices, the stiffness method won the upper hand, since it involves no decisions whatsoever in the choice of redundant quantities, unlike the flexibility method. Understandably, the stiffness method is more amenable to automatic computation.[17] There are also some theoretical reasons why the battle should have been won by the actual victor. These reasons became more evident when the FEM was being developed. Indeed, the natural nodal unknowns in the weak formulation of a problem in elasticity turn out to be the displacements, not the stresses. Moreover, even if that were not the case, the number of degrees of freedom involved in some of the most mundane problems solved by the FEM can easily run in the hundreds of thousands, so that an automated procedure is of the essence.

Our presentation of the FEM so far has been fully implemented only for the case of a single ODE of the second order. We will later show how this idea can be extended to more general problems in two or three dimensions. At any rate, the procedure commenced with a translation of the ODE and the boundary conditions into the language of the weak formulation, where only first derivatives appear in integral expressions. We should point out in passing that if the problem is formulated as a variational problem, whose Euler–Lagrange equation is the above mentioned ODE, as is sometimes possible, then the

---

[16] Technically, the closer a structure is to being statically determinate, the smaller the number of equations needed in the flexibility method of analysis, whereas the number required by the stiffness method is smaller the more statically indeterminate the structure is (i.e., the more supports it has).

[17] If you solve a simple truss by the flexibility method, since it is a statically determinate structure, there are no large systems of equations to solve. When solving the same problem by the stiffness method, the number of equations is proportional to the number of joints, which can be very large. Nevertheless, the computer power available is such that this is hardly an inconvenience when considering that for the computer every truss (or frame, for that matter) is solved by exactly the same completely automated procedure.

weak formulation is obtained directly as the first variation of the functional followed by an integration by parts. The second step in our procedure consisted of applying the Galerkin method to the resulting equations. If the problem can be formulated as the stationary value of an integral, this method coincides with the Rayleigh–Ritz method. Finally, for the procedure to be legitimately called a finite-element method, the domain of interest was subdivided into a number of finite (as opposed to infinitesimal) pieces, called finite elements, and then a number of shape functions equal to the number of free nodes was defined with a minimal degree of smoothness. These shape functions have the property of vanishing except at a small neighbourhood of a node, as shown in Figure 8.10.

What we have not shown is how this procedure can be made as routine and automatic as possible. There are two items to emphasize in this respect, the second being more important than the first. The first point attempts to rid us of the tyranny of the special shape function that we have called ϕ_0, a function whose only justification is to guarantee the satisfaction of the non-homogeneous essential boundary conditions by all possible choices of the unknown coefficients A_i. Clearly, the presence of this special function is bothersome for two reasons. One reason is that this function has to be chosen, and an intelligent choice can only be made by a human, not by a computer. But even forgetting this debatable reason, we see that, with its coefficient immutably set to 1, this function introduces an asymmetry in the treatment.

This asymmetry can be easily remedied as follows. You may recall that, with the presence of ϕ_0, we had to eliminate the shape functions associated with nodes on which essential boundary conditions were stipulated. For instance, in the last example, we had to eliminate the functions and ϕ_1 and ϕ_6. But let us allow these functions and their coefficients to stay in the game, on an equal footing with all the other coefficients, while getting rid of the function ϕ_0. Thus, we write instead of equation (8.66), the following approximation:

$$y(x) \approx A_1\phi_1(x) + A_2\phi_2(x) + \ldots + A_n\phi_n(x). \tag{8.115}$$

In order to satisfy the essential boundary conditions, however, we need to add extra relations between the coefficients. For example, if one such boundary condition is that the unknown function attains the value y_0 at the origin, we demand that

$$A_1\phi_1(0) + A_2\phi_2(0) + \ldots + A_n\phi_n(0) = y_0. \tag{8.116}$$

If you recall the type of shape functions we are using in the FEM (see Figure 8.10), you understand how easily this extra condition can be satisfied. Indeed, the function $\phi_1(x)$, which we would have eliminated in the previous treatment, is the only function that does not vanish at the left end. By keeping it, we restore the desired symmetry, in the sense that all nodes are treated equally, without causing any damage to the system of equations, except that a simple equation such as (8.116) needs to be added to it. As we have just mentioned, in our example this equation simply establishes that the new coefficient is equal to the desired boundary value. In our example, two such equations need to be added, one for A_1 and another for A_6. This is a very small price to pay for having a symmetric, automated treatment of all nodes.

The second, and more important, item to be considered towards a full automation of the procedure consists of looking at the integral of the weak formulation as made up of a sum of integrals, one over each element. This is certainly permitted, since integration is

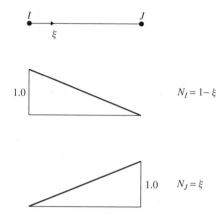

Figure 8.11 Linear element shape functions

a linear operation with respect to subdivisions of the domain of integration. So, instead of looking at *what each weight function contributes* to the total expression, we look at *what each element contributes* to it. This slight change of perspective constitutes a giant step towards automation. Why? If you look at what happens within each element in terms of how the unknown function is interpolated within it, you immediately realize, always looking at our example, that what happens is controlled by two and just two parameters. And these parameters, for the element number i (counting from the left end), are precisely the coefficients A_i and A_{i+1}. The meaning of these coefficients is clear: they are the values that the unknown field $y(x)$ attains at the nodes of this element in the approximate solution to the problem.

How do these two numbers control the field within the element? Given the shape functions that we have chosen, it is clear that, as far as this particular element is concerned, A_i governs a descending straight line and A_{i+1} an ascending one, as shown in Figure 8.11. These functions are called *element shape functions* and are usually denoted by $N_I(\xi)$ and $N_J(\xi)$. The reason for using the variable name ξ, rather than x, is that x is a *global coordinate* for the whole domain, whereas within each element we may prefer to use a *local coordinate*, with its origin within the element. We may also want to render the local coordinate non-dimensional, so as to make the result of the calculations independent of the size of the element. For instance, we may choose ξ as 0 at the left end (I) and as 1 at the right end (J) of the element. Notice that from this perspective, all elements look exactly the same and, therefore, their contributions will be identical (or, perhaps, if the element sizes are different, they will be proportional to each other).

The interpolated field within the element is given by

$$y(\xi) = A_I \, N_I(\xi) + A_J \, N_J(\xi). \tag{8.117}$$

We want to assess the contribution of each particular element to the weak form of the problem, which in our previous example is given by equation (8.100), that is,

$$\int_0^{\pi/2} (-y'\delta y' + y\delta y) \, dx = 0. \tag{8.118}$$

This integral can be broken up into elemental contributions:

$$\int_0^{\pi/2} (-y'\delta y' + y\delta y)\, dx = \sum_e \int_e (-y'\delta y' + y\delta y)\, dx = 0, \qquad (8.119)$$

where e denotes a generic element. The summation extends over all elements into which the domain has been subdivided. If we concentrate our attention on one particular element and make use of equation (8.117), we obtain

$$\int_e (-y'\delta y' + y\delta y)\, dx = \int_0^1 \left[-\left(A_I \frac{dN_I}{d\xi} + A_J \frac{dN_J}{d\xi} \right) \left(\delta A_I \frac{dN_I}{d\xi} + \delta A_J \frac{dN_J}{d\xi} \right) \left(\frac{d\xi}{dx} \right)^2 \right.$$

$$\left. + \left(A_I N_I + A_J N_J \right) \left(\delta A_I N_I + \delta A_J N_J \right) \right] \frac{dx}{d\xi}\, d\xi. \qquad (8.120)$$

The derivative $\frac{dx}{d\xi}$ and its inverse appear only because we have decided to use a non-dimensional coordinate along the element. In this example, if we call L_e the length of the element, we have that

$$\frac{dx}{d\xi} = L_e. \qquad (8.121)$$

Notice that by choosing a non-dimensional coordinate with its origin at the left end of each element, the range of integration is $[0, 1]$ for all elements. We can now calculate, once and for all, the following integrals:

$$\int_0^1 N_I N_I\, d\xi = \frac{1}{3}, \quad \int_0^1 N_I N_J\, d\xi = \frac{1}{6}, \quad \int_0^1 N_J N_J\, d\xi = \frac{1}{3}, \qquad (8.122)$$

and

$$\int_0^1 \frac{dN_I}{d\xi} \frac{dN_I}{d\xi}\, d\xi = 1, \quad \int_0^1 \frac{dN_I}{d\xi} \frac{dN_J}{d\xi}\, d\xi = -1, \quad \int_0^1 \frac{dN_J}{d\xi} \frac{dN_J}{d\xi}\, d\xi = 1 \qquad (8.123)$$

where we have used the formulas of Figure 8.11. Plugging these values back into equation (8.120), we obtain

$$\int_e (-y'\delta y' + y\delta y)\, dx = \left\{ [-A_I \delta A_I + A_I \delta A_J + A_J \delta A_I - A_J \delta A_J] \frac{1}{L_e^2} \right.$$

$$\left. + \left[\frac{1}{3} A_I \delta A_I + \frac{1}{6} A_I \delta A_J + \frac{1}{6} A_J \delta A_I + \frac{1}{3} \delta A_J \delta A_J \right] \right\} L_e, \qquad (8.124)$$

which can also be written in matrix notation as

$$\int_e (-y'\delta y' + y\delta y)\, dx = \langle\, \delta A_I \;\; \delta A_J \,\rangle \begin{bmatrix} -\frac{1}{L_e} + \frac{L_e}{3} & \frac{1}{L_e} + \frac{L_e}{6} \\[2mm] \frac{1}{L_e} + \frac{L_e}{6} & -\frac{1}{L_e} + \frac{L_e}{3} \end{bmatrix} \begin{Bmatrix} A_I \\ A_J \end{Bmatrix}. \qquad (8.125)$$

If you are wondering why is it that the units do not seem to match, the reason is that in our differential equation (8.97) we used inconsistent units.

Observe that, in this example, the contributions of all the elements are given by the same expression. This constitutes a significant simplification both in the understanding and in the coding of the finite-element procedure. All we need to do now is add together all the contributions of the individual elements and we are done. This addition, however, involves a logistic subtlety because, for each element, the meanings of the indices I and J, when seen from the vantage point of the whole domain, are different. For example, for the first element counting from the left we have $I = 1$ and $J = 2$; for the second element, $I = 2$ and $J = 3$, and so on. Expressed differently, there exists a mapping between the *local numbering* (at the element level) and the *global numbering* (at the whole domain level). But this is the kind of operation that computers can do well. Indeed, when we define for the computer the subdivision into elements (and even this so-called *meshing* is handled automatically these days by most finite-element codes), the computer stores this topological information for later use ('such and such element connects between such and such nodes').

Before proceeding to show how the addition of the individual element contributions is achieved, it is worth pointing out that equation (8.124) could be expressed as the matrix equation (8.125) only because our differential equation in this example is linear. The result of this fortunate circumstance is that the integral of the weak formulation turns out to be linear both in the coefficients of the candidate function and in the coefficients of the test functions. If you were to apply this procedure to a problem such as that described by equation (8.88), which is non-linear, then the linearity in the coefficients would be lost. But the linearity in the coefficient variations is never lost. What this means is that the integral in equation (8.82) will be always approximated (after substitution of the approximations for the function and its variation) by an expression of the form

$$\int (\ldots)\, dx = [\ldots]_1\, \delta A_1 + [\ldots]_2\, \delta A_2 + [\ldots]_3\, \delta A_3 + \ldots + [\ldots]_n\, \delta A_n. \quad (8.126)$$

In this expression, the contents of the square brackets are functions of the unknown coefficients A_i only. If the original differential equation happens to be linear, then these expressions are also linear. Otherwise, they will be non-linear. And what are the algebraic equations that we need to solve in order to obtain the values of the unknown coefficients A_i? If you go back and glance at equation (8.88), you will notice that the only thing we have left out is the consideration of the non-essential boundary condition. But the expression for it is clearly also linear in the coefficients. So, assuming that we have consolidated the whole of equation (8.88) into an approximation of the form (8.126), we conclude that the algebraic equations to be solved consist precisely of the vanishing of each of the expressions within the square brackets in equation (8.126). This conclusion clearly follows from the mutual independence and arbitrariness of the variations δA_i.

Let us consider the peculiar form of the shape functions in the FEM. Recall that there is one such function associated with each node, and that the function associated with a specific node is non-zero *only on all those elements that converge at that node*. In other words, if an element does not own that node, then it does not contribute at all to the corresponding coefficient within the square brackets. What this means is that, when calculating the value of one specific expression within brackets, we only need to look at the contributions of the elements that converge at the corresponding node. These are usually very few in number. For example, in our one-dimensional situation there are at most two elements converging at any specific node! Let us consider, for example, node number 3. It involves just the second and third elements (counting from the left). And how much does element number 2 contribute to the corresponding (third) bracketed expression in equation (8.126)? We have just seen (e.g., in equation (8.124)) that each element contributes to the total integral an expression of the form

$$\int_e \{\ldots\} \, d\xi = [\ldots]_{e,\,I} \, \delta A_I + [\ldots]_{e,\,J} \, \delta A_J. \tag{8.127}$$

Element number 2 has node number 3 as its J node (namely, the one to the right), while element number 3 has it as its I node. We conclude, therefore, that the total value of the coefficient of δA_3 in equation (8.126) is given by the sum of just two terms,

$$[\ldots]_3 = [\ldots]_{2,\,J} + [\ldots]_{3,\,J}. \tag{8.128}$$

It is that simple. In two- and three-dimensional problems, of course, there are usually more than two elements converging at each node, but the idea is exactly the same. And this is precisely the type of procedure that is quite amenable to automation. This procedure is known as *assembly*.

In the case of a linear differential equation, for which the bracketed expressions can be expressed as a product of a constant square matrix times the vector of the coefficients A_i, the procedure of assembly is even easier and more elegant. Historically, within the context of the analysis of linear elastic structures, the FEM emerged in the minds of structural engineers as a glorified version of the stiffness method of analysis of linear elastic frames, for which the assembly procedure had been understood and codified in previous generations.

Exercise 8.5.4   Solve the example of equations (8.97) and (8.98) using the assembly procedure just explained. Use a division into five elements. Compare your results with the ones obtained before in equation (8.112). Note that now you will have six shape functions (rather than just four), but your procedure will be more template-like.

8.6   The FEM in Solid Mechanics

8.6.1   *The Principle of Virtual Work*

In Section 6.6 we summarized the formulation of an initial and boundary value problem in thermoelasticity using the strong formulation, namely, in terms of PDEs. For these equations of motion to be valid, we need at least twice-differentiable displacements. Concentrated forces are out of the question, and so are knife-edge supports. The main numerical technique for solving these equations with appropriate initial and boundary conditions is the method of finite differences.

Our objective in this section is to provide the weak formulation counterpart in a manner amenable to treatment by the direct methods of the calculus of variations and, in particular, by the FEM. For simplicity, we will confine ourselves to the purely mechanical problem, disregarding thermal effects. Figure 8.12 is a schematic representation of a typical boundary value problem in solid mechanics in an Eulerian setting (in terms of the Cauchy stress). The Lagrangian setting is similar (in terms of the Piola stress). We are identifying the referential and spatial coordinate systems.

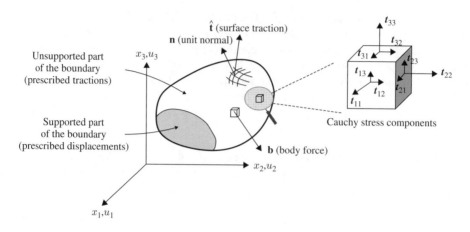

Figure 8.12 Boundary value problem in solid mechanics

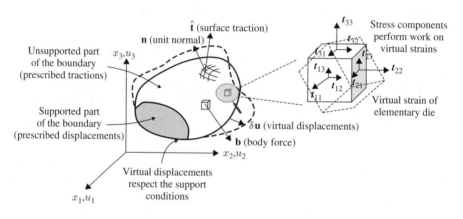

Figure 8.13 A virtual displacement field

Although the weak formulation can be arrived at by purely mathematical means, it is refreshing to know that in fact it has had a life of its own since antiquity. Archimedes of Siracuse (d. 212 BC) and Heron of Alexandria (first century) already understood the basic idea that in levers and pulleys what is gained in force is lost in displacement. This notion, formalized as the *principle of virtual work* or *principle of virtual power*, would much later be generalized to deformable bodies and shown to correspond to the weak formulation of the balance of linear momentum.

A *virtual displacement* is a small displacement field $\delta\mathbf{u}$ whose components vanish in that part of the boundary where the corresponding components of the displacement have been prescribed.[18] In physical terms, a virtual displacement field must be small and respect the support conditions. A convenient way to think of a virtual displacement is as some kind of *small perturbation* of a given configuration of the body, indicated by the thick broken line in Figure 8.13.

---

[18] Notice how this idea corresponds to the notion of test function.

When a virtual displacement field is superimposed on a given configuration, the external forces acting on the body perform a (small) amount of work EVW, called the *external virtual work*. By definition, the work of a force over a small displacement equals the dot (inner) product of the force times the displacement. Virtual displacements are assumed to take place instantaneously, so that everything is instantaneously frozen (forces, stresses), while the only change taking place is in the displacements themselves. Notice that, by the very definition of virtual displacement, the reactions at the supports perform no virtual work.

In the calculation of the external virtual work we must include the body forces, the surface tractions acting on the unsupported part of the boundary and, if we are dealing with dynamics, the inertia forces (according to *d'Alembert's principle*). More explicitly, the external virtual work is given by the expression

$$EVW = \int_v (\mathbf{b} - \rho \ddot{\mathbf{u}}) \cdot \delta\mathbf{u}\, dv + \int_{\partial_1 v} \hat{\mathbf{t}} \cdot \delta\mathbf{u}\, da. \tag{8.129}$$

In this expression, v represents the volume currently occupied by the body and $\partial_1 v$ is the unsupported part of the boundary, where the boundary conditions of traction are specified. It is sometimes useful to write equation (8.129) in matrix notation. For this purpose, we introduce the following matrices:

$$\{\mathbf{u}\} = \begin{Bmatrix} u_1 \\ u_2 \\ u_3 \end{Bmatrix}, \quad \{\delta\mathbf{u}\} = \begin{Bmatrix} \delta u_1 \\ \delta u_2 \\ \delta u_3 \end{Bmatrix}, \quad \{\mathbf{b}\} = \begin{Bmatrix} b_1 \\ b_2 \\ b_3 \end{Bmatrix}, \quad \{\hat{\mathbf{t}}\} = \begin{Bmatrix} \hat{t}_1 \\ \hat{t}_2 \\ \hat{t}_3 \end{Bmatrix}. \tag{8.130}$$

Accordingly, we can write equation (8.129) as

$$EVW = \int_v \{\delta\mathbf{u}\}^T (\{\mathbf{b}\} - \rho\{\ddot{\mathbf{u}}\})\, dv + \int_{\partial_1 v} \{\delta\mathbf{u}\}^T \{\hat{\mathbf{t}}\}\, da. \tag{8.131}$$

For a rigid body, the virtual displacements have to respect the rigidity condition. In this case, the principle of virtual work states that the actual configuration corresponds to the vanishing of the external virtual work. As we learn in high-school physics courses, this criterion correctly delivers the laws of equilibrium of the lever and other elementary mechanisms. In the case of a deformable body, on the other hand, the internal stresses consume some internal virtual work (IVW) as a result of the small changes in the strain components brought about by the virtual displacements. Physically, a small perturbation of the displacement field produces a small perturbation of the strain field.

To motivate the exact expression for the IVW, we recall that, when formulating the law of energy balance in Section 6.2.2, we obtained, in equation (6.57), the following expression for the power of the internal forces:

$$W_{\text{int}} = -\int_v \text{trace } (\mathbf{tD})\, dv. \tag{8.132}$$

We recall that, as already remarked in Section 6.2.2, the minus sign is indicative of the fact that, when defining the flux operator through Cauchy's tetrahedron argument, we made a choice of sign that, in the case of the stress, corresponds to the reaction, rather than the action, in the discrete analogue. We also recall that, if we multiply the tensor $\mathbf{D}$

(the symmetric part of the velocity gradient) by a time increment dt, we obtain the strain of the infinitesimal theory, given by equation (5.190), as expected from the assumption that the virtual displacements are small.[19] In view of these considerations, we arrive at the conclusion that the internal virtual work can be defined as

$$IVW = \int_v \frac{1}{2} t^{ij} \left(\delta u_{i,j} + \delta u_{j,i} \right) dv, \tag{8.133}$$

where the summation convention is in force.

Equation (8.133) can also be expressed in matrix notation. Recalling the symmetry of the Cauchy stress and introducing the six-dimensional column vectors

$$\{t\} = \begin{Bmatrix} t^{11} \\ t^{22} \\ t^{33} \\ t^{23} \\ t^{31} \\ t^{12} \end{Bmatrix}, \quad \{\delta \varepsilon\} = \begin{Bmatrix} \delta u_{1,1} \\ \delta u_{2,2} \\ \delta u_{3,3} \\ \frac{1}{2}(\delta u_{2,3} + \delta u_{3,2}) \\ \frac{1}{2}(\delta u_{3,1} + \delta u_{1,3}) \\ \frac{1}{2}(\delta u_{1,2} + \delta u_{2,1}) \end{Bmatrix}, \tag{8.134}$$

we can write

$$IVW = \int_v \{\delta \varepsilon\}^T \{t\} \, dv. \tag{8.135}$$

The principle of virtual work for deformable bodies states that the actual configuration of the system at each instant is such that the external virtual work is equal to the internal virtual work *identically for all virtual displacement fields*:

$$EVW \equiv IVW. \tag{8.136}$$

Recall that, by definition, the virtual displacements must respect the support conditions. The principle of virtual work states that the actual configuration of the system is completely characterized by the peculiar way it behaves under small perturbations that respect the support conditions. That peculiarity consists of the fact that the work of the external forces is exactly matched by that of the internal forces, regardless of the perturbation given. For all other configurations, one may find perturbations for which this is no longer true. This principle is a remarkable conceptual paradigm that differs radically from the Newtonian cause–effect paradigm ('forces cause acceleration').

Substituting equations (8.131) and (8.133) into (8.136), we obtain the explicit form of the principle of virtual work as

$$\int_v \frac{1}{2} t^{ij} \left(\delta u_{i,j} + \delta u_{j,i} \right) dv \equiv \int_v (b^i - \rho \ddot{u}^i) \delta u_i \, dv + \int_{\partial_1 v} \hat{t}^i \delta u_i \, da. \tag{8.137}$$

This statement has the form of a weak formulation. We still need to show that, to make physical sense, the principle of virtual work is equivalent to the law of balance of linear momentum whenever the displacement fields are at least twice differentiable. We start by

---

[19] See, in particular, Exercise 5.5.25.

noticing that, in fact, according to equation (6.56) (before even enforcing the symmetry of the Cauchy stress), equation (8.137) can be written as

$$\int_v t^{ij} \delta u_{i,\,j}\, dv \equiv \int_v (b^i - \rho \ddot{u}^i) \delta u_i\, dv + \int_{\partial_1 v} \hat{t}^i \delta u_i\, da. \tag{8.138}$$

Using the divergence theorem component by component, the left-hand side of equation (8.138) can be further developed as

$$\int_v t^{ij} \delta u_{i,\,j}\, dv = \int_v \left((t^{ij} \delta u_i)_{,\,j} - t^{ij}_{\,,\,j} \delta u_i \right) dv = \int_{\partial v} t^{ij} \delta u_i n_j\, da - \int_v t^{ij}_{\,,\,j} \delta u_i\, dv, \tag{8.139}$$

where the n_j are the components of the exterior unit normal to the boundary. Introducing this result into equation (8.138) yields:

$$\int_{\partial v} t^{ij} \delta u_i n_j\, da - \int_v t^{ij}_{\,,\,j} \delta u_i\, dv \equiv \int_v (b^i - \rho \ddot{u}^i) \delta u_i\, dv + \int_{\partial_1 v} \hat{t}^i \delta u_i\, da. \tag{8.140}$$

Since this is an identity to be satisfied *for all virtual displacement fields*, we can certainly choose arbitrary virtual displacements that happen to vanish over the whole boundary of the domain, thus eliminating the two surface integrals. The remaining two (volume) integrals must now be identical for all such virtual displacements. But, by the fundamental lemma of the calculus of variations, this is impossible unless the integrands are identical. We obtain therefore, as our first conclusion from identity (8.140), that

$$-t^{ij}_{\,,\,j} = b^i - \rho \ddot{u}^i. \tag{8.141}$$

Comparing this result with equation (6.42), we ascertain that indeed the weak formulation implies the validity of the equation of balance of linear momentum of the strong formulation (provided the divergence theorem can be applied, that is, provided we have enough smoothness). But there is more.

We can keep exploiting identity (8.140) by working on the remaining two terms, now that the volume integrals have cancelled each other out. Notice that, although the two surface integrals seem to extend over different domains, in fact the vanishing of the admissible virtual displacements over the supported part of the boundary renders the domains of integration the same. Since over the unsupported part we can prescribe arbitrary boundary virtual displacements, a new application of the fundamental lemma tells us that the integrands of the surface integrals must equal each other. That is,

$$t^{ij} n_j = \hat{t}^i. \tag{8.142}$$

This is precisely the boundary condition of traction!

We have, therefore, shown that the principle of virtual work, in the case of smooth enough fields, yields not only the PDEs of the strong formulation but also the (non-essential) boundary conditions. To prove that the converse is true, namely, that the strong formulation implies the virtual work identity, all we have to do is multiply equation (6.42) by δu_i (this is, of course, a dot product of vectors) and retrace our steps.

Exercise 8.6.1   Carry out the procedure just described to show that the strong formulation implies its weak counterpart.

Exercise 8.6.2 Lagrangian version of the principle of virtual work. Show that the Lagrangian version of the principle of virtual work is given by the identity:

$$\int_V T^{iJ} \delta F_{i,\,J} \; dV \equiv \int_V (B^i - \rho_0 \ddot{u}^i) \delta u_i \; dV + \int_{\partial_1 V} \hat{T}^i \delta u_i \; dA. \tag{8.143}$$

What we have achieved is a generalization of the purely mechanical equations of solid mechanics to a weak (integral) version that is amenable to treatment by the Galerkin method and, more generally, by the FEM. This being so, we can afford the luxury of accepting other (less smooth) contributions to the external virtual work. For example, if we have a concentrated force acting at a point of the body, its contribution to the external virtual work can be added to the right-hand side of (8.137) by taking the dot product of the force times the virtual displacement evaluated at the point of application. It is interesting to point out that the single *scalar* identity (8.137) yields the *vector* equation (6.42), which consists, in any coordinate system, of three scalar differential equations.

8.6.2 The Principle of Stationary Potential Energy

The preceding derivations are completely independent of the particular constitutive equation of the material, which means that the principle of virtual work is valid for any material response (elasticity, plasticity, etc.). If, on the other hand, the material happens to be hyperelastic, we may write, according to equation (6.123),

$$\mathbf{T} = \rho_0 \, \frac{\partial \psi(\mathbf{F})}{\partial \mathbf{F}}. \tag{8.144}$$

In this case, the internal virtual work is given by

$$IVW = \int_V T^{iJ} \delta F_{i,\,J} \; dV = \int_V \rho_0 \frac{\partial \psi}{\partial F_{i,\,J}} \, \delta F_{i,\,J} \; dV = \int_V \rho_0 \delta \psi \; dV = \delta \int_V \rho_0 \psi \; dV, \tag{8.145}$$

where the strain energy density ψ is a function of the deformation gradient alone. The internal virtual work can be regarded, therefore, as the variation of a functional, which we call the *elastic potential energy* of the body:

$$U_{\text{int}} = \int_V \rho_0 \psi \; dV. \tag{8.146}$$

It represents the recoverable mechanical energy stored by an elastic body in the form of interatomic potential energy.

Let us assume, moreover, that the external body forces and the surface tractions are *dead loads*. This means that these forces are independent of the deformation.[20] We define the *potential energy of the external forces* as

$$U_{\text{ext}} = - \int_V \mathbf{B} \cdot \mathbf{u} \; dV - \int_{\partial_1 V} \hat{\mathbf{T}} \cdot \mathbf{u} \; dA. \tag{8.147}$$

---

[20] External forces that depend on the deformation are sometimes called *follower forces*, a common example being the forces involved in the inflation of a balloon.

Moreover, we will consider problems of static equilibrium only, so that there are no inertia forces. The external virtual work can, therefore, be written as[21]

$$EVW = -\delta U_{\text{ext}}. \tag{8.148}$$

If we define the *total potential energy* as

$$U = U_{\text{int}} + U_{\text{ext}}, \tag{8.149}$$

the principle of virtual work is equivalent to the identity

$$\delta U \equiv 0, \tag{8.150}$$

for all virtual displacement fields. Thus, for a hyperelastic body in the presence of dead load only, an equilibrium configuration renders the total potential energy stationary (in the sense of the calculus of variations). This is the principle of stationary potential energy. When this principle is available, the Galerkin procedure reduces to the Rayleigh–Ritz method. We have already exploited these ideas for the particular case of elastic beams, in which we assumed the principle of stationary energy to be valid without proof. The availability of a true variational principle, as opposed to a mere weak formulation, is an important mathematical bonus. It facilitates proofs of convergence and stability.

8.7   Finite-Element Implementation

8.7.1   General Considerations

The essence of the computer implementation of the FEM resides in the coding of a routine to evaluate the internal virtual work within a single element. The other routines dance around it in a seemingly interminable *danse macabre* of calls and counter-calls. This is particularly true when the programmer, rather than being guided by efficiency in the coding process, attempts to follow the spirit and the letter of the Galerkin method. The result of this theoretically driven approach can be a surprisingly compact program.

A first choice the programmer must make is whether to use the Lagrangian or the Eulerian formulation.[22] In solid mechanics the preferred formulation is the Lagrangian one, in which the element mesh is attached to the material particles as opposed to being fixed in space. Nevertheless, within the Lagrangian setting, it is still possible to adhere to a strict Lagrangian formulation of the balance equations in a fixed reference configuration or to choose a variable reference configuration which is constantly updated in time or with load step, in such a way that the equations of balance correspond to successive regimes of small deformations. In the first option, the equations are fully non-linear, and this is the choice that we will make for our presentation. The discretized (algebraic) non-linear equations will be solved by the Newton–Raphson method, which essentially proceeds by a succession of linearizations of the mathematical equations, rather than by a succession of physical linearizations of the deformation itself.

Having made the choice of the strictest Lagrangian formulation, we must decide on the type of element geometry and of shape functions to be used. To be sure, in large-scale

---

[21] The scope of this expression can be extended to the case of special follower loads that derive from a potential function of the displacements.

[22] Another possibility is known as the arbitrary Lagrangian–Eulerian (ALE) method.

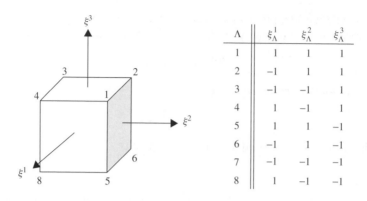

Figure 8.14 Ideal element

commercial codes there is a wide choice of element types. This *element library* is usually very well documented so that the user can decide which elements are most suitable for the particular kind of problem to be solved. Our presentation will be based on a single element type, namely, the eight-node isoparametric brick, which we will explain in greater detail below.

Large-scale multi-purpose programs offer also a wide range of analysis types (statics, dynamics, modal analysis, buckling, thermomechanical coupling, electromagnetic interactions, and so on) and of material behaviour (elasticity, plasticity, viscoelasticity, poroelasticity and many other types). Moreover, commercial codes are equipped with powerful graphic pre- and post-processors that facilitate the input of the geometric data and the interpretation of the final results. One of many other features offered by these programs is the automatic meshing algorithms that, with the help of a few parameters provided by the user, generate optimal element meshes.

No single individual can compete with the commercial codes, except in one particular aspect: the flexibility afforded by the intimate knowledge and the free accessibility of one's own source code. In spite of claims of extreme generality, it is often the case that what appears to require minor improvements to the existing capabilities of a program cannot actually be implemented. Many instances of this kind occur, particularly in biomechanical applications. The activation of striated muscle, for example, with its typical fibre directionality and with its possible coupling to control mechanisms, let alone chemical reactions, is most probably not available in even the most general and friendly multi-purpose commercial codes.

8.7.2 An Ideal Element

We imagine an ideal world, a copy of $\mathbb{R}^3$, which is home to a single perfect cube, as shown in Figure 8.14. The natural coordinates are denoted by ξ^α, $\alpha = 1, 2, 3$. The vertices of this *ideal element* occupy the eight points with coordinates that have unit absolute value. These vertices are, by definition, the *element nodes*.

Remark 8.7.1 The adjective 'ideal' is not meant to convey any notion of higher quality, but only to suggest that ideal elements are defined in a world of their own and only later mapped into the body itself, thus becoming 'real' elements. In principle, an ideal element can have more nodes than just the vertices. For instance, the midpoints of the sides or of the faces could be additional nodes. These extra nodes, if included in the definition of the element, render the element more flexible and improve the quality of the approximation of fields defined within the element.

To complete the characterization of this ideal element, in addition to its geometry and its nodes, we need to specify its *shape functions*, whose role is to provide for the interpolation of fields. The number of shape functions is equal to the number of nodes (eight in our case). We will denote the shape functions by

$$N^\Lambda = N^\Lambda(\xi^1, \xi^2, \xi^3), \quad \Lambda = 1, \ldots, 8. \tag{8.151}$$

Any suggested shape function will be assumed to satisfy the following condition:

Condition 8.7.2 Nodal consistency. The shape function N^Λ attains a unit value at node Λ and vanishes at all the other nodes.

To appreciate the usefulness of this condition, consider a field ψ and assume that all the information we have about this field consists of the values of ψ at the nodes, values that we denote as ψ_Λ, $\Lambda = 1, \ldots, 8$. Then the field within the element can be approximated by the function

$$\psi(\xi^1, \xi^2, \xi^3) = \sum_{\Lambda=1}^{8} \psi_\Lambda \, N^\Lambda(\xi^1, \xi^2, \xi^3). \tag{8.152}$$

Condition 8.7.2 guarantees that the interpolated field does indeed attain the known values at the nodes. The nodal consistency condition can be expressed more compactly as

$$N^\Lambda(\xi^1_\Delta, \xi^2_\Delta, \xi^3_\Delta) = \delta^\Lambda_\Delta, \tag{8.153}$$

where ξ^α_Δ designates the coordinate number α of the node number Δ, and δ is the Kronecker symbol.

The following shape functions will be adopted:

$$N^\Lambda = \frac{1}{8}(1 + \xi^1_\Lambda \xi^1)(1 + \xi^2_\Lambda \xi^2)(1 + \xi^3_\Lambda \xi^3). \tag{8.154}$$

These functions satisfy the nodal consistency condition, as can be verified directly. They also satisfy many other desirable conditions, on some of which we will comment later. At this point, it is worth remarking that these cubic functions are *trilinear*, which means that, keeping two coordinates fixed, the dependence on the remaining coordinate is linear. Moreover, we notice that the values of the interpolated field on an edge or a face depend only on the nodal values pertaining to that edge or face. This is an important condition that will later guarantee the inter-element continuity of the field in contiguous real elements.

Exercise 8.7.3   Starting your own FEM code. The array *CORNERS* (8, 3) is a direct copy of the table of nodal coordinates shown in Figure 8.14. This array is available to all the routines of the FEM code that we intend to write. Let the vector xi (3) contain specific values of the coordinates ξ^1, ξ^2, ξ^3. In the computer language of your choice, and with an obvious notation, write a function $NS\,(Lambda, xi\,())$ that returns the value of the shape function $N^\Lambda(\xi^1, \xi^2, \xi^3)$. In the same spirit, write a function $dNSdxi\,(Lambda, alpha, xi\,())$ that returns the value of the derivative $\partial N^\Lambda / \partial \xi^\alpha$ at the point ξ^1, ξ^2, ξ^3.

8.7.3 Meshing, Insertion Maps and the Isoparametric Idea

We turn our attention to the body itself in a given reference configuration and consider the task of subdividing it by means of a finite-element *mesh*.

Remark 8.7.4 The meshing process, akin to the triangulation of the Earth's surface, will in general leave some small parts near the body boundary unaccounted for. This will be the case, for example, if the body boundary is curved while the element boundaries are flat, as shown in Figure 8.15. As the elements become smaller and more numerous (in other words, as the mesh is *refined*), the uncovered areas too become smaller. Thus, the process of approximating the fields undergoes improvement at the same time as the domain itself is better approximated.

We will consider each element of the mesh as a smooth transformation of the ideal element, which is inserted into the reference configuration by means of an *insertion map* ϕ, as shown in Figure 8.16. In coordinates, this map translates into three smooth, and smoothly invertible, functions,

$$X^I = X^I(\xi^1, \xi^2, \xi^3), \quad I = 1, 2, 3. \tag{8.155}$$

In principle, any functions may be chosen to perform the insertion. As a result, we will obtain some kind of curved brick, with identifiable corners and edges (since the insertion functions are smooth). Nevertheless, since the shape functions have already been chosen, a good idea is to use them also as insertion functions. What we mean by this is that, given the desired locations X^I_Λ of the nodes of the inserted element, we construct the insertion map *in the same way* as we would interpolate any other field, namely

$$X^I(\xi^1, \xi^2, \xi^3) = \sum_{\Lambda=1}^{8} X^I_\Lambda \, N^\Lambda(\xi^1, \xi^2, \xi^3). \tag{8.156}$$

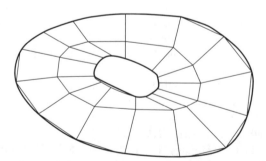

Figure 8.15   A coarse two-dimensional mesh

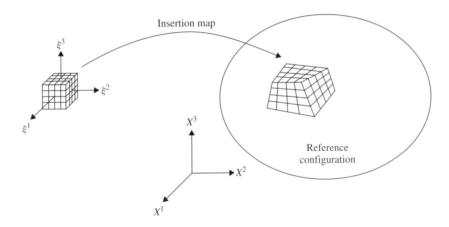

Figure 8.16   An insertion map

This technique of insertion gives rise to the so-called *isoparametric elements*, the name suggesting the origin of the concept.[23]

In the particular case of our shape functions (8.154), if two adjacent elements share four nodes, they will share the corresponding entire face, so that no gaps exist between contiguous elements. The meshing process can be imagined as a constellation of body points that, in an orderly fashion, are used in groups of eight as nodes of inserted elements.

If $\psi = \psi(X^1, X^2, X^3)$ is some (Lagrangian) field defined over the body in its reference configuration, and if ψ_Λ, $\Lambda = 1, \ldots, 8$, are the values of this field at the nodes of a particular element (as inserted in the body), then the interpolated field within this element is obtained by means of equation (8.152), just as in the ideal element, except that now the coordinates ξ^α are seen as curvilinear coordinates within the element, as indicated by the coordinate transformation (8.156). This transformation is, of course, invertible: in principle, if needed, the coordinates ξ^α can be expressed as functions of the global referential coordinates X^I.

8.7.4   The Contractibility Condition and its Consequences

The collection of permitted insertions of the ideal isoparametric element is completely determined by the choice of shape functions. One may, therefore, ask the following question: does this collection enjoy certain desirable properties, and does it include certain desirable types of insertions? This section, which may be skipped without further consequence, is included only to explore a few of those desiderata and their physical interpretation.

Let us assume that the coordinates X^I_Λ of the nodes of the inserted element get arbitrarily close to each other, to the extent that all nodes almost coalesce to a single point in the reference configuration. In such a case, we obviously would expect that the image of the whole inserted element also shrinks to a point. Nevertheless, if the shape functions are specified arbitrarily, there is no reason why this should be the case. We, therefore, impose the following condition.

[23] Irons and Zinkiewicz (1968)

Condition 8.7.5 Contractibility. If the images of all the nodal points coincide, the image of the entire inserted element must degenerate to a single point.

It is remarkable that this condition is equivalent to a somewhat unexpected result, which allows us to check the satisfaction of the contractibility condition in a simple manner, as established in the following theorem.

Theorem 8.7.6 Partition of unity. *The element shape functions satisfy the contractibility condition if, and only if, their sum is identically 1.*

Proof. If the contractibility condition is satisfied, given any three constants C^I, $I = 1, 2, 3$, representing the target point where all M nodes coincide, we must have

$$C^I = X^I(\xi^1, \xi^2, \xi^3) = \sum_{\Lambda=1}^{M} X_{\Lambda}^I \, N^{\Lambda}(\xi^1, \xi^2, \xi^3) = C^I \sum_{\Lambda=1}^{M} N^{\Lambda}(\xi^1, \xi^2, \xi^3) \qquad (8.157)$$

identically for all ξ^α within the element. Since the constants C^I are arbitrary, we conclude that

$$\sum_{\Lambda=1}^{M} N^{\Lambda}(\xi^1, \xi^2, \xi^3) = 1. \qquad (8.158)$$

Conversely, if equation (8.158) is satisfied, and if the nodal coordinates coincide with some fixed values C^I, we have

$$X^I(\xi^1, \xi^2, \xi^3) = \sum_{\Lambda=1}^{M} X_{\Lambda}^I \, N^{\Lambda}(\xi^1, \xi^2, \xi^3) = C^I \sum_{\Lambda=1}^{M} N^{\Lambda}(\xi^1, \xi^2, \xi^3) = C^I. \qquad (8.159)$$

∎

Exercise 8.7.7 Show that the isoparametric shape functions (8.154) satisfy the contractibility condition by checking that their sum is 1 everywhere in the element.

The contractibility condition is of paramount importance since one would like to see that, as the inserted elements become smaller and smaller by mesh refinement, the approximate fields become closer and closer to their exact counterparts. For this reason, it is important to avoid any pathological behaviour in the process of shrinking the elements to a point. It is remarkable, however, that the contractibility condition alone implies yet another, quite unexpected, desirable property of the insertions, as stated in the following theorem.

Theorem 8.7.8 Affine insertions. *The shape functions satisfy the contractibility condition if, and only if, any affine transformation of an admissible insertion is also admissible.*

The proof of this important theorem is left to the reader. By an affine transformation we mean a transformation of the reference space of the form

$$\{Y\} = [A]\{X\} + \{B\}, \qquad (8.160)$$

where $[\mathbf{A}]$ and $\{\mathbf{B}\}$ are constant matrices. A particular case of an affine transformation is a rigid motion, namely, a translation followed by a rotation. This happens when the matrix $[\mathbf{A}]$ is orthogonal.

The importance of the theorem of affine insertions in the case of isoparametric elements can be understood when one realizes that the displacements themselves will be interpolated by the same shape functions as the insertions. In other words, the deformed elements can themselves be seen as permissible insertions. As a consequence, the isoparametric elements are consistent with rigid body motions and with constant strain states, a condition whose physical desirability cannot be questioned. Indeed, if rigid-body motions were not available, any attempt to move the nodes rigidly would result in a state of non-zero strain within the element. In that case, intended rigid-body motions would cause internal stresses.

There are many other desirable conditions that one would like to impose on the shape functions, such as the condition of *edge and face autonomy* establishing that the restriction of an insertion map to an edge or face must be completely determined by the placement of the nodes of that edge or face alone. Clearly, this condition is partially responsible for the fact that we want the faces and edges determined by common nodes with adjacent elements to agree, so that there are no gaps left. The isoparametric elements based on polynomial shape functions can be shown to be optimal with respect to the satisfaction of these desirable conditions. We will not pursue this topic any further.

8.7.5 The Element IVW Routine

Suppose we have, for one particular element in the meshed reference configuration, three arrays, $X(8,3)$, $U(8,3)$ and $DU(8,3)$, containing the ordered values of the nodal global coordinates X^I_Λ, given nodal displacement components u^i_Λ and nodal virtual displacement components δu^i_Λ, respectively. These components are understood in the global spatial coordinate axes which, as the use of displacements indicates, are identified with the referential coordinate axes. Our first objective is to calculate the deformation gradient at an arbitrary point within the element.

As pointed out in the previous section, points within an element can be conveniently addressed by their coordinates ξ^α, induced by the insertion map. The advantage of this way of addressing element points stems from the fact that it represents the *relative location* of the point within the element. For example, the point with ξ-coordinates $(0,0,0)$ is at the 'centre' of the element. This kind of address is precisely what is needed to carry out a procedure of numerical integration, which will be needed to evaluate the internal virtual work contribution of an element.

In accordance with the general interpolation formula (8.152), the displacement field within the element can be calculated as

$$u^i(\xi^1,\xi^2,\xi^3) = \sum_{\Lambda=1}^{8} u^i_\Lambda \, N^\Lambda(\xi^1,\xi^2,\xi^3), \tag{8.161}$$

which can be written more compactly as

$$u^i = u^i_\Lambda \, N^\Lambda, \tag{8.162}$$

provided we extend the summation convention to include upper-case Greek indices, whose range is $\Lambda = 1,\ldots,8$. The virtual displacement field is given analogously by

$$\delta u^i = \delta u^i_\Lambda \, N^\Lambda. \tag{8.163}$$

The deformation gradient global components can be calculated as

$$F_I^i(\xi^1,\xi^2,\xi^3) = \delta_I^i + \frac{\partial u^i}{\partial X^I} = \delta_I^i + \frac{\partial u^i}{\partial \xi^\alpha}\,\frac{\partial \xi^\alpha}{\partial X^I}, \tag{8.164}$$

where the (usual) summation convention is used.

At this junction, we seem to have run into a problem. Equation (8.155) provides us with the derivatives

$$\frac{\partial X^I}{\partial \xi^\alpha} = X_\Lambda^I \sum_{\Lambda=1}^{8} \frac{\partial N^\Lambda(\xi^1,\xi^2,\xi^3)}{\partial \xi^\alpha}, \tag{8.165}$$

which we can write more compactly as

$$\frac{\partial X^I}{\partial \xi^\alpha} = X_\Lambda^I\,\frac{\partial N^\Lambda}{\partial \xi^\alpha}. \tag{8.166}$$

Observe that, for our particular shape functions (8.154), these derivatives can be calculated explicitly. The apparent problem is that we need the derivatives of the inverse map. According to a well-known theorem of calculus in several variables, however, the Jacobian matrix of the inverse map is precisely the inverse matrix of the Jacobian of the original map, namely

$$\left[\frac{\partial \xi^\alpha}{\partial X^I}\right] = \left[\frac{\partial X^I}{\partial \xi^\alpha}\right]^{-1}. \tag{8.167}$$

This is a 3×3 matrix, so that its inverse can be calculated explicitly at an arbitrary point with coordinates ξ^α, $\alpha = 1,2,3$.

Emulating equation (8.164), we can also calculate the variation of the deformation gradient associated with the given variations of the displacements as

$$\delta F_I^i(\xi^1,\xi^2,\xi^3) = \delta_I^i + \frac{\partial \delta u^i}{\partial X^I} = \delta_I^i + \frac{\partial \delta u^i}{\partial \xi^\alpha}\,\frac{\partial \xi^\alpha}{\partial X^I}. \tag{8.168}$$

Exercise 8.7.9 Jacobian of the insertion map. Given the global nodal coordinates $X(8,3)$ of an inserted element and the element coordinates $xi(3)$ of a point within the element, write a subroutine that returns the Jacobian matrix $dXdxi(3,3)$ of the insertion map according to equation (8.165), and its inverse $dxidX(3,3)$. Use the results of Exercise 8.7.3.

Exercise 8.7.10 Deformation gradient. Given the global nodal coordinates $X(8,3)$ of an inserted element and the arrays of nodal displacements $U(8,3)$ and nodal virtual displacements $DU(8,3)$ for that element, write a subroutine that returns the deformation gradient $F(3,3)$ and its variation $DF(3,3)$ at a point within the element with element coordinates $xi(3)$. Use the results of Exercise 8.7.9.

In conclusion, the coding of the deformation-gradient field **F** within an element and of its variation δ**F** is a straightforward programming task. Given an elastic constitutive law by some formula of the type

$$\mathbf{T} = \mathbf{T}(\mathbf{F}), \tag{8.169}$$

we have at our disposal the stress field as well. The internal virtual work contributed by the element is

$$IVW_e = \int_{V_e} \text{trace} \ (\mathbf{T}^T \delta \mathbf{F}) \, dV_e = \int_{-1}^{1} \int_{-1}^{1} \int_{-1}^{1} \text{trace} \ (\mathbf{T}^T \delta \mathbf{F}) \det \left[\frac{\partial X^I}{\partial \xi^\alpha} \right] d\xi^1 \, d\xi^2 \, d\xi^3,$$

(8.170)

where V_e is the volume occupied by the element in the reference configuration. The determinant of the Jacobian of the insertion map appears because of the volume element correction between the ideal and the real element.

Even if the integral in (8.170) were available analytically, it would be pointless to insist on its exact evaluation, since the FEM is already an approximate method. What we need is some numerical integration technique that is in some sense consistent with the error introduced by the approximation of the fields brought about by the assumed shape functions. For our trilinear shape functions, it is sufficient to use an eight-point Gauss integration procedure, which requires the evaluation of the integrand at just eight judiciously located points. The exact location of these points is given by the eight combinations of ξ-coordinates with absolute value equal to $\sqrt{3}/3$.

Exercise 8.7.11   Gauss integration. Let a function $\psi = \psi(\xi^1, \xi^2, \xi^3)$ be available for all values of the element coordinates ξ^α within the ideal element. Write a computer function that delivers the approximate value of the integral

$$\int_e \psi \, dV_\xi = \int_{-1}^{1} \int_{-1}^{1} \int_{-1}^{1} \psi(\xi^1, \xi^2, \xi^3) \, d\xi^1 \, d\xi^2 \, d\xi^3,$$

(8.171)

by an eight-point Gauss integration procedure. The integral is approximated by the following sum:

$$\int_e \psi \, dV_\xi \approx \sum_{\Lambda=1}^{8} \psi \left(\frac{\sqrt{3}}{3} \xi_\Lambda^1, \frac{\sqrt{3}}{3} \xi_\Lambda^2, \frac{\sqrt{3}}{3} \xi_\Lambda^3 \right).$$

(8.172)

Recall that the values ξ_Λ^α are available in the array $CORNERS\,(8,3)$.

Exercise 8.7.12   Internal virtual work routine. Given the global nodal coordinates $X(8,3)$ of an inserted element and the arrays of nodal displacements $U(8,3)$ and nodal virtual displacements $DU(8,3)$ for that element, write a computer function that returns the value of the internal virtual work of that element. For specificity, use the constitutive law

$$\mathbf{T} = G \left(\mathbf{F} - \mathbf{F}^{-T} \right),$$

(8.173)

where G is a material modulus, or any other elastic constitutive law of your choice, as long as the material is not subject to any geometric constraint, such as incompressibility.

Summing up, we may say that the central routine of the FEM is a rather straightforward programming task. It requires knowledge of the values of the three arrays $X(8,3)$, $U(8,3)$ and $DU(8,3)$, the shape functions, the constitutive equation and nothing else. Since the (scalar) function to be integrated needs to be calculated at eight points only, the number of operations involved is relatively small.

8.7.6   The Element EVW Routine

There are at least three different types of contributions to the external virtual work. The
first kind consists of forces directly applied in correspondence with the nodal displacements. This is the case of a concentrated force acting directly at a node. Its *EVW* is easily
calculated by taking its dot product with the virtual displacement at that node. We will
deal with the implementation of these loads later. At this point we are concerned with
loads at the element level. These can be actual *element loads* (such as body forces, surface
tractions, concentrated forces applied at points other than the nodes) or *inertia loads*.

The treatment of the forces of inertia, which are essentially body forces (unless there are
also concentrated masses), necessitates the implementation of a numerical procedure in the
space-time domain. Among these techniques, we may mention implicit time-integration
methods, which combine the usual finite-element method with a finite-difference algorithm
in the time domain. In this way, the dynamic analysis is resolved into a discrete succession
of static solutions, each one of them subjected to additional fictitious forces arising from
the available solution at a number (usually three) of previous time-steps.[24] For this reason,
we will confine ourselves to the discussion of static problems.

We are left with the evaluation of the *EVW* of element loads. One way to deal with
these forces is by a technique known as *inconsistent lumping* at the nodes. In this method,
the element loads are intuitively apportioned to the element nodes. For example, the
distributed weight of an element (a body force) can be totalled and the total weight
divided into eight equal parts, each of which is assigned to a different node. The claim
can be made that this technique becomes more and more accurate as the mesh is refined.
A more rational technique is the *consistent lumping*, which we will now exhibit for a body
force, but which can also be used for other types of element loads. Consistent lumping
is nothing but the evaluation of the *EVW* via the interpolated displacement field, just as
was the case for the *IVW*. We have

$$EVW_e = \int_{V_e} \mathbf{B} \cdot \delta \mathbf{u} \, dV_e, \tag{8.174}$$

where $\delta\mathbf{u}$ is given in components by equation (8.163). The body force itself, $\mathbf{B}$, must be
given. The most common case is the force of gravity.

In conclusion, given the external element loads and the arrays $X(8,3)$ and $DU(8,3)$,
the integral (8.174) can be evaluated by Gauss quadrature. Follower loads can be included
in the same way, but involving also the displacement array $U(8,3)$.

8.7.7   Assembly and Solution

The total number of possible degrees of freedom (i.e., nodal displacements) of a meshed
body is equal to $3N$, where N is the total number of nodes. We have two different
numbering systems at play for these degrees of freedom. The first system consists of
naming (or numbering) the elements and, within each element, numbering the nodes from
1 to 8 in the order induced by the insertion map and, finally, the components (from 1
to 3). An example of this system would read as follows: 'element number 154, node
number 7, component number 2'. Since most nodes are shared by several elements this
numbering contains a built-in redundancy.

---

[24] The first procedure of this kind is due to Houbolt (1950).

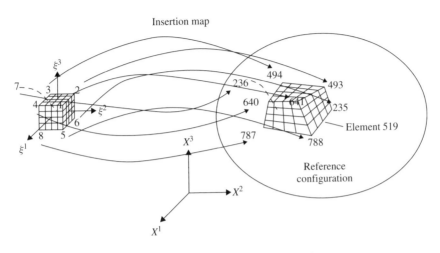

Insertion map

Reference
configuration

Figure 8.17   Local to global node numbers

The second numbering system is purely global. Assume that the nodes are numbered sequentially starting from 1 and following some numbering pattern.[25] For each node thus numbered, the displacement components are ordered sequentially form 1 to 3. All these triads are then placed in a large vector $UUU\,(3N)$ following the nodal numbering. An example of the address of a degree of freedom in this numbering system could be: degree of freedom number 5672. Dividing $5672 - 1$ by 3, we obtain a quotient of 1890 and a remainder of 1. This means that by the number 5672 we are addressing the second component of the displacement of node number 1891.

The *assembly* procedure consists largely of a proper administration of the logistics of the passage from one numbering system to the other. It is accomplished by means of a table that can be established at the time of meshing the domain. It contains for each element the numbers of the nodes it possesses in the order prescribed by the insertion map of that element. Calling this array $ETABLE\,(N,8)$, a typical row (row number 519 for the example depicted in Figure 8.17) will look as follows:

$$\text{Row number } 519 \rightarrow \langle 641\ 493\ 494\ 640\ 788\ 235\ 236\ 787\rangle. \qquad (8.175)$$

This information is sufficient to accomplish our goal. Given an address in the large global array UUU, we subtract 1 and divide by 3 to obtain the node number and the component direction. By scanning the array $ETABLE$ (or by other more sophisticated means) we determine all the (few) elements that share that node and which specific position the node occupies (from 1 to 8) in each of the sharing elements. In this way, we find for each element which of the available 24 degrees of freedom is affected. This is easier to program than to explain.

As we have learned from the theoretical presentation in Section 8.5.3, the equations to be solved consist of the vanishing of the coefficients of the variations, as implied in equation (8.126). In the particular case of solid mechanics, those expressions arise

---

[25] The numbering of the nodes in a mesh can be optimized according to the following criterion: try to minimize the difference between the numbers assigned to pairs of nodes that belong to the same element. The optimal numbering minimizes the number of algebraic operations in the solution procedure by avoiding 'trapped zeros' in the solution matrices.

from the contribution of the *IVW* minus the contribution of the *EVW*. To isolate one particular coefficient, we choose a variation of unit value for that particular degree of freedom and zero for all others. We then perform the logistic exercise explained above and focus only on those elements that own that particular degree of freedom. The routine for the element *IVW* is called once for each of those few elements, each time with the corresponding degree of freedom at the element level. The corresponding contributions to the coefficient of that variation are precisely the values returned by the *IVW* routine (which is a function; that is, it returns a single value). A similar procedure is followed for the external virtual work. Finally, we still need to add the *EVW* of the load (if any) applied directly in correspondence with the degree of freedom under treatment. But this contribution is simply equal to minus the force (since the virtual displacement has been assumed of unit value). As a result of this exercise, the value of the (equilibrium) equation associated with the particular degree of freedom selected is obtained. This value clearly depends only on the present content of the vector *UUU*.

To implement a support condition, all that has to be done is eliminate the equation associated with the restrained degree of freedom and replace it with the constraint itself (usually, equating to zero the value of the degree of freedom), as we discussed in the general presentation of the Galerkin method.

Window 8.1   Review of the Newton–Raphson method

Given m differentiable functions f_i of m variables z_i, we seek an approximate solution of the system of equations

$$f_i(z_1, z_2, \ldots, z_m) = 0, \quad i = 1, \ldots, m. \tag{8.176}$$

The search process gets started with an *initial guess* $z_1^0, z_2^0, \ldots, z_m^0$. The first-order approximation to the given functions around this initial guess is used to obtain the next approximation $(z_1^1, z_2^1, \ldots, z_m^1)$ according to

$$f_i(z_1^0, z_2^0, \ldots, z_m^0) + \sum_{k=1}^{m} \left. \frac{\partial f_i}{\partial z_k} \right|_{z_1^0, \ldots, z_m^0} (z_k^1 - z_k^0) = 0, \quad i = 1, \ldots, m.$$

This is a system of *linear* algebraic equations. The procedure is repeated recursively with $z_1^1, z_2^1, \ldots, z_m^1$ as the new initial guess until convergence is achieved according to some suitable convergence criterion.

Remark 8.7.13   Numerical evaluation of the derivatives. When analytic expressions for the functions f_i are not available, the entries in the coefficient matrix can be approximated as

$$\left. \frac{\partial f_i}{\partial z_k} \right|_{z_1^0, \ldots, z_m^0} \approx \frac{f_i(z_1^0, z_2^0, \ldots, z_k^0 + h, \ldots, z_m^0) - f_i(z_1^0, z_2^0, \ldots, z_k^0, \ldots, z_m^0)}{h}, \tag{8.177}$$

where h is a small number. This policy must be followed in our FEM program, since the values of the functions f_i are available only through a lengthy process involving, among other factors, numerical integrations.

Let us assess the situation. We have at our disposal a method to obtain the numerical value of $3N$ (in general non-linear) functions of $3N$ variables (the totality of nodal displacement components) for any value of these variables. When equated to zero, these functions constitute a system of simultaneous algebraic equations. To solve this system, we can resort to the classical Newton–Raphson method (see Window 8.1). It requires, in addition to the value of the functions (which we already have), the evaluation of the partial derivatives of each of the $3N$ functions with respect to each of the $3N$ variables. Because of the fact that each node is connected to only a small number of other nodes, only a few of these partial derivatives do not vanish. Since the functions are available only numerically, we can resort to evaluating those derivatives numerically. This takes two evaluations: one is the one we already have (the value of the function for the present value of the displacements) and the other is obtained by varying just one of the displacements by a small increment, going through the evaluation process again, taking the difference and dividing by the increment. The Newton–Raphson method is then left on automatic pilot until (one hopes) it converges to a solution. This concludes the whole endeavour, since, once the nodal displacement vector UUU becomes available, one can calculate interpolated deformation gradients and stresses at any desired point of the body.

Exercise 8.7.14   The Newton–Raphson routine. Let m functions be available via a computer function named $f(i, z())$. Its arguments are the function number ($i = 1, \ldots, m$) and the array $z(m)$ of independent variables. Write a subroutine to solve approximately the system of equations (8.176). When calling this subroutine, the array $z()$ contains the initial guess. Upon completion, the converged result is placed in this same array. You will need to define extra arrays within the subroutine. As a criterion, using the notation of Window 8.1, a good possibility is:

$$\varepsilon = \frac{|\{\mathbf{z}\}^s - \{\mathbf{z}\}^{s-1}|}{|\{\mathbf{z}^s\}|} \le 10^{-6}, \quad s = 1, 2, \ldots. \tag{8.178}$$

In this expression, s is an index denoting the iteration step, and $|\cdot|$ is the Cartesian length defined in equation (5.46). Since, in general, the derivatives of the functions may not be explicitly available, you will need to write and check a numerical differentiation routine that implements the formula (8.177). Moreover, you need a linear equation solver.

Exercise 8.7.15   Final project: a FEM program. On the basis of the computer codes developed in previous exercises, write a complete finite-element program for the static analysis of elastic media in a regime of large deformations. The Newton–Raphson routine acts as the main program. The array of unknown quantities is a long vector $UUU(m)$ whose dimension m is 3 times the number of nodes. The function number i is obtained as follows: with the current value of $UUU()$, define a vector $DUUU()$ with vanishing entries, except for the single entry $UUU(i)$, which is set equal to 1. The degree of freedom number i is owned by a small number of elements, which can be found by scanning the array $ETABLE$, as explained above. Accordingly, for an unconstrained degree of freedom, the function number i is constructed by calling the internal virtual work function as many times as needed and adding the results. In a similar way, the external virtual work (if any) is subtracted. For constrained degrees of freedom, the corresponding function is just the expression of the constraint (usually, the vanishing of the corresponding degree of freedom). Material properties, loads and constraints are part of the input data. The only part of the program that we have not explained is the mesh generation. You can either enter the mesh by hand or generate it as the outcome of any of the meshing programs available on the market. At any rate, in addition to the array *ETABLE*, you need your meshing procedure to produce an array containing the global coordinates of the nodes of the mesh.

References

Houbolt, J.C. (1950) A recurrence matrix solution for the dynamic response of elastic aircraft. *Journal of Aeronautical Science* **17**, 540–550.

Irons, B.M. and Zinkiewicz, O.C. (1968) The isoparametric finite element system – a new concept in finite element analysis. In *Proceedings of the Conference on Recent Advances in Stress Analysis*, Royal Aeronautical Society, London.

Timoshenko, S.P. (1983) *History of Strength of Materials*. New York: Dover.

Index